Heat Convection

Latif M. Jiji

Heat Convection

Second Edition

 Springer

Prof. Latif M. Jiji
Department of Mechanical Engineering
Grove School of Engineering
The City College of
The City University of New York
New York, New York 10031
USA
E-mail: jiji@ccny.cuny.edu

ISBN 978-3-642-44763-1   ISBN 978-3-642-02971-4 (eBook)

DOI 10.1007/978-3-642-02971-4

© 2009 Springer-Verlag Berlin Heidelberg

Typesetting: Data supplied by the authors

Production: Scientific Publishing Services Pvt. Ltd., Chennai, India

Cover Design: WMXDesign GmbH, Heidelberg

Printed in acid-free paper

9 8 7 6 5 4 3 2 1

springer.com

To my sister Sophie and brother Fouad
for their enduring love and affection

PREFACE

Why have I chosen to write a book on convection heat transfer when several already exist? Although I appreciate the available publications, in recent years I have not used a textbook to teach our graduate course in convection. Instead, I have relied on my own notes, not because existing textbooks are unsatisfactory, but because I preferred to select and organize the subject matter to cover the most basic and essential topics and to strike a balance between physical description and mathematical requirements. As I developed my material, I began to distribute lecture notes to students, abandon blackboard use, and rely instead on PowerPoint presentations. I found that PowerPoint lecturing works most effectively when the presented material follows a textbook very closely, thus eliminating the need for students to take notes. Time saved by this format is used to raise questions, engage students, and gauge their comprehension of the subject. This book evolved out of my success with this approach.

This book is designed to:

- Provide students with the fundamentals and tools needed to model, analyze, and solve a wide range of engineering applications involving convection heat transfer.

- Present a comprehensive introduction to the important new topic of convection in microchannels.

- Present textbook material in an efficient and concise manner to be covered in its entirety in a one semester graduate course.

- Liberate students from the task of copying material from the blackboard and free the instructor from the need to prepare extensive notes.

- Drill students in a systematic problem solving methodology with emphasis on thought process, logic, reasoning, and verification.

- Take advantage of internet technology to teach the course online by posting ancillary teaching materials and solutions to assigned problems.

This edition adds two new chapters on turbulent convection. I am fortunate that Professor Glen E. Thorncroft of California Polytechnic State University, San Luis Obispo, California, agreed to take on this responsibility and prepared all the material for chapters 8 and 9.

Hard as it is to leave out any of the topics usually covered in classic texts, cuts have been made so that the remaining material can be taught in one semester. To illustrate the application of principles and the construction of solutions, examples have been carefully selected, and the approach to solutions follows an orderly method used throughout. Detailed solution to all end of chapter problems follow the same format. They are prepared for posting electronically.

This work owes a great deal to published literature on heat transfer. As I developed my notes, I used examples and problems taken from published work on the subject. Since I did not always record references in my early years of teaching, I have tried to eliminate any that I knew were not my own. I would like to express regret if a few have been unintentionally included.

Latif M. Jiji
Department of Mechanical Engineering
The City College
of the City University of New York
New York, New York
May 2009

Contents

1

Basic Concepts

1.1 Convection Heat Transfer

In general, convection heat transfer deals with thermal interaction between a surface and an adjacent moving fluid. Examples include the flow of fluid over a cylinder, inside a tube and between parallel plates. Convection also includes the study of thermal interaction between fluids. An example is a jet issuing into a medium of the same or a different fluid.

1.2 Important Factors in Convection Heat Transfer

Consider the case of the electric bulb shown in Fig. 1.1. Surface temperature and heat flux are T_s and q_s'', respectively. The ambient fluid temperature is T_∞. Electrical energy is dissipated into heat at a fixed rate determined by the capacity of the bulb. Neglecting radiation, the dissipated energy is transferred by convection from the surface to the ambient fluid. Suppose that the resulting surface temperature is too high and that we wish to lower it. What are our options?

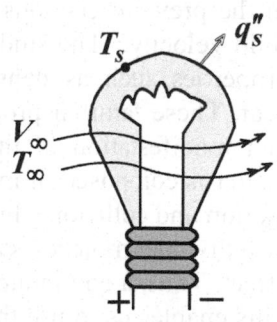

Fig. 1.1

(1) Place a fan in front of the bulb and force the ambient fluid to flow over the bulb.

(2) Change the fluid, say, from air to a non-conducting liquid.

(3) Increase the surface area by redesigning the bulb geometry.

We conclude that three factors play major roles in convection heat transfer: (i) *fluid motion,* (ii) *fluid nature, and* (iii) *surface geometry.*

Other common examples of the role of fluid motion in convection are:

- Fanning to feel cool.

- Stirring a mixture of ice and water.

- Blowing on the surface of coffee in a cup.

- Orienting a car radiator to face air flow.

Common to all these examples is a moving fluid which is exchanging heat with an adjacent surface.

1.3 Focal Point in Convection Heat Transfer

Of interest in convection heat transfer problems is the determination of surface heat transfer rate and/or surface temperature. These important engineering factors are established once the temperature distribution in the moving fluid is determined. Thus the focal point in convection heat transfer is *the determination of the temperature distribution in a moving fluid.* In Cartesian coordinates this is expressed as

$$T = T(x, y, z, t).$$ (1.1)

1.4 The Continuum and Thermodynamic Equilibrium Concepts

In the previous sections we have invoked the concept of temperature and fluid velocity. The study of convection heat transfer depends on material properties such as density, pressure, thermal conductivity, and specific heat. These familiar properties which we can quantify and measure are in fact manifestation of the molecular nature and activity of material. All matter is composed of molecules which are in a continuous state of random motion and collisions. In the *continuum* model we ignore the characteristics of individual molecules and instead deal with their average or macroscopic effect. Thus, a continuum is assumed to be composed of continuous matter. This enables us to use the powerful tools of calculus to model and analyze physical phenomena. However, there are conditions under which the continuum assumption breaks down. It is valid as long as there is sufficiently large number of molecules in a given volume to make the statistical average of their activities meaningful. A measure of the validity of the continuum assumption is the molecular-mean-free path λ relative to the characteristic dimension of the system under consideration. The mean-free-path is the average distance traveled by molecules before they collide. The ratio of these two length scales is called the *Knudson number, Kn,* defined as

$$Kn = \frac{\lambda}{D_e},$$ (1.2)

where D_e is the characteristic length, such as the equivalent diameter or the spacing between parallel plates. The criterion for the validity of the continuum assumption is [1]

$$Kn < 10^{-1}.$$ (1.3a)

Thus this assumption begins to break down, for example, in modeling convection heat transfer in very small channels.

Thermodynamic equilibrium depends on the collisions frequency of molecules with an adjacent surface. At thermodynamic equilibrium the fluid and the adjacent surface have the same velocity and temperature. This is called the *no-velocity slip* and *no-temperature jump*, respectively. The condition for thermodynamic equilibrium is

$$Kn < 10^{-3}.$$ (1.3b)

The continuum and thermodynamic equilibrium assumptions will be invoked throughout this book. Chapter 11, Convection in Microchannels, deals with applications where the assumption of thermodynamic equilibrium breaks down.

1.5 Fourier's Law of Conduction

Our experience shows that if one end of a metal bar is heated, its temperature at the other end will eventually begin to rise. This transfer of energy is due to molecular activity. Molecules at the hot end exchange their kinetic and vibrational energies with neighboring layers through random motion and collisions. A temperature gradient, or slope, is established with energy continuously being transported in the direction of decreasing temperature. This mode of energy transfer is called *conduction*. The same mechanism takes place in fluids, whether they are stationary or moving. It is important to recognize that the mechanism for energy interchange at the interface between a fluid and a surface is conduction. However, energy transport throughout a moving fluid is by conduction and convection.

We now turn our attention to formulating a law that will help us determine the rate of heat transfer by conduction. Consider the wall shown in Fig.1.2. The temperature of one surface ($x = 0$) is T_{si} and of the other surface ($x = L$) is T_{so}. The wall thickness is L and its surface area is A. The remaining four surfaces are well insulated and thus heat is transferred in the x-direction only. Assume steady state and let q_x be the rate of heat transfer in the x-direction. Experiments have shown

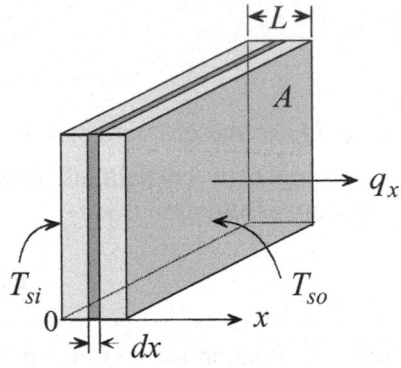

Fig. 1.2

that q_x is directly proportional to A and $(T_{si} - T_{so})$ and inversely proportional to L. That is

$$q_x \propto \frac{A\,(T_{si} - T_{so})}{L}.$$

Introducing a proportionality constant k, we obtain

$$q_x = k\frac{A\,(T_{si} - T_{so})}{L}, \tag{1.4}$$

where k is a property of material called *thermal conductivity*. We must keep in mind that (1.4) is valid for: (i) steady state, (ii) constant k and (iii) one-dimensional conduction. These limitations suggest that a re-formulation is in order. Applying (1.4) to the element dx shown in Fig.1.2 and noting that $T_{si} \rightarrow T(x)$, $T_{so} \rightarrow T(x+dx)$, and L is replaced by dx, we obtain

$$q_x = k\,A\frac{T(x) - T(x+dx)}{dx} = -k\,A\frac{T(x+dx) - T(x)}{dx}.$$

Since $T(x+dx) - T(x) = dT$, the above gives

$$q_x = -k\,A\frac{dT}{dx}. \tag{1.5}$$

It is useful to introduce the term *heat flux* q_x'', which is defined as the heat flow rate per unit surface area normal to x. Thus,

$$q_x'' = \frac{q_x}{A}. \tag{1.6}$$

Therefore, in terms of heat flux, (1.5) becomes

$$q_x'' = -k\frac{dT}{dx}. \tag{1.7}$$

Although (1.7) is based on one-dimensional conduction, it can be generalized to three-dimensional and transient conditions by noting that heat flow is a vector quantity. Thus, the temperature derivative in (1.7) is changed to partial derivative and adjusted to reflect the direction of heat flow as follows:

$$q_x'' = -k\frac{\partial T}{\partial x}\ , \quad q_y'' = -k\frac{\partial T}{\partial y}\ , \quad q_z'' = -k\frac{\partial T}{\partial z}, \tag{1.8}$$

where x, y, and z are the rectangular coordinates. Equation (1.8) is known as *Fourier's law of conduction*. Four observations are worth making: (i) The negative sign means that when the gradient is negative, heat flow is in the positive direction, i.e., towards the direction of decreasing temperature, as dictated by the second law of thermodynamics. (ii) The conductivity k need not be uniform since (1.8) applies at a point in the material and not to a finite region. In reality thermal conductivity varies with temperature. However, (1.8) is limited to *isotropic material*, i.e., k is invariant with direction. (iii) Returning to our previous observation that the focal point in heat transfer is the determination of temperature distribution, we now recognize that once $T(x,y,z,t)$ is known, the heat flux in any direction can be easily determined by simply differentiating the function T and using (1.8). (iv) By manipulating fluid motion, temperature distribution can be altered. This results in a change in heat transfer rate, as indicated in (1.8).

1.6 Newton's Law of Cooling

An alternate approach to determining heat transfer rate between a surface and an adjacent fluid in motion is based on *Newton's law of cooling*. Using experimental observations by Isaac Newton, it is postulated that surface flux in convection is directly proportional to the difference in temperature between the surface and the streaming fluid. That is

$$q_s'' \propto (T_s - T_\infty),$$

where q_s'' is surface flux, T_s is surface temperature and T_∞ is the fluid temperature far away from the surface. Introducing a proportionality constant to express this relationship as equality, we obtain

$$q_s'' = h(T_s - T_\infty).\tag{1.9}$$

This result is known as *Newton's law of cooling*. The constant of proportionality h is referred to as the *heat transfer coefficient*. This simple result is very important, deserving special attention and will be examined in more detail in the following section.

1.7 The Heat Transfer Coefficient h

The heat transfer coefficient plays a major role in convection heat transfer. We make the following observations regarding h:

(1) Equation (1.9) is a definition of h and not a phenomenological law.

(2) Unlike thermal conductivity k, the heat transfer coefficient is not a material property. Rather it depends on geometry, fluid properties, motion, and in some cases temperature difference, $\Delta T = (T_s - T_\infty)$. That is

$$h = f \text{ (geometry, fluid motion, fluid properties, } \Delta T).\tag{1.10}$$

(3) Although no temperature distribution is explicitly indicated in (1.9), the analytical determination of h requires knowledge of temperature distribution in a moving fluid. This becomes evident when both Fourier's law and Newton's law are combined. Application of Fourier's law in the y-direction for the surface shown in Fig. 1.3 gives

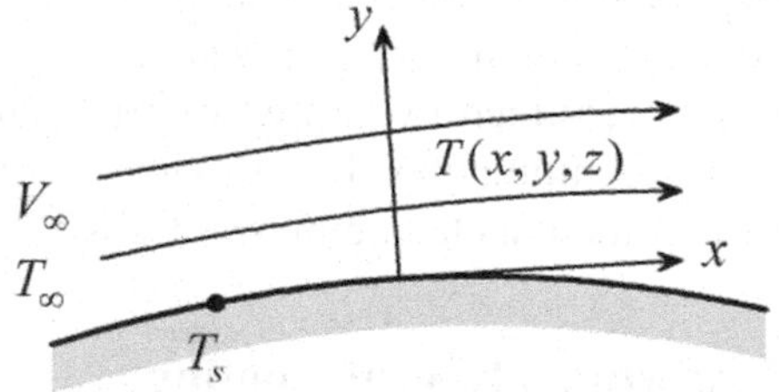

$$q_s'' = -k\frac{\partial T(x,0,z)}{\partial y},\tag{1.11}$$

Fig. 1.3

where y is normal to the surface, $\partial T(x,0,z)/\partial y$ is temperature gradient in the *fluid* at the interface, and k is the thermal conductivity of the *fluid*. Combining (1.9) and (1.11) and solving for h, gives

$$h = -k \frac{\dfrac{\partial T(x,0,z)}{\partial y}}{(T_s - T_\infty)}. \qquad (1.12)$$

This result shows that to determine h analytically one must determine temperature distribution.

(4) Since both Fourier's law and Newton's law give surface heat flux, what is the advantage of introducing Newton's law? In some applications the analytical determination of the temperature distribution may not be a simple task, for example, turbulent flow over a complex geometry. In such cases one uses equation (1.9) to determine h experimentally by measuring q_s'', T_s and T_∞ and constructing an empirical equation to correlate experimental data. This eliminates the need for the determination of temperature distribution.

(5) We return now to the bulb shown in Fig. 1.1. Applying Newton's law (1.9) and solving for surface temperature T_s, we obtain

$$T_s = T_\infty + \frac{q_s''}{h}. \qquad (1.13)$$

For specified q_s'' and T_∞, surface temperature T_s can be altered by changing h. This can be done by changing the fluid, surface geometry and/or fluid motion. On the other hand, for specified surface temperature T_s and ambient temperature T_∞, equation (1.9) shows that surface flux can be altered by changing h.

(6) One of the major objectives of convection is the determination of h.

(7) Since h is not a property, its values cannot be tabulated as is the case with thermal conductivity, enthalpy, density, etc. Nevertheless, it is useful to have a rough idea of its magnitude for common processes and fluids. Table 1.1 gives the approximate range of h for various conditions.

Table 1.1 Typical values of h	
Process	$h\,(\text{W/m}^2 - {}^\circ\text{C})$
Free convection Gases Liquids	 5-30 20-1000
Forced convection Gases Liquids Liquid metals	 20-300 50-20,000 5,000-50,000
Phase change Boiling Condensation	 2,000-100,000 5,000-100,000

1.8 Radiation: Stefan-Boltzmann Law

Radiation energy exchange between two surfaces depends on the geometry, shape, area, orientation, and emissivity of the two surfaces. In addition, it depends on the *absorptivity* α of each surface. Absorptivity is a surface property defined as the fraction of radiation energy incident on a surface which is absorbed by the surface. Although the determination of the net heat exchange by radiation between two surfaces, q_{12}, can be complex, the analysis is simplified for an ideal model for which the absorptivity α is equal to the emissivity ε. Such an ideal surface is called a *gray* surface. For the special case of a gray surface which is completely enclosed by a much larger surface, q_{12} is given by Stefan-Boltzmann radiation law

$$q_{12} = \varepsilon_1 \sigma A_1 (T_1^4 - T_2^4),\qquad(1.14)$$

where ε_1 is the emissivity of the small surface, A_1 its area, T_1 its absolute temperature, and T_2 is the absolute temperature of the surrounding surface. Note that for this special case neither the area A_2 of the large surface nor its emissivity ε_2 affect the result.

1.9 Differential Formulation of Basic Laws

The analysis of convection heat transfer relies on the application of the three basic laws: conservation of mass, momentum, and energy. In addition, Fourier's conduction law and Newton's law of cooling are also applied. Since the focal point is the determination of temperature distribution, the three basic laws must be cast in an appropriate form that lends itself to the determination of temperature distribution. This casting process is called *formulation*. Various formulation procedures are available. They include *differential, integral, variational,* and *finite difference* formulation. This section deals with differential formulation. Integral formulation is presented in Chapter 5.

Differential formulation is based on the key assumption of continuum. This assumption ignores the molecular structure of material and focuses on the gross effect of molecular activity. Based on this assumption, fluids are modeled as continuous matter. This makes it possible to treat variables such as temperature, pressure, and velocity as continuous function in the domain of interest.

1.10 Mathematical Background

We review the following mathematical definitions which are needed in the differential formulation of the basic laws.

(a) Velocity Vector $\vec{V}$. Let u, v, and w be the velocity components in the x, y and z directions, respectively. The vector $\vec{V}$ is given by

$$\vec{V} = ui + vj + wk . \qquad (1.15a)$$

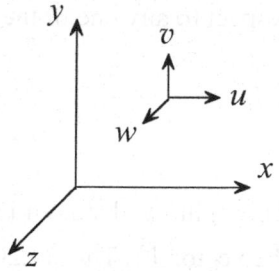

Fig. 1.4

(b) Velocity Derivative. The derivative of the velocity vector with respect to any one of the three independent variables is given by

$$\frac{\partial \vec{V}}{\partial x} = \frac{\partial u}{\partial x}i + \frac{\partial v}{\partial x}j + \frac{\partial w}{\partial x}k . \qquad (1.15b)$$

(c) The Operator ∇. In Cartesian coordinates the operator ∇ is a vector defined as

$$\nabla \equiv \frac{\partial}{\partial x}i + \frac{\partial}{\partial y}j + \frac{\partial}{\partial z}k . \qquad (1.16)$$

In cylindrical coordinates this operator takes the following form

$$\nabla \equiv \frac{\partial}{\partial r}i_r + \frac{1}{r}\frac{\partial}{\partial \theta}i_\theta + \frac{\partial}{\partial z}i_z . \qquad (1.17)$$

Similarly, the form in spherical coordinate is

$$\nabla \equiv \frac{\partial}{\partial r}i_r + \frac{1}{r}\frac{\partial}{\partial \theta}i_\theta + \frac{1}{r\sin\theta}\frac{\partial}{\partial \phi}i_\phi . \qquad (1.18)$$

(d) Divergence of a Vector. The divergence of a vector $\vec{V}$ is a scalar defined as

$$div.\vec{V} \equiv \nabla \cdot \vec{V} = \frac{\partial u}{\partial x} + \frac{\partial v}{\partial y} + \frac{\partial w}{\partial z} . \qquad (1.19)$$

(e) Derivative of the Divergence. The derivative of the divergence with respect to any one of the three independent variables is given by

$$\frac{\partial}{\partial x}\left(\nabla \cdot \vec{V}\right) = \frac{\partial}{\partial x}\left(\frac{\partial u}{\partial x} + \frac{\partial v}{\partial y} + \frac{\partial w}{\partial z}\right). \tag{1.20}$$

The right hand side of (1.20) represents the divergence of the derivative of the vector $\vec{V}$. Thus (1.20) can be rewritten as

$$\frac{\partial}{\partial x}\left(\nabla \cdot \vec{V}\right) = \nabla \cdot \frac{\partial}{\partial x}\left(u\,i + v\,j + w\,k\right),$$

or

$$\frac{\partial}{\partial x}\left(\nabla \cdot \vec{V}\right) = \nabla \cdot \frac{\partial \vec{V}}{\partial x}. \tag{1.21}$$

(f) Gradient of a Scalar. The gradient of a scalar, such as temperature T, is a vector given by

$$Grad\,T \equiv \nabla \cdot T = \frac{\partial T}{\partial x}i + \frac{\partial T}{\partial y}j + \frac{\partial T}{\partial z}k. \tag{1.22}$$

(g) Total Differential and Total Derivative. We consider a variable of the flow field designated by the symbol f. This is a scalar quantity such as temperature T, pressure p, density ρ, or velocity component u. In general this quantity is a function of the four independent variables x, y, z and t. Thus in Cartesian coordinates we write

$$f = f(x, y, z, t). \tag{a}$$

The *total differential* of f is the total change in f resulting from changes in x, y, z and t. Thus, using (a)

$$df = \frac{\partial f}{\partial x}dx + \frac{\partial f}{\partial y}dy + \frac{\partial f}{\partial z}dz + \frac{\partial f}{\partial t}dt.$$

Dividing through by dt

$$\frac{df}{dt} \equiv \frac{Df}{Dt} = \frac{\partial f}{\partial x}\frac{dx}{dt} + \frac{\partial f}{\partial y}\frac{dy}{dt} + \frac{\partial f}{\partial z}\frac{dz}{dt} + \frac{\partial f}{\partial t} \ . \tag{b}$$

However

$$\frac{dx}{dt} = u\ , \quad \frac{dy}{dt} = v\ , \quad \frac{dz}{dt} = w\ . \tag{c}$$

Substituting (c) into (b)

$$\frac{df}{dt} = \frac{Df}{Dt} = u\frac{\partial f}{\partial x} + v\frac{\partial f}{\partial y} + w\frac{\partial f}{\partial z} + \frac{\partial f}{\partial t} \ . \tag{1.23}$$

df / dt in the above is called the *total derivative*. It is also written as Df / Dt, to emphasize that it represents the change in f which results from changes in the four independent variables. It is also referred to as the *substantial derivative*. Note that the first three terms on the right hand side are associated with motion and are referred to as the *convective derivative*. The last term represents changes in f with respect to time and is called the *local derivative*. Thus

$$u\frac{\partial f}{\partial x} + v\frac{\partial f}{\partial y} + w\frac{\partial f}{\partial z} = \text{convective derivative}, \tag{d}$$

$$\frac{\partial f}{\partial t} = \text{local derivative}. \tag{e}$$

To appreciate the physical significance of (1.23), we apply it to the velocity component u. Setting $f = u$ in (1.23) gives

$$\frac{du}{dt} = \frac{Du}{Dt} = u\frac{\partial u}{\partial x} + v\frac{\partial u}{\partial y} + w\frac{\partial u}{\partial z} + \frac{\partial u}{\partial t} \ . \tag{1.24}$$

Following (d) and (e) format, (1.24) represents

$$u\frac{\partial u}{\partial x} + v\frac{\partial u}{\partial y} + w\frac{\partial w}{\partial z} = \text{convective acceleration in the } x\text{-direction}, \tag{f}$$

$$\frac{\partial u}{\partial t} = \text{local acceleration}. \tag{g}$$

Similarly, (1.23) can be applied to the y and z directions to obtain the corresponding total acceleration in these directions.

The three components of the total acceleration in the cylindrical coordinates r, θ, z are

$$\frac{dv_r}{dt} = \frac{Dv_r}{Dt} = v_r \frac{\partial v_r}{\partial r} + \frac{v_\theta}{r} \frac{\partial v_r}{\partial \theta} - \frac{v_\theta^2}{r} + v_z \frac{\partial v_r}{\partial z} + \frac{\partial v_r}{\partial t}, \tag{1.25a}$$

$$\frac{dv_\theta}{dt} = \frac{Dv_\theta}{Dt} = v_r \frac{\partial v_\theta}{\partial r} + \frac{v_\theta}{r} \frac{\partial v_\theta}{\partial \theta} + \frac{v_r v_\theta}{r} + v_z \frac{\partial v_\theta}{\partial z} + \frac{\partial v_\theta}{\partial t}, \tag{1.25b}$$

$$\frac{dv_z}{dt} = \frac{Dv_z}{Dt} = v_r \frac{\partial v_z}{\partial r} + \frac{v_\theta}{r} \frac{\partial v_z}{\partial \theta} + v_z \frac{\partial v_z}{\partial z} + \frac{\partial v_z}{\partial t}. \tag{1.25c}$$

Another example of total derivative is obtained by setting $f = T$ in (1.23) to obtain the total temperature derivative

$$\frac{dT}{dt} = \frac{DT}{Dt} = u \frac{\partial T}{\partial x} + v \frac{\partial T}{\partial y} + w \frac{\partial T}{\partial z} + \frac{\partial T}{\partial t}. \tag{1.26}$$

1.11 Units

SI units are used throughout this text. The basic units in this system are:

Length (L): meter (m).
Time (t): second (s).
Mass (m): kilogram (kg).
Temperature (T): kelvin (K).

Temperature on the *Celsius* scale is related to the *kelvin* scale by

$$T(^\circ C) = T(K) - 273.15. \tag{1.27}$$

Note that temperature difference on the two scales is identical. Thus, a change of one kelvin is equal to a change of one Celsius. This means that quantities that are expressed per unit kelvin, such as thermal conductivity, heat transfer coefficient, and specific heat, are numerically the same as per degree Celsius. That is, $W/m^2\text{-}K = W/m^2\text{-}°C$.

The basic units are used to derive units for other quantities. Force is measured in *newtons* (N). One newton is the force needed to accelerate a mass of one kilogram one meter per second per second:

$$\text{Force} = \text{mass} \times \text{acceleration},$$

$$N = kg - m/s^2.$$

Energy is measured in *joules* (J). One joule is the energy associated with a force of one newton moving a distance of one meter.

$$J = N \times m = kg - m^2/s^2.$$

Power is measured in *watts* (W). One watt is energy rate of one joule per second.

$$W = J/s = N \times m/s = kg - m^2/s^3.$$

1.12 Problem Solving Format

Convection problems lend themselves to a systematic solution procedure. The following basic format which builds on the work of Ver Planck and Teare [2] is used throughout the text.

(1) Observations. Study the situation, operation, process, design, etc. under consideration. Read the problem statement very carefully and note essential facts and features. Identify cueing information in the problem statement. Show a schematic diagram describing the situation. Where appropriate show the origin and coordinate axes.

(2) Problem Definition. Identify the key factors which must be determined so that a solution can be constructed. Distinguish between the question asked and the problem to be solved. Look for cues in the problem statement to construct a problem definition that cues a solution plan.

(3) Solution Plan. Identify the problem's basic laws and concepts.

(4) Plan Execution. This stage is carried out in four steps.

   (i) Assumptions. Model the problem by making simplifications and approximations. List all assumptions.

(ii) **Analysis.** Apply the basic laws identified in the solution plan. Carry out analysis in terms of symbols representing variables, parameters and constants rather than numerical values. Define all terms and give their units.

(iii) **Computations.** Execute the necessary computations and calculations to generate the desired numerical results.

(iv) **Checking.** Check each step of the solution as you proceed. Apply dimensional checks and examine limiting cases.

(5) **Comments.** Review your solution and comment on such things as the role of assumptions, the form of the solution, the number of governing parameters, etc.

Example 1.1: Heat Loss from Identical Triangles

Consider two identical triangles drawn on the surface of a flat plate as shown. The plate, which is maintained at uniform surface temperature T_S, is cooled by forced convection. The free stream temperature is T_∞. Under certain conditions the heat transfer coefficient varies with distance x from the leading edge of the plate according to

$$h(x) = \frac{C}{\sqrt{x}},$$

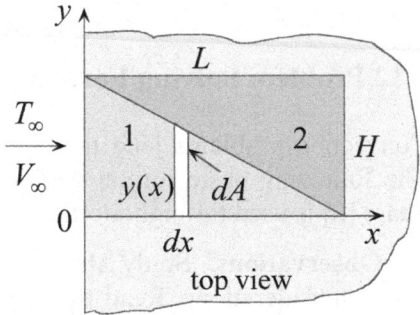

top view

where C is constant. Determine the ratio of the heat transfer rate from the two triangles, q_1/q_2.

(1) **Observations.** (i) Convection heat transfer from a surface can be determined using Newton's law of cooling. (ii) The local heat transfer coefficient varies along the plate. (iii) For each triangle the area of an element dx varies with distance along the plate. (iv) The total heat transfer rate can be determined by integration along the length of each triangle.

(2) **Problem Definition.** Determine the heat rate by convection from an element dx of each triangle.

(3) **Solution Plan.** Apply Newton's law of cooling to an element of each triangle and integrate over the area.

(4) **Plan Execution.**

(i) **Assumptions.** (1) Steady state, (2) one-dimensional variation of heat transfer coefficient, (3) uniform free stream temperature, (4) uniform surface temperature, and (5) negligible radiation.

(ii) **Analysis.** Of interest is the ratio of the total heat transfer rate from triangle 1 to that of triangle 2. Since both the heat transfer coefficient and area vary along each triangle, it follows that Newton's law of cooling should be applied to an element dA at a distance x from the leading edge:

$$dq = h(x)(T_s - T_\infty)dA,$$ (a)

where

dA = area of element, m^2
$h(x)$ = local heat transfer coefficient,
dq = rate of heat transfer from the element, W
T_s = surface temperature, $^\circ C$
T_∞ = free stream temperature, $^\circ C$
x = distance along plate, m

The local heat transfer coefficient is given by

$$h = \frac{C}{\sqrt{x}}.$$ (b)

Using the subscripts 1 and 2 to refer to triangles 1 and 2, respectively, the infinitesimal area dA for each triangle is given by

$$dA_1 = y_1(x)dx,$$ (c)

and

$$dA_2 = y_2(x)dx,$$ (d)

where

$y_1(x)$ = side of element in triangle 1, m
$y_2(x)$ = side of element in triangle 2, m

Similarity of triangles gives

$$y_1(x) = \frac{H}{L}(L - x),$$ (e)

$$y_2(x) = \frac{H}{L}x. \tag{f}$$

Substituting (e) into (c) and (f) into (d) gives

$$dA_1 = \frac{H}{L}(L - x)dx, \tag{g}$$

$$dA_2 = \frac{H}{L}xdx, \tag{h}$$

where

H = base of triangle, m
L = length of triangle, m

Substituting (b) and (g) into (a) and integrating from $x = 0$ to $x = L$, gives

$$q_1 = \int dq_1 = \int_0^L C(T_s - T_\infty)\frac{H}{L}\frac{L-x}{x^{1/2}}dx = C(T_s - T_\infty)\frac{H}{L}\int_0^L \frac{L-x}{x^{1/2}}dx.$$

Carrying out the integration yields

$$q_1 = (4/3)C(T_s - T_\infty)HL^{1/2}. \tag{i}$$

Similarly, substituting (b) and (h) into (a) and integrating from $x = 0$ to $x = L$ gives

$$q_2 = \int dq_2 = \int_0^L C(T_s - T_\infty)\frac{H}{L}\frac{x}{x^{1/2}}dx = C(T_s - T_\infty)\frac{H}{L}\int_0^L x^{1/2}dx.$$

Carrying out the integration yields

$$q_2 = (2/3)C(T_s - T_\infty)HL^{1/2}. \tag{j}$$

Taking the ratio of (i) and (j)

$$\frac{q_1}{q_2} = 2. \tag{k}$$

(iii) Checking. *Dimensional check*: Units of q_1 in equation (i) should be W. First, units of C are

$$C = \text{W/m}^{3/2} - {}^\circ\text{C}$$

Thus units of q_1 are

$$q_1 = C \, (\text{W/m}^{3/2}\text{-}{}^\circ\text{C})(T_s - T_\infty)({}^\circ\text{C})H(\text{m})L^{1/2} \, (\text{m}^{1/2}) = \text{W}$$

Since q_2 has the same form as q_1, it follows that units of q_2 in equation (j) are also correct.

Qualitative check: The result shows that the rate of heat transfer from triangle 1 is greater than that from triangle 2. This is expected since the heat transfer coefficient increases as the distance from the leading edge is decreased and triangle 1 has its base at $x = 0$ where h is maximum. According to (b), the heat transfer coefficient is infinite at $x = 0$.

(5) Comments. (i) Although the two triangles have the same area, the rate of heat transfer from triangle 1 is double that from triangle 2. Thus, orientation and proximity to the leading edge of a flat plate play an important role in determining the rate of heat transfer.

(ii) The same approach can be used to determine heat transfer for configurations other than rectangles, such as circles and ellipses.

REFERENCES

[1] Gad-el-Hak, M., Flow Physics, in *The MEMS Handbook*, M. Gad-el-Hak, ed., CRC Press, 2005.

[2] Ver Planck, D.W., and Teare Jr., B.R., *Engineering Analysis: An Introduction to Professional Method*, Wiley, New York, 1952.

PROBLEMS

1.1   Heat is removed from a rectangular surface by convection to an ambient fluid at T_∞. The heat transfer coefficient is h. Surface temperature is given by

$$T_s = \frac{A}{x^{1/2}},$$

where A is constant. Determine the steady state heat transfer rate from the plate.

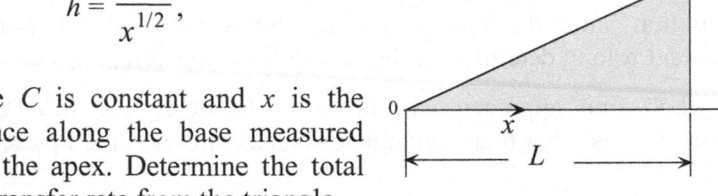

1.2   A right angle triangle is at a uniform surface temperature T_s. Heat is removed by convection to an ambient fluid at T_∞. The heat transfer coefficient h varies along the surface according to

$$h = \frac{C}{x^{1/2}},$$

where C is constant and x is the distance along the base measured from the apex. Determine the total heat transfer rate from the triangle.

1.3   A high intensity light bulb with surface heat flux $(q/A)_s$ is cooled by a fluid at T_∞. Sketch the fluid temperature profiles for three values of the heat transfer coefficient: h_1, h_2, and h_3, where $h_1 < h_2 < h_3$.

1.4   Explain why fanning gives a cool sensation.

1.5   A block of ice is submerged in water above the melting temperature. Explain why stirring the water accelerates the melting rate.

1.6   Consider steady state, incompressible, axisymmetric parallel flow in a tube of radius r_o. The axial velocity distribution for this flow is given by

$$u = 2\bar{u}\left(1 - \frac{r^2}{r_o^2}\right),$$

where $\bar{u}$ is the mean or average axial velocity. Determine the three components of the total acceleration for this flow.

1.7   Consider transient flow in the neighborhood of a vortex line where the velocity is in the tangential direction, given by

$$V(r,t) = \frac{\Gamma_o}{2\pi r}\left[1 - \exp\left(-\frac{r^2}{4vt}\right)\right].$$

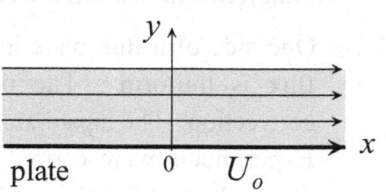

Here r is the radial coordinate, t is time, Γ_o is circulation (constant), and v is kinematic viscosity. Determine the three components of the total acceleration.

1.8   An infinitely large plate is suddenly moved parallel to its surface with a velocity U_o. The resulting transient velocity distribution of the surrounding fluid is given by

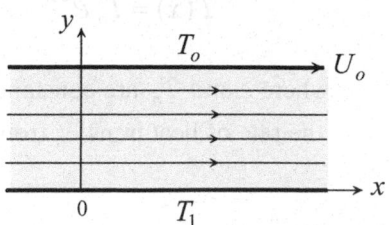

$$u = U_o\left[1 - (2/\sqrt{\pi})\int_0^{\eta} \exp(-\eta^2)d\eta\right],$$

where the variable η is defined as

$$\eta(x,t) = \frac{y}{2\sqrt{vt}}.$$

Here t is time, y is the vertical coordinate and v is kinematic viscosity. Note that streamlines for this flow are parallel to the plate. Determine the three components of the total acceleration.

1.9   Consider two parallel plates with the lower plate stationary and the upper plate moving with a velocity U_o. The lower plate is maintained at temperature T_1 and the upper plate at T_o. The axial velocity of the fluid for steady state and parallel streamlines is given by

$$u = U_o \frac{y}{H},$$

where H is the distance between the two plates. Temperature distribution is given by

$$T = \frac{\mu U_o^2}{2kH} \left[y - \frac{y^2}{H} \right] + (T_o - T_1) \frac{y}{H} + T_1,$$

where k is thermal conductivity and μ is viscosity. Determine the total temperature derivative.

1.10 One side of a thin plate is heated electrically such that surface heat flux is uniform. The opposite side of the plate is cooled by convection. The upstream velocity is V_∞ and temperature is T_∞. Experiments were carried out at two upstream velocities, $V_{\infty 1}$ and $V_{\infty 2}$ where $V_{\infty 2} > V_{\infty 1}$. All other conditions were unchanged. The heat transfer coefficient was found to increase as the free stream velocity is increased. Sketch the temperature profile $T(y)$ of the fluid corresponding to the two velocities.

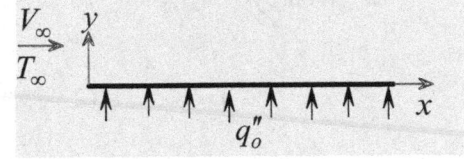

1.11 Heat is removed from an L-shaped area by convection. The heat transfer coefficient is h and the ambient temperature is T_∞. Surface temperature varies according to

$$T(x) = T_o e^{cx},$$

where c and T_o are constants. Determine the rate of heat transfer from the area.

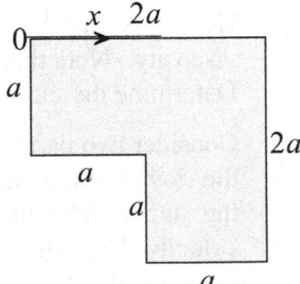

2

Differential Formulation of the Basic Laws

2.1 Introduction

In a moving fluid the three fundamental laws, conservation of mass, momentum, and energy, must be satisfied at *every point* in the domain. Thus the first step is to formulate (cast) the three laws in a form that satisfies this condition. This is accomplished by applying each law to a differential (infinitesimal) element. Following this approach, each law is described by a partial differential equation. Differential formulation of the three laws will be presented using rectangular coordinates. The corresponding forms in cylindrical and spherical coordinates will be stated without details.

2.2 Flow Generation

Since fluid motion is central to convection heat transfer we will be concerned with two common flow classifications:

(a) *Forced convection*. Fluid motion is generated mechanically through the use of a fan, blower, nozzle, jet, etc.. Fluid motion relative to a surface can also be obtained by moving an object, such as a missile, through a fluid.

(b) *Free (natural) convection*. Fluid motion is generated by gravitational field. However, the presence of a gravitational field is not sufficient to set a fluid in motion. Fluid density change is also required for free convection to occur. In free convection, density variation is primarily due to temperature changes.

2.3 Laminar vs. Turbulent Flow

One classification of fluid flow and convection heat transfer is based on certain flow characteristics. If the flow is characterized by random fluctuations in quantities such as velocity, temperature, pressure, and density, it is referred to as *turbulent*. On the other hand, in the absence of such fluctuations the flow is called *laminar*. These two basic flow patterns are illustrated in Fig.2.1. Since flow and heat transfer characteristics differ significantly for these two modes, it is essential to establish if a flow is laminar, turbulent, or mixed. Transition from laminar to turbulent flow takes place at an experimentally determined value of the Reynolds number known as the *transition Reynolds number*, Re_t. The magnitude of this number depends primarily on flow geometry but can be influenced by surface roughness, pressure gradient and other factors. For uniform flow over a semi-infinite flat plate $Re_t = V_\infty x_t / v \approx 500{,}000$, where V_∞ is the free stream velocity, x_t is the distance along the plate, measured from the leading edge to where transition occurs, and v is the kinematic viscosity. On the other hand, for flow through tubes $Re_t = \bar{u}D/v \approx 2300$, where D is tube diameter and $\bar{u}$ is the mean fluid velocity.

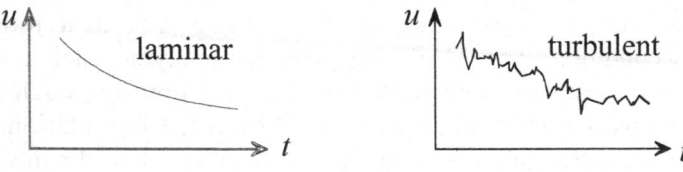

Fig. 2.1

2.4   Conservation of Mass: The Continuity Equation

2.4.1 Cartesian Coordinates

Consider an element $dxdydz$ as a control volume in the flow field of Fig.2.2a. For simplicity, the z-direction is not shown. The element is enlarged in Fig. 2.2b showing the flow of mass through it. Conservation of mass, applied to the element, states that

> Rate of mass added to element - Rate of mass removed from element = Rate of mass change within element

$$(2.1)$$

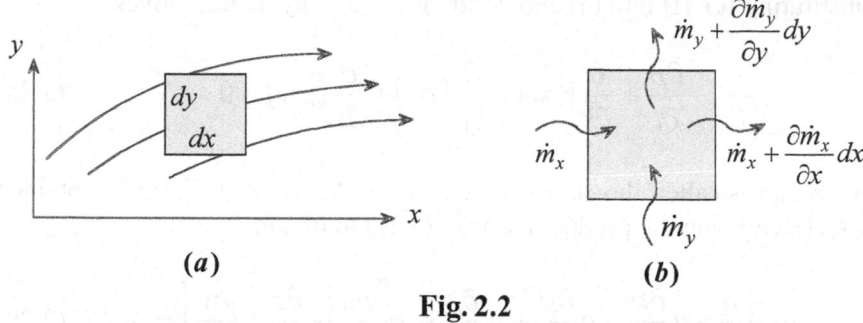

Fig. 2.2

Assuming continuum and using the notation of Fig. 2.2b, equation (2.1) is expressed as

$$\delta \dot{m}_x + \delta \dot{m}_y + \delta \dot{m}_z - \left[\delta \dot{m}_x + \frac{\partial (\delta \dot{m}_x)}{\partial x} dx \right]$$

$$- \left[\delta \dot{m}_y + \frac{\partial (\delta \dot{m}_y)}{\partial y} dy \right] - \left[\delta \dot{m}_z + \frac{\partial (\delta \dot{m}_z)}{\partial z} dz \right] = \frac{\partial (\delta m)}{\partial t},$$

(a)

where

$\delta \dot{m}_x$ = mass flow rate entering element in the x-direction
$\delta \dot{m}_y$ = mass flow rate entering element in the y-direction
$\delta \dot{m}_z$ = mass flow rate entering element in the z-direction
δm = mass within element

To express (a) in terms of fluid density and velocity, we utilize the one-dimensional flow rate equation

$$m = \rho V A,$$

(b)

where V is the velocity normal to the flow area A, and ρ is density. It should be emphasized that in this form both ρ and V must be uniform over the flow area A. Applying (b) to the element, gives

$$\delta \dot{m}_x = \rho u dy dz,$$

(c)

$$\delta \dot{m}_y = \rho v dx dz,$$

(d)

$$\delta \dot{m}_z = \rho w dx dy,$$

(e)

where u, v and w are the velocity components in the x, y and z-direction, respectively. The mass, δm, within the element is given by

$$\delta m = \rho \, dx dy dz.$$

(f)

Substituting (c)–(f) into (a) and dividing through by *dxdydz*, gives

$$\frac{\partial \rho}{\partial t} + \frac{\partial}{\partial x}(\rho u) + \frac{\partial}{\partial y}(\rho v) + \frac{\partial}{\partial z}(\rho w) = 0. \tag{2.2a}$$

This result is called the *continuity equation*. An alternate form is obtained by differentiating the product terms in (2.2a) to obtain

$$\frac{\partial \rho}{\partial t} + u\frac{\partial \rho}{\partial x} + v\frac{\partial \rho}{\partial y} + w\frac{\partial \rho}{\partial z} + \rho\left[\frac{\partial u}{\partial x} + \frac{\partial v}{\partial y} + \frac{\partial w}{\partial z}\right] = 0. \tag{2.2b}$$

Note that the first four terms in (2.2b) represent the total derivative of ρ and the last three terms represent the divergence of the velocity vector $\vec{V}$. Thus, (2.2b) is rewritten as

$$\frac{D\rho}{Dt} + \rho \nabla \cdot \vec{V} = 0. \tag{2.2c}$$

An alternate form of (2.2c) is

$$\frac{\partial \rho}{\partial t} + \nabla \cdot \rho\vec{V} = 0. \tag{2.2d}$$

For constant density (incompressible fluid) the total derivative in (2.2d) vanishes. That is

$$\frac{D\rho}{Dt} = 0.$$

Substituting into (2.2d) gives

$$\nabla \cdot \vec{V} = 0. \tag{2.3}$$

Equation (2.3) is the continuity equation for incmopressible fluid.

2.4.2 Cylindrical Coordinates

Applying (2.1) to an infinitesimal element $rd\theta\,drdz$ in the cylindrical coordinates shown in Fig. 2.3, gives

$$\frac{\partial \rho}{\partial t} + \frac{1}{r}\frac{\partial}{\partial r}(\rho r v_r) + \frac{1}{r}\frac{\partial}{\partial \theta}(\rho v_\theta) + \frac{\partial}{\partial z}(\rho v_z) = 0,$$

(2.4)

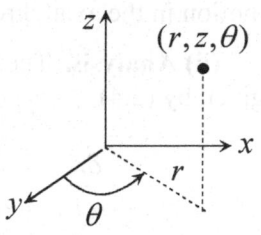

Fig. 2.3

where v_r, v_θ and v_z are the velocity components in r, θ and z-direction, respectively.

2.4.3 Spherical Coordinates

Applying (2.1) to an infinitesimal element $r d\theta r d\phi dr$ in the spherical coordinates shown in Fig. 2.4 and following the procedure of Section 2.4.1, gives

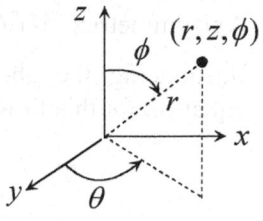

$$\frac{\partial \rho}{\partial t} + \frac{1}{r^2}\frac{\partial}{\partial r}(\rho r^2 v_r) + \frac{1}{r \sin \theta}\frac{\partial}{\partial \theta}(\rho v_\theta \sin \theta)$$

$$+ \frac{1}{r \sin \theta}\frac{\partial}{\partial \phi}(\rho v_\phi) = 0.$$

(2.5)

Fig. 2.4

Example 2.1: Fluid in Angular Motion

A shaft rotates concentrically inside a tube. The annular space between the shaft and the tube is filled with incompressible fluid. Neglecting fluid motion in the axial direction z, write the continuity equation for this case.

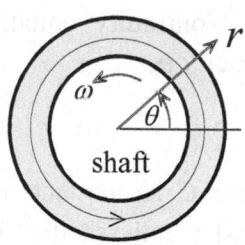

(1) Observations. (i) Use cylindrical coordinates. (ii) No variation in the axial and angular directions. (iii) The fluid is incompressible (constant density).

(2) Problem Definition.   Simplify the three-dimensional continuity equation for this flow.

(3) Solution Plan. Apply the continuity in cylindrical coordinates.

(4) Plan Execution.

   (i) Assumptions. (1) Continuum, (2) incompressible fluid, (3) no motion in the axial direction, and (4) shaft and tube are concentric.

(ii) Analysis. The continuity equation in cylindrical coordinates is given by (2.4)

$$\frac{\partial \rho}{\partial t} + \frac{1}{r}\frac{\partial}{\partial r}\left(\rho r v_r\right) + \frac{1}{r}\frac{\partial}{\partial \theta}\left(\rho v_\theta\right) + \frac{\partial}{\partial z}\left(\rho v_z\right) = 0. \tag{2.4}$$

This equation is simplified based on:

Incompressible fluid: ρ is constant, $\partial \rho / dt = 0$.
No axial velocity: $v_z = 0$.
Axisymmetric: $\partial / \partial \theta = 0$.

Introducing the above simplifications into (2.4), gives the continuity equation for this flow

$$\frac{\partial}{\partial r}\left(r v_r\right) = 0. \tag{a}$$

(iii) Checking. *Dimensional check*: Each term in (2.4) has units of density per unit time.

(5) Comments. (i) Equations (a) and (d) are valid for transient as well as steady state as long as the fluid is incompressible.

(ii) Continuity equation (a) can be integrated to give the radial velocity v_r

$$r v_r = C, \tag{b}$$

where C is constant or a function of θ. Since the radial velocity v_r vanishes at the shaft's surface, if follows from (b) that

$$C = 0. \tag{c}$$

Equation (b) gives

$$v_r = 0. \tag{d}$$

(iii) Since $v_r = 0$ everywhere in the flow field, it follows that the streamlines are concentric circles.

2.5 Conservation of Momentum: The Navier-Stokes Equations of Motion

2.5.1 Cartesian Coordinates

We note first that momentum is a vector quantity. Thus conservation of momentum (Newton's law of motion) provides three equations, one in each of the three coordinates. Application of Newton's law of motion to the element shown in Fig. 2.5, gives

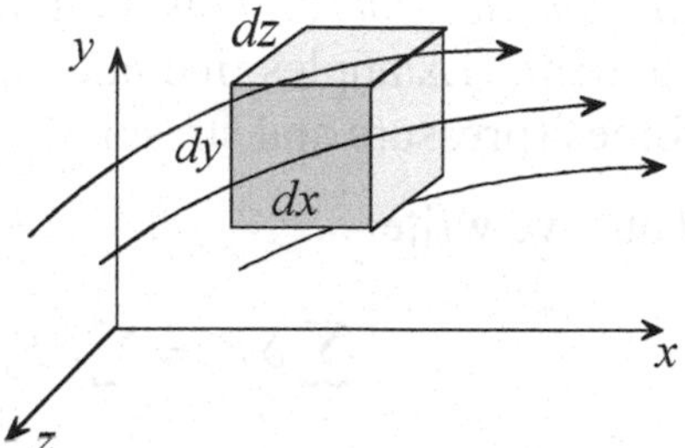

$$\sum \delta \vec{F} = (\delta m)\vec{a}, \qquad (a)$$

Fig. 2.5

where

$\vec{a}$ = acceleration of the element

$\delta \vec{F}$ = external force acting on the element

δm = mass of the element

Application of (a) in the x-direction, gives

$$\sum \delta F_x = (\delta m)a_x. \qquad (b)$$

The mass of the element is

$$\delta m = \rho dxdydz. \qquad (c)$$

Based on the assumption of continuum, the total acceleration of the element in the x-direction, a_x, is

$$a_x = \frac{du}{dt} = \frac{Du}{Dt} = u\frac{\partial u}{\partial x} + v\frac{\partial u}{\partial y} + w\frac{\partial u}{\partial z} + \frac{\partial u}{\partial t}. \qquad (d)$$

Substituting (c) and (d) into (b)

$$\sum \delta F_x = \rho \frac{Du}{Dt} dxdydz. \qquad (e)$$

Next we determine the sum of all external forces acting on the element in the x-direction. We classify external forces as:

(i) *Body force*. This is a force that acts on every particle of the material or element. Examples include gravity and magnetic forces.

(ii) *Surface force*. This is a force that acts on the surface of the element. Examples include tangential forces (shear) and normal forces (pressure and stress).

Thus we write

$$\sum \delta F_x = \sum \delta F_x\big)_{body} + \sum \delta F_x\big)_{surface}. \qquad (f)$$

We consider gravity as the only body force acting on the element. The x-component of this force is

$$\sum \delta F_x\big)_{body} = \rho g_x dx dy dz, \qquad (g)$$

where g_x is gravitational acceleration component in the plus x-direction.

Next we formulate an expression for the surface forces in the x-direction. These forces are shown in Fig. 2.6. They are:

σ_{xx} = normal stress on surface $dydz$
τ_{yx} = shearing (tangential) stress on surface $dxdz$
τ_{zx} = shearing (tangential) stress on surface $dxdy$

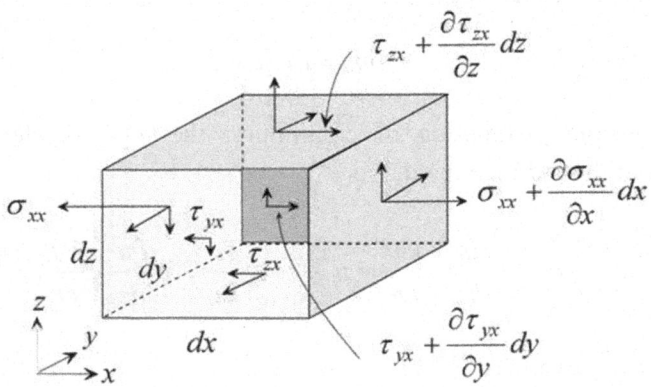

Fig. 2.6

Summing up all the x-component forces shown in Fig. 2.6 gives

$$\sum \delta F_x \Big)_{surface} = \left(\frac{\partial \sigma_{xx}}{\partial x} + \frac{\partial \tau_{yx}}{\partial y} + \frac{\partial \tau_{zx}}{\partial z} \right) dx dy dz .$$ (h)

Similar expressions are obtained for surface forces in the y and z-directions. Substituting (f), (g) and (h) into (e), gives the x-direction equation

$$\rho \frac{Du}{Dt} = \rho g_x + \frac{\partial \sigma_{xx}}{\partial x} + \frac{\partial \tau_{yx}}{\partial y} + \frac{\partial \tau_{zx}}{\partial z} .$$ (2.6a)

Similarly, applying Newton's law of motion in the y and z-directions gives the two corresponding momentum equations. By analogy with (2.6a), these equations are

$$\rho \frac{Dv}{Dt} = \rho g_y + \frac{\partial \tau_{xy}}{\partial x} + \frac{\partial \sigma_{yy}}{\partial y} + \frac{\partial \tau_{zy}}{\partial z} ,$$ (2.6b)

and

$$\rho \frac{Dw}{Dt} = \rho g_z + \frac{\partial \tau_{xz}}{\partial x} + \frac{\partial \tau_{yz}}{\partial y} + \frac{\partial \sigma_{zz}}{\partial z} .$$ (2.6c)

Equations (2.6a), (2.6b), and (2.6c) are general in nature since they are based on fundamental laws of motion. The only restriction is the assumption of continuum. Examination of these equations shows that they contain 13 unknowns: u, v, w, ρ, σ_{xx}, σ_{yy}, σ_{zz}, τ_{xy}, τ_{yx}, τ_{xz}, τ_{zx}, τ_{yz} and τ_{zy}. Application of the moment of momentum principle to a differential element gives

$$\tau_{xy} = \tau_{yx} , \quad \tau_{xz} = \tau_{zx} , \quad \tau_{yz} = \tau_{zy} .$$ (i)

To further reduce the number of unknown variables, an important restriction is introduced. The basic idea is to relate normal and shearing stresses to the velocity field. This is accomplished through the introduction of experimentally based relations known as *constitutive equations*. These equations are [1]:

$$\tau_{xy} = \tau_{yx} = \mu \left(\frac{\partial v}{\partial x} + \frac{\partial u}{\partial y} \right) ,$$ (2.7a)

$$\tau_{xz} = \tau_{zx} = \mu\left(\frac{\partial w}{\partial x} + \frac{\partial u}{\partial z}\right), \qquad\qquad (2.7b)$$

$$\tau_{yz} = \tau_{zy} = \mu\left(\frac{\partial v}{\partial z} + \frac{\partial w}{\partial y}\right), \qquad\qquad (2.7c)$$

$$\sigma_{xx} = -p + 2\mu\frac{\partial u}{\partial x} - \frac{2}{3}\mu\nabla\cdot\vec{V}, \qquad\qquad (2.7d)$$

$$\sigma_{yy} = -p + 2\mu\frac{\partial v}{\partial y} - \frac{2}{3}\mu\nabla\cdot\vec{V}, \qquad\qquad (2.7e)$$

$$\sigma_{zz} = -p + 2\mu\frac{\partial w}{\partial z} - \frac{2}{3}\mu\nabla\cdot\vec{V}. \qquad\qquad (2.7f)$$

where μ is a property called viscosity and p is the hydrostatic pressure. A fluid that obeys (2.7) is referred to as *Newtonian* fluid. Examples of Newtonian fluids include air, water and most oils. Fluids such as tar, honey and polymers are called non-Newtonian. Substituting (2.7) into (2.6), we obtain

$$\rho\frac{Du}{Dt} = \rho g_x - \frac{\partial p}{\partial x} + \frac{\partial}{\partial x}\left[\mu\left(2\frac{\partial u}{\partial x} - \frac{2}{3}\nabla\cdot\vec{V}\right)\right]$$

$$+ \frac{\partial}{\partial y}\left[\mu\left(\frac{\partial u}{\partial y} + \frac{\partial v}{\partial x}\right)\right] + \frac{\partial}{\partial z}\left[\mu\left(\frac{\partial w}{\partial x} + \frac{\partial u}{\partial z}\right)\right], \qquad (2.8x)$$

$$\rho\frac{Dv}{Dt} = \rho g_y - \frac{\partial p}{\partial y} + \frac{\partial}{\partial y}\left[\mu\left(2\frac{\partial v}{\partial y} - \frac{2}{3}\nabla\cdot\vec{V}\right)\right]$$

$$+ \frac{\partial}{\partial z}\left[\mu\left(\frac{\partial v}{\partial z} + \frac{\partial w}{\partial y}\right)\right] + \frac{\partial}{\partial x}\left[\mu\left(\frac{\partial u}{\partial y} + \frac{\partial v}{\partial x}\right)\right], \qquad (2.8y)$$

$$\rho \frac{Dw}{Dt} = \rho g_z - \frac{\partial p}{\partial z} + \frac{\partial}{\partial z}\left[\mu\left(2\frac{\partial w}{\partial z} - \frac{2}{3}\nabla \cdot \vec{V}\right)\right]$$

$$+ \frac{\partial}{\partial x}\left[\mu\left(\frac{\partial w}{\partial x} + \frac{\partial u}{\partial z}\right)\right] + \frac{\partial}{\partial y}\left[\mu\left(\frac{\partial \upsilon}{\partial z} + \frac{\partial w}{\partial y}\right)\right].$$

$$(2.8z)$$

The following observations are made regarding (2.8):

(1) These equations are known as the *Navier-Stokes equations of motion*. They apply to Newtonian fluids.

(2) The number of unknowns in the three equations are 6: u, υ, w, p, ρ, and μ.

(3) The assumptions leading to (2.8) are: continuum and Newtonian fluid.

Expressing equations (2.8x), (2.8y) and (2.8z) in a vector form, gives

$$\rho \frac{D\vec{V}}{Dt} = \rho\vec{g} - \nabla p + \frac{4}{3}\nabla\left(\mu\nabla \cdot \vec{V}\right) + \nabla\left(\vec{V} \cdot \nabla\mu\right)$$

$$- \vec{V}\nabla^2\mu + \nabla\mu \times \left(\nabla \times \vec{V}\right) - \left(\nabla \cdot \vec{V}\right)\nabla\mu - \nabla \times \left(\nabla \times \mu\vec{V}\right).$$

$$(2.8)$$

Equation (2.8) is now applied to two simplified cases:

(i) Constant viscosity. For this case

$$\nabla\mu = 0, \tag{j}$$

and

$$\nabla \times \left(\nabla \times \mu\vec{V}\right) = \nabla\left(\nabla \cdot \mu\vec{V}\right) - \nabla \cdot \nabla\mu\vec{V} = \mu\nabla\left(\nabla \cdot \vec{V}\right) - \mu\nabla^2\vec{V}. \tag{k}$$

Substituting (j) and (k) into (2.8)

$$\rho \frac{D\vec{V}}{Dt} = \rho\vec{g} - \nabla p + \frac{1}{3}\mu\nabla\left(\nabla \cdot \vec{V}\right) + \mu\nabla^2\vec{V}. \tag{2.9}$$

Thus (2.9) is valid for: (1) continuum, (2) Newtonian fluid, and (3) constant viscosity.

(ii) Constant viscosity and density. The continuity equation for constant density fluid is given by equation (2.3)

$$\nabla \cdot \vec{V} = 0 .$$ (2.3)

Substituting (2.3) into (2.9) gives

$$\rho \frac{D\vec{V}}{Dt} = \rho \vec{g} - \nabla \vec{p} + \mu \nabla^2 \vec{V} .$$ (2.10)

Equation (2.10) is valid for: (1) continuum, (2) Newtonian fluid, (3) constant viscosity and (4) constant density. Note that this vector equation represents the three components of the Navier-Stokes equations of motion. These three x, y, and z components are

$$\rho \left(\frac{\partial u}{\partial t} + u \frac{\partial u}{\partial x} + v \frac{\partial u}{\partial y} + w \frac{\partial u}{\partial z} \right) = \rho g_x - \frac{\partial p}{\partial x} + \mu \left(\frac{\partial^2 u}{\partial x^2} + \frac{\partial^2 u}{\partial y^2} + \frac{\partial^2 u}{\partial z^2} \right),$$

(2.10x)

$$\rho \left(\frac{\partial v}{\partial t} + u \frac{\partial v}{\partial x} + v \frac{\partial v}{\partial y} + w \frac{\partial v}{\partial z} \right) = \rho g_y - \frac{\partial p}{\partial y} + \mu \left(\frac{\partial^2 v}{\partial x^2} + \frac{\partial^2 v}{\partial y^2} + \frac{\partial^2 v}{\partial z^2} \right),$$

(2.10y)

$$\rho \left(\frac{\partial w}{\partial t} + u \frac{\partial w}{\partial x} + v \frac{\partial w}{\partial y} + w \frac{\partial w}{\partial z} \right) = \rho g_z - \frac{\partial p}{\partial z} + \mu \left(\frac{\partial^2 w}{\partial x^2} + \frac{\partial^2 w}{\partial y^2} + \frac{\partial^2 w}{\partial z^2} \right).$$

(2.10z)

2.5.2 Cylindrical Coordinates

Applying Newton's law of motion to an infinitesimal element $r d\theta \, drdz$ in the cylindrical coordinates shown in Fig. 2.3 and following the procedure of Section 2.5.1, gives the three Navier-Stokes equations in cylindrical coordinates. We limit the result to the following case:

(1) Continuum, (2) Newtonian fluid, (3) constant viscosity, and (4) constant density. The r, θ, and z components for this case are

$$\rho\left(v_r\frac{\partial v_r}{\partial r}+\frac{v_\theta}{r}\frac{\partial v_r}{\partial\theta}-\frac{v_\theta^2}{r}+v_z\frac{\partial v_r}{\partial z}+\frac{\partial v_r}{\partial t}\right)=$$

$$\rho g_r-\frac{\partial p}{\partial r}+\mu\left[\frac{\partial}{\partial r}\left(\frac{1}{r}\frac{\partial}{\partial r}(rv_r)\right)+\frac{1}{r^2}\frac{\partial^2 v_r}{\partial\theta^2}-\frac{2}{r^2}\frac{\partial v_\theta}{\partial\theta}+\frac{\partial^2 v_r}{\partial z^2}\right],$$

(2.11r)

$$\rho\left(v_r\frac{\partial v_\theta}{\partial r}+\frac{v_\theta}{r}\frac{\partial v_\theta}{\partial\theta}+\frac{v_r v_\theta}{r}+v_z\frac{\partial v_\theta}{\partial z}+\frac{\partial v_\theta}{\partial t}\right)=$$

$$\rho g_\theta-\frac{1}{r}\frac{\partial p}{\partial\theta}+\mu\left[\frac{\partial}{\partial r}\left(\frac{1}{r}\frac{\partial}{\partial r}(rv_\theta)\right)+\frac{1}{r^2}\frac{\partial^2 v_\theta}{\partial\theta^2}+\frac{2}{r^2}\frac{\partial v_r}{\partial\theta}+\frac{\partial^2 v_\theta}{\partial z^2}\right],$$

(2.11θ)

$$\rho\left(v_r\frac{\partial v_z}{\partial r}+\frac{v_\theta}{r}\frac{\partial v_z}{\partial\theta}+v_z\frac{\partial v_z}{\partial z}+\frac{\partial v_z}{\partial t}\right)=$$

$$\rho g_z-\frac{\partial p}{\partial z}+\mu\left[\frac{1}{r}\frac{\partial}{\partial r}\left(r\frac{\partial v_z}{\partial r}\right)+\frac{1}{r^2}\frac{\partial^2 v_z}{\partial\theta^2}+\frac{\partial^2 v_z}{\partial z^2}\right].$$

(2.11z)

2.5.3 Spherical Coordinates

Applying Newton's law of motion to an infinitesimal element $rd\theta rd\phi dr$ in the spherical coordinates shown in Fig. 2.4 and following the procedure of Section 2.5.1, gives the three Navier-Stokes equations in spherical coordinates. We limit the result to the following case:

Continuum, (2) Newtonian fluid, (3) constant viscosity, and (4) constant density. The r, θ, and ϕ components for this case are

$$\rho\left(v_r\frac{\partial v_r}{\partial r}+\frac{v_\theta}{r}\frac{\partial v_r}{\partial\theta}+\frac{v_\phi}{r\sin\theta}\frac{\partial v_r}{\partial\phi}-\frac{v_\theta^2+v_\phi^2}{r}+\frac{\partial v_r}{\partial t}\right)=$$

$$\rho g_r-\frac{\partial p}{\partial r}+\mu\left(\nabla^2 v_r-\frac{2v_r}{r^2}-\frac{2}{r^2}\frac{\partial v_\theta}{\partial\theta}-\frac{2v_\theta\cot\theta}{r^2}-\frac{2}{r^2\sin\theta}\frac{\partial v_\phi}{\partial\phi}\right),$$

(2.12r)

$$\rho \left(v_r \frac{\partial v_\theta}{\partial r} + \frac{v_\theta}{r} \frac{\partial v_\theta}{\partial \theta} + \frac{v_\phi}{r \sin \theta} \frac{\partial v_\theta}{\partial \phi} + \frac{v_r v_\theta}{r} - \frac{v_\phi^2 \cot \theta}{r} + \frac{\partial v_\theta}{\partial t} \right) =$$

$$\rho g_\theta - \frac{1}{r} \frac{\partial p}{\partial \theta} + \mu \left(\nabla^2 v_\theta + \frac{2}{r^2} \frac{\partial v_r}{\partial \theta} - \frac{v_\theta}{r^2 \sin^2 \theta} - \frac{2 \cos \theta}{r^2 \sin^2 \theta} \frac{\partial v_\phi}{\partial \phi} \right),$$

$$(2.12\theta)$$

$$\rho \left(v_r \frac{\partial v_\phi}{\partial r} + \frac{v_\theta}{r} \frac{\partial v_\phi}{\partial \theta} + \frac{v_\phi}{r \sin \theta} \frac{\partial v_\phi}{\partial \phi} + \frac{v_\phi v_r}{r} + \frac{v_\theta v_\phi}{r} \cot \theta + \frac{\partial v_\theta}{\partial t} \right) = \rho g_\phi$$

$$- \frac{1}{r \sin \theta} \frac{\partial p}{\partial \phi} + \mu \left(\nabla^2 v_\phi - \frac{v_\phi}{r^2 \sin^2 \theta} + \frac{2}{r^2 \sin^2 \theta} \frac{\partial v_r}{\partial \phi} + \frac{2 \cos \theta}{r^2 \sin^2 \theta} \frac{\partial v_\theta}{\partial \phi} \right).$$

$$(2.12\phi)$$

Note that in equations (2.12) the operator ∇ in spherical coordinates is defined as

$$\nabla^2 = \frac{1}{r^2} \frac{\partial}{\partial r} \left(r^2 \frac{\partial}{\partial r} \right) + \frac{1}{r^2 \sin \theta} \frac{\partial}{\partial \theta} \left(\sin \theta \frac{\partial}{\partial \theta} \right) + \frac{1}{r^2 \sin^2 \theta} \frac{\partial^2}{\partial \phi^2}. \quad (2.13)$$

Example 2.2: Thin Liquid Film Flow over an Inclined Surface

A thin liquid film flows axially down an inclined plane. Consider the example of incompressible, steady flow with parallel streamlines. Write the Navier-Stokes equations of motion for this flow.

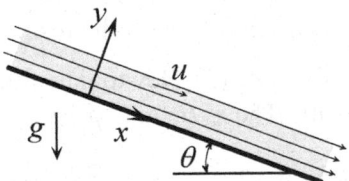

(1) Observations. (i) The flow is due to gravity. (ii) For parallel streamlines the vertical component $v = 0$. (iii) Pressure at the free surface is uniform (atmospheric). (iv) The component of gravity in the direction tangent to the surface causes the fluid to flow downwards. (v) The geometry is Cartesian.

(2) Problem Definition. Determine the x and y components of the Navier-Stokes equations of motion for the flow under consideration.

(3) Solution Plan. Start with the Navier-Stokes equations of motion in Cartesian coordinates and simplify them for this special case.

(4) Plan Execution.

(i) **Assumptions.** (1) Continuum, (2) Newtonian fluid, (3) steady, (4) flow is in the x-direction, (5) constant properties, (6) uniform ambient pressure, and (7) parallel streamlines.

(ii) **Analysis.** Start with the Navier Stokes equations of motion in Cartesian coordinates for constant properties, equations (2.10x) and (2.10y)

$$\rho\left(\frac{\partial u}{\partial t} + u\frac{\partial u}{\partial x} + v\frac{\partial u}{\partial y} + w\frac{\partial u}{\partial z}\right) = \rho g_x - \frac{\partial p}{\partial x} + \mu\left(\frac{\partial^2 u}{\partial x^2} + \frac{\partial^2 u}{\partial y^2} + \frac{\partial^2 u}{\partial z^2}\right),$$

$$(2.10x)$$

$$\rho\left(\frac{\partial v}{\partial t} + u\frac{\partial v}{\partial x} + v\frac{\partial v}{\partial y} + w\frac{\partial v}{\partial z}\right) = \rho g_y - \frac{\partial p}{\partial y} + \mu\left(\frac{\partial^2 v}{\partial x^2} + \frac{\partial^2 v}{\partial y^2} + \frac{\partial^2 v}{\partial z^2}\right).$$

$$(2.10y)$$

The two gravitational components are

$$g_x = g\sin\theta, \quad g_y = -g\cos\theta. \tag{a}$$

Based on the above assumptions, these equations are simplified as follows:

Steady state:
$$\frac{\partial u}{\partial t} = \frac{\partial v}{\partial t} = 0. \tag{b}$$

Axial flow (x-direction only):
$$w = \frac{\partial}{\partial z} = 0. \tag{c}$$

Parallel flow:
$$v = 0. \tag{d}$$

Substituting (a)-(d) into (2.10x) and (2.10y), gives

$$\rho u\frac{\partial u}{\partial x} = \rho g\sin\theta - \frac{\partial p}{\partial x} + \mu\left(\frac{\partial^2 u}{\partial x^2} + \frac{\partial^2 u}{\partial y^2}\right), \tag{e}$$

and

$$0 = -\rho g \cos\theta - \frac{\partial p}{\partial y} \; . \tag{f}$$

The x-component (e) can be simplified further using the continuity equation for incompressible flow, equation (2.3)

$$\nabla \cdot \vec{V} = \frac{\partial u}{\partial x} + \frac{\partial v}{\partial y} + \frac{\partial w}{\partial z} = 0 \; . \tag{g}$$

Substituting (c) and (d) into (g), gives

$$\frac{\partial u}{\partial x} = 0 \; . \tag{h}$$

Using (h) into (e) gives the x-component

$$\rho g \sin\theta - \frac{\partial p}{\partial x} + \mu \frac{\partial^2 u}{\partial y^2} = 0. \tag{i}$$

Integrating (f) with respect to y

$$p = -(\rho g \cos\theta) y + f(x), \tag{j}$$

where $f(x)$ is constant of integration.   At the free surface, $y = H$, the pressure is uniform equal to p_∞. Therefore, setting $y = H$ in (j) gives

$$f(x) = p_\infty + \rho g H \cos\theta. \tag{k}$$

Substituting (k) into (j) gives the pressure solution

$$p = \rho g (H - y)\cos\theta + p_\infty. \tag{l}$$

Differentiating (l) with respect to x gives

$$\frac{\partial p}{\partial x} = 0. \tag{m}$$

Substituting (m) into (i) gives the x-component of the Navier-Stokes equations

$$\rho g \sin\theta + \mu \frac{d^2 u}{dy^2} = 0. \tag{n}$$

(iii) Checking. *Dimensional check*: Each term of the y-component equation (f) must have the same units:

$$\rho g \cos\theta = (kg/m^3)(m/s^2) = kg/m^2\text{-}s^2.$$

$$\frac{\partial p}{\partial y} = (N/m^2)/m = N/m^3 = (kg - m/s^2)/m^3 = kg/m^2\text{-}s^2.$$

Similarly, units of the x-component equation (n) must also be consistent

$$\rho g \sin\theta = kg/m^2\text{-}s^2.$$

$$\mu \frac{d^2 u}{dy^2} = (kg/m - s)(m/s)/(m^2) = kg/m^2\text{-}s^2.$$

Limiting check: For the special case of zero gravity the fluid will not flow. That is, $u = 0$. Setting $g = 0$ in (n) gives

$$\frac{d^2 u}{dy^2} = 0. \tag{o}$$

It can be shown that the solution to (o) gives $u = 0$.

(5) Comments. (i) For two-dimensional incompressible parallel flow, the momentum equations are significantly simplified because the vertical velocity v, vanishes.

(ii) The flow is one-dimensional since u does not change with x and is a function of y only.

2.6 Conservation of Energy: The Energy Equation

2.6.1 Formulation: Cartesian Coordinates

Consider an element *dxdydz* as a control volume in the flow field of Fig. 2.7. Fluid enters and leaves the element through its six surfaces. We introduce the principle of conservation of energy (first law of thermodynamics). We begin with the statement

Energy cannot be created or destroyed

This statement is not very useful in solving heat transfer problems. We rewrite it as an equation and apply it to the element:

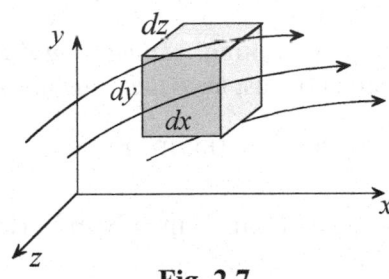

Fig. 2.7

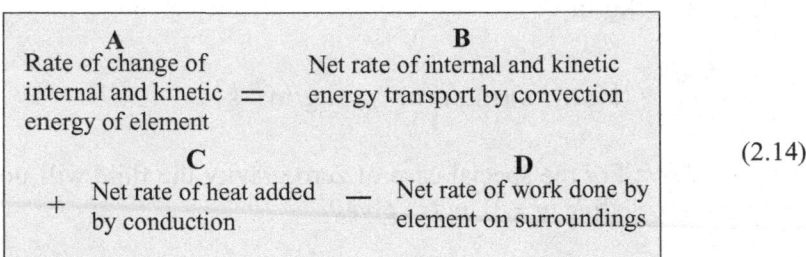

A Rate of change of internal and kinetic $=$ energy of element	B Net rate of internal and kinetic energy transport by convection
C $+$ Net rate of heat added $-$ by conduction	D Net rate of work done by element on surroundings

$$(2.14)$$

Note that *net rate* in equation (2.14) refers to rate of energy added minus rate of energy removed. The objective is to express each term in equation (2.14) in terms of temperature to obtain what is known as the *energy equation*. This formulation is detailed in Appendix A. In this section we will explain the physical significance of each term in equation (2.14) and its relation to temperature. The resulting energy equation will be presented in various forms. The formulation assumes: (1) continuum, (2) Newtonian fluid, and (3) negligible nuclear, electromagnetic and radiation energy transfer.

(1) A = Rate of change of internal and kinetic energy of element

The material inside the element has internal and kinetic energy. Internal energy can be expressed in terms of temperature using thermodynamic relations. Kinetic energy depends on the flow field.

(2) B = Net rate of internal and kinetic energy transport by convection

Mass flow through the element transports kinetic and thermal energy. Energy convected through each side of the element in Fig. 2.7 depends on mass flow rate and internal and kinetic energy per unit mass. Mass flow rate depends on density and velocity. Thus this component of energy balance can be expressed in terms of temperature and the velocity fields.

(3) C = Net rate of heat addition by conduction

Energy is conducted through each side of the element in Fig. 2.7. Using Fourier's law this component of energy can be expressed in terms of temperature gradient using equation (1.6).

(4) D = Net rate of work done by the element on the surroundings

The starting point in formulating this term is the observation that a moving force by the element on the surrounding represents work done or energy supplied by the element. That is

$$\textit{Rate of work = force} \times \textit{velocity}$$

Thus we must account for all surface forces acting on each side of the element as well as on the mass of the element (body forces). Examination of Fig. 2.6 shows that there are three forces on each side for a total of 18 forces. Each force moves with its own velocity. Body forces act on the mass of the element. Here the only body force considered is gravity. Accounting for all the forces and their respective velocities determines the net work done by the element on the surroundings.

Formulation of the four terms A, B, C and D and substitution into (2.14) give the following energy equation (See Appendix A)

$$\rho c_p \frac{DT}{Dt} = \nabla \cdot k\nabla T + \beta T \frac{Dp}{Dt} + \mu\Phi, \tag{2.15}$$

where

c_p = specific heat at constant pressure
k = thermal conductivity
p = pressure
β = coefficient of thermal expansion (compressibility)
Φ = dissipation function

The coefficient of thermal expansion β is a property of material defined as

$$\beta = -\frac{1}{\rho}\left[\frac{\partial \rho}{\partial T}\right]_p . \tag{2.16}$$

The dissipation function Φ is associated with energy dissipation due to friction. It is important in high speed flow and for very viscous fluids. In Cartesian coordinates Φ is given by

$$\Phi = 2\left[\left(\frac{\partial u}{\partial x}\right)^2 + \left(\frac{\partial v}{\partial y}\right)^2 + \left(\frac{\partial w}{\partial z}\right)^2\right]$$

$$+ \left[\left(\frac{\partial u}{\partial y} + \frac{\partial v}{\partial x}\right)^2 + \left(\frac{\partial v}{\partial z} + \frac{\partial w}{\partial y}\right)^2 + \left(\frac{\partial w}{\partial x} + \frac{\partial u}{\partial z}\right)^2\right] - \frac{2}{3}\left(\frac{\partial u}{\partial x} + \frac{\partial v}{\partial y} + \frac{\partial w}{\partial z}\right)^2 .$$

$$\tag{2.17}$$

2.6.2 Simplified Form of the Energy Equation

Equation (2.15) is based on the following assumptions: (1) continuum, (2) Newtonian fluid, and (3) negligible nuclear, electromagnetic and radiation energy transfer. It can be simplified under certain conditions. Three cases are considered.

(1) Incompressible fluid. According to (2.16), $\beta = 0$ for incompressible fluid. In addition, thermodynamic relations show that

$$c_p = c_v = c ,$$

where c_v is specific heat at constant volume. Equation (2.15) becomes

$$\rho c_p \frac{DT}{Dt} = \nabla \cdot k\nabla T + \mu \Phi . \tag{2.18}$$

(2) Incompressible constant conductivity fluid

Equation (2.18) is simplified further if the conductivity k is assumed constant. The result is

$$\rho c_p \frac{DT}{Dt} = k\nabla^2 T + \mu \Phi . \tag{2.19a}$$

Using the definition of total derivative and operator ∇, this equation is expressed as

$$\rho c_p \left(\frac{\partial T}{\partial t} + u \frac{\partial T}{\partial x} + v \frac{\partial T}{\partial y} + w \frac{\partial T}{\partial z} \right) = k \left(\frac{\partial^2 T}{\partial x^2} + \frac{\partial^2 T}{\partial y^2} + \frac{\partial^2 T}{\partial z^2} \right) + \mu \Phi. \quad (2.19b)$$

Note that for incompressible fluid, the last term in the dissipation function, equation (2.17), vanishes. Furthermore, if dissipation is negligible equation (2.19b) is simplified by setting $\Phi = 0$.

(3) Ideal gas. The ideal gas law gives

$$\rho = \frac{p}{RT}. \quad (2.20)$$

Substituting into (2.16)

$$\beta = -\frac{1}{\rho} \left(\frac{\partial \rho}{\partial T} \right)_p = \frac{1}{\rho} \frac{p}{RT^2} = \frac{1}{T}. \quad (2.21)$$

Equation (2.21) into (2.15), gives

$$\rho c_p \frac{DT}{Dt} = \nabla \cdot k \nabla T + \frac{Dp}{Dt} + \mu \Phi. \quad (2.22)$$

This result can be expressed in terms of c_v using continuity (2.2c) and the ideal gas law (2.20)

$$\rho c_v \frac{DT}{Dt} = \nabla \cdot k \nabla T - p \nabla \cdot \vec{V} + \mu \Phi. \quad (2.23)$$

2.6.3 Cylindrical Coordinates

The energy equation in cylindrical coordinates will be presented for the simplified case based on the following assumptions:

(1) Continuum, (2) Newtonian fluid, (3) negligible nuclear, electromagnetic and radiation energy transfer, (4) incompressible fluid, and (5) constant conductivity. The energy equation for this case is

$$\rho c_p \left(\frac{\partial T}{\partial t} + v_r \frac{\partial T}{\partial r} + \frac{v_\theta}{r} \frac{\partial T}{\partial \theta} + v_z \frac{\partial T}{\partial z} \right) =$$

$$k \left[\frac{1}{r} \frac{\partial}{\partial r} \left(r \frac{\partial T}{\partial r} \right) + \frac{1}{r^2} \frac{\partial^2 T}{\partial \theta^2} + \frac{\partial^2 T}{\partial z^2} \right] + \mu\Phi .$$

(2.24)

The dissipation function in cylindrical coordinates for incompressible fluid is given by

$$\Phi = 2\left(\frac{\partial v_r}{\partial r} \right)^2 + 2\left(\frac{1}{r} \frac{\partial v_\theta}{\partial \theta} + \frac{v_r}{r} \right)^2 + 2\left(\frac{\partial v_z}{\partial z} \right)^2 + \left(\frac{\partial v_\theta}{\partial r} - \frac{v_\theta}{r} + \frac{1}{r} \frac{\partial v_r}{\partial \theta} \right)^2$$

$$+ \left(\frac{1}{r} \frac{\partial v_z}{\partial \theta} + \frac{\partial v_\theta}{\partial z} \right)^2 + \left(\frac{\partial v_r}{\partial z} + \frac{\partial v_z}{\partial r} \right)^2 .$$

(2.25)

2.6.4 Spherical Coordinates

The energy equation in spherical coordinates will be presented for the simplified case based on the following assumptions:

(1) Continuum, (2) Newtonian fluid, (3) negligible nuclear, electromagnetic and radiation energy transfer, (4) incompressible fluid, and (5) constant conductivity. The energy equation for this case is

$$\rho c_p \left(\frac{\partial T}{\partial t} + v_r \frac{\partial T}{\partial r} + \frac{v_\phi}{r} \frac{\partial T}{\partial \phi} + \frac{v_\theta}{r \sin \phi} \frac{\partial T}{\partial \theta} \right) =$$

$$k \left[\frac{1}{r^2} \frac{\partial}{\partial r} \left(r^2 \frac{\partial T}{\partial r} \right) + \frac{1}{r^2 \sin \phi} \frac{\partial}{\partial \phi} \left(\sin \phi \frac{\partial T}{\partial \phi} \right) + \frac{1}{r^2 \sin^2 \phi} \frac{\partial^2 T}{\partial \theta^2} \right] + \mu\Phi.$$

(2.26)

The dissipation function in cylindrical coordinates for incompressible fluid is given by

$$\Phi = 2\left[\left(\frac{\partial v_r}{\partial r}\right)^2 + \left(\frac{1}{r}\frac{\partial v_\phi}{\partial \phi} + \frac{v_r}{r}\right)^2 + \left(\frac{1}{r\sin\phi}\frac{\partial v_\theta}{\partial \theta} + \frac{v_r}{r} + \frac{v_\phi \cot\phi}{r}\right)^2\right]$$

$$+\left[r\frac{\partial}{\partial r}\left(\frac{v_\phi}{r}\right) + \frac{1}{r}\frac{\partial v_r}{\partial \phi}\right]^2 + \left[\frac{\sin\phi}{r}\frac{\partial}{\partial \phi}\left(\frac{v_\theta}{r\sin\phi}\right) + \frac{1}{r\sin\phi}\frac{\partial v_\theta}{\partial \theta}\right]^2$$

$$+\left[\frac{1}{r\sin\phi}\frac{\partial v_r}{\partial \phi} + r\frac{\partial}{\partial r}\left(\frac{v_\theta}{r}\right)\right]^2. \tag{2.27}$$

Example 2.3: Flow between Parallel Plates

A fluid flows axially (x-direction) between parallel plates. Assume: Newtonian fluid, steady state, constant density and conductivity, and parallel streamlines. Taking dissipation into consideration, write the energy equation for this flow.

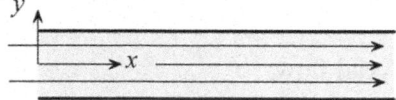

(1) Observations. (i) For parallel streamlines the vertical component $v = 0.$ (ii) Density and thermal conductivity are constant. (iii) Dissipation must be included in the energy equation. (iv) The geometry is Cartesian.

(2) Problem Definition. Determine the energy equation for parallel flow.

(3) Solution Plan. Start with the energy equation in Cartesian coordinates for constant density and conductivity and simplify it for this special case.

(4) Plan Execution.

(i) **Assumptions.** (1) Continuum, (2) Newtonian fluid, (3) steady state, (4) axial flow, (5) constant density and conductivity, (6) negligible nuclear, electromagnetic and radiation energy transfer, and (7) parallel streamlines.

(ii) **Analysis.** The energy equation in Cartesian coordinates for incompressible constant conductivity fluid is given by equation (2.19b)

$$\rho c_p\left(\frac{\partial T}{\partial t} + u\frac{\partial T}{\partial x} + v\frac{\partial T}{\partial y} + w\frac{\partial T}{\partial z}\right) = k\left(\frac{\partial^2 T}{\partial x^2} + \frac{\partial^2 T}{\partial y^2} + \frac{\partial^2 T}{\partial z^2}\right) + \mu\Phi, \tag{2.19b}$$

where the dissipation function in Cartesian coordinates is given by equation (2.17)

$$\Phi = 2\left[\left(\frac{\partial u}{\partial x}\right)^2 + \left(\frac{\partial v}{\partial y}\right)^2 + \left(\frac{\partial w}{\partial z}\right)^2\right]$$

$$+\left[\left(\frac{\partial u}{\partial y} + \frac{\partial v}{\partial x}\right)^2 + \left(\frac{\partial v}{\partial z} + \frac{\partial w}{\partial y}\right)^2 + \left(\frac{\partial w}{\partial x} + \frac{\partial u}{\partial z}\right)^2\right] - \frac{2}{3}\left(\frac{\partial u}{\partial x} + \frac{\partial v}{\partial y} + \frac{\partial w}{\partial z}\right)^2.$$

$$(2.17)$$

Based on the above assumptions, these equations are simplified as follows:

Steady state:
$$\frac{\partial T}{\partial t} = 0. \tag{a}$$

Axial flow:
$$w = \frac{\partial}{\partial z} = 0. \tag{b}$$

Parallel flow:
$$v = 0. \tag{c}$$

Substituting (a)-(c) into (2.19b), gives

$$\rho c_p u \frac{\partial T}{\partial x} = k\left(\frac{\partial^2 T}{\partial x^2} + \frac{\partial^2 T}{\partial y^2}\right) + \mu\Phi. \tag{d}$$

The dissipation function (2.17) is simplified using (b) and (c)

$$\Phi = 2\left(\frac{\partial u}{\partial x}\right)^2 + \left(\frac{\partial u}{\partial y}\right)^2 - \frac{2}{3}\left(\frac{\partial u}{\partial x}\right)^2. \tag{e}$$

Continuity equation (2.3) gives

$$\frac{\partial u}{\partial x} = 0. \tag{f}$$

Using (f) into (e) gives

$$\Phi = \left(\frac{\partial u}{\partial y}\right)^2. \tag{g}$$

Substituting (g) into (d) gives the energy equation

$$\rho c_p u \frac{\partial T}{\partial x} = k \left(\frac{\partial^2 T}{\partial x^2} + \frac{\partial^2 T}{\partial y^2} \right) + \mu \left(\frac{\partial u}{\partial y} \right)^2 . \tag{h}$$

(iii) Checking. *Dimensional check*: Each term in (h) has units of W/m^3.

Limiting check: If the fluid is not moving, the energy equation should reduce to pure conduction. Setting $u = 0$ in (h) gives

$$\frac{\partial^2 T}{\partial x^2} + \frac{\partial^2 T}{\partial y^2} = 0 .$$

This is the correct equation for this limiting case.

(5) Comments. In energy equation (h), properties c_p, k, ρ and μ represent fluid nature. The velocity u represents fluid motion. This confirms the observation made in Chapter 1 that fluid motion and nature play a role in convection heat transfer (temperature distribution).

2.7 Solutions to the Temperature Distribution

Having formulated the three basic laws, continuity (2.2), momentum (2.8) and energy (2.15), we examine the mathematical consequence of these equations with regard to obtaining solutions to the temperature distribution. Table 2.1 lists the governing equations and the unknown variables.

TABLE 2.1

Basic law	No. of Equations	Unknowns							
Energy	1	T	u	v	w		ρ	μ	k
Continuity	1		u	v	w		ρ		
Momentum	3		u	v	w	p	ρ	μ	
Equation of State	1	T				p	ρ		
Viscosity relation $\mu = \mu(p,T)$	1	T				p		μ	
Conductivity relation $k = k(p,T)$	1	T				p			k

The following observations are made regarding Table 2.1:

(1) Although specific heats c_p and c_v, and the coefficient of thermal expansion β appear in the energy equation, they are not listed in Table 2.1 as unknown. These properties are determined once the equation of state is specified.

(2) For the general case of variable properties, the total number of unknowns is 8: T, u, v, w, p, ρ, μ, and k. To determine the temperature distribution, the eight equations must be solved simultaneously for the eight unknowns. Thus the velocity and temperature fields are *coupled*.

(3) For the special case of constant conductivity and viscosity the number of unknowns is reduced to six: T, u, v, w, p and ρ, Thus the six equations, energy, continuity, momentum and state must be solved simultaneously to determine the temperature distribution. This case is defined by the largest dashed rectangle in Table 2.1.

(4) For the important case of constant density (incompressible fluid), viscosity and conductivity, the number of unknowns is reduced to five: T, u, v, w, p. This case is defined by the second largest dashed rectangle in Table 2.1. However a significant simplification takes place: the four equations, continuity, and momentum, contain four unknowns: u, v, w and p, as defined by the smallest rectangle in Table 2.1. Thus the velocity and temperature fields are *uncoupled*. This means that the velocity field can be determined first by solving the continuity and momentum equations without using the energy equation. Once the velocity field is determined, it is substituted into the energy equation and the resulting equation is solved for the temperature distribution.

2.8 The Boussinesq Approximation

Fluid motion in free convection is driven by density change and gravity. Thus the assumption of constant density cannot be made in the analysis of free convection problems. Instead an alternate simplification called the *Boussinesq approximation* is made. The basic approach in this approximation is to treat the density as constant in the continuity equation and the inertia term of the momentum equation, but allow it to change with temperature in the gravity term. We begin with the momentum equation for constant viscosity

$$\rho \frac{D\vec{V}}{Dt} = \rho \vec{g} - \nabla p + \frac{1}{3} \mu \nabla (\nabla \cdot \vec{V}) + \mu \nabla^2 \vec{V}.$$ (2.9)

This equation is valid for variable density ρ. However, we will assume that ρ is constant in the inertia (first) term but not in the gravity term $\rho \vec{g}$. Thus (2.9) is rewritten as

$$\rho_\infty \frac{D\vec{V}}{Dt} = \rho \vec{g} - \nabla p + \mu \nabla^2 \vec{V},$$ (a)

where ρ_∞ is fluid density at some reference state, such as far away from an object where the temperature is uniform and the fluid is either stationary or moving with uniform velocity. Thus at the reference state we have

$$\frac{D\vec{V}_\infty}{Dt} = \nabla^2 \vec{V}_\infty = 0.$$ (b)

Applying (a) at the reference state ∞ and using (b), gives

$$\rho_\infty \vec{g} - \nabla p_\infty = 0.$$ (c)

Subtracting (c) from (a)

$$\rho \frac{D\vec{V}}{Dt} = (\rho - \rho_\infty) \vec{g} - \nabla (p - p_\infty) + \mu \nabla^2 \vec{V}.$$ (d)

The objective of the next step is to eliminate the $(\rho - \rho_\infty)$ term in (d) and express it in terms of temperature difference. This is accomplished through the introduction of the coefficient of thermal expansion β, defined as

$$\beta = -\frac{1}{\rho} \left[\frac{\partial \rho}{\partial T} \right]_p.$$ (2.16)

Pressure variation in free convection is usually small and in addition, the effect of pressure on β is also small. In other words, in free convection β can be assumed independent of p. Thus we rewrite (2.16) as

$$\beta \approx -\frac{1}{\rho_\infty}\frac{d\rho}{dT}\ .\qquad\qquad(e)$$

We further note that over a small change in temperature the change in density is approximately linear. Thus we rewrite (e) as

$$\beta \approx -\frac{1}{\rho_\infty}\frac{\rho-\rho_\infty}{T-T_\infty}\ .\qquad\qquad(f)$$

This result gives

$$\rho-\rho_\infty = -\beta\rho_\infty(T-T_\infty)\ .\qquad\qquad(2.28)$$

Equation (2.28) relates density change to temperature change. Substituting (2.28) into (d)

$$\frac{D\vec{V}}{Dt} = -\beta\,\vec{g}(T-T_\infty)-\frac{1}{\rho_\infty}\nabla(p-p_\infty)+\nu\nabla^2\vec{V}\ .\qquad(2.29)$$

The simplification leading to (2.29) is known as the *Boussinesq* approximation. The importance of this approximation lies in the elimination of density as a variable in the analysis of free convection problems. However, the momentum and energy equations remain coupled.

2.9 Boundary Conditions

To obtain solutions to the flow and temperature fields, boundary conditions must be formulated and specified. Boundary conditions are mathematical equations describing what takes place physically at a boundary. In convection heat transfer it is necessary to specify boundary conditions on the velocity and temperature. The following are commonly encountered conditions.

(1) No-slip condition. Fluid velocity vanishes at a stationary boundary such as the wall of a tube, surface of a plate, cylinder, or sphere. Thus all three velocity components must vanish. For example, for the flow over a surface located at $y = 0$ in Cartesian coordinates, this condition is expressed mathematically as

$$\vec{V}(x,0,z,t) = 0\ ,\qquad\qquad(2.30a)$$

It follows from the above that

$$u(x,0,z,t) = v(x,0,z,t) = w(x,0,z,t) = 0. \qquad (2.30b)$$

Equation (2.30) is referred to as the *no-slip condition*.

(2) Free stream condition. Far away from an object it is common to assume a uniform or zero velocity. For example, a uniform x-component velocity at $y = \infty$ is expressed as

$$u(x,\infty,z,t) = V_\infty . \qquad (2.31)$$

Similarly, uniform temperature far away from an object is expressed as

$$T(x,\infty,z,t) = T_\infty . \qquad (2.32)$$

(3) Surface thermal conditions. Two common surface thermal conditions are used in the analysis of convection problems. They are:

(i) Specified surface temperature. This condition is written as

$$T(x,0,z,t) = T_s . \qquad (2.33)$$

Note that surface temperature T_s need not be uniform or constant. It can vary with location x and z as well as time.

(ii) Specified surface heat flux. The boundary condition for a surface which is heated or cooled at a specified flux is expressed as

$$-k\frac{\partial T(x,0,z,t)}{\partial y} = \pm q_o'' . \qquad (2.34)$$

Note that in (2.34) the plus sign is selected if the heat flux q_o'' points in the positive y-direction. The flux need not be uniform or constant. It can vary with location x and z as well as time.

Example 2.4: Heated Thin Liquid Film Flow over an Inclined Surface

A thin liquid film flows axially down an inclined plate. The film thickness H is uniform. The plate is maintained at uniform temperature T_o and the

free surface is heated with a flux q_o''.
Write the velocity and thermal boundary
conditions at these two surfaces.

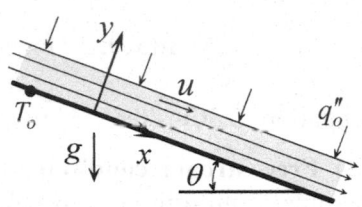

(1) Observations. (i) The free surface is
parallel to the inclined plate. (ii) The no-
slip condition applies at the inclined
surface. (iii) The temperature is specified
at the plate. The flux is specified at the
free surface. (iv) Cartesian geometry.

(2) Problem Definition. Write the boundary conditions at the two surfaces
for the velocity components u and v and for the thermal field.

(3) Solution Plan. Select an origin and coordinate axes. Identify the
physical flow and thermal conditions at the two surfaces and express them
mathematically.

(4) Plan Execution.

  (i) Assumptions. (1) Continuum, (2) Newtonian fluid, (3) negligible
shearing stress at the free surface, and (4) constant film thickness.

  (ii) Analysis. The origin and Cartesian coordinate axes are selected as
shown. The velocity and thermal boundary conditions at the two surfaces
are:

(1) No-slip condition at the inclined surface:

$$u(x,0) = 0, \tag{a}$$

$$v(x,0) = 0. \tag{b}$$

(2) Free surface is parallel to inclined plate:

$$v(x, H) = 0. \tag{c}$$

(3) Negligible shear at the free surface: Shearing stress for a Newtonian
fluid is given by equation (2.7a)

$$\tau_{yx} = \mu\left(\frac{\partial v}{\partial x} + \frac{\partial u}{\partial y}\right). \tag{2.7a}$$

Applying (2.7a) at the free surface and using (c), gives

$$\frac{\partial u(x,H)}{\partial y} = 0.$$
(d)

(4) Specified temperature at the inclined surface:

$$T(x,0) = T_o.$$
(e)

(5) Specified heat flux at the free surface. Application of equation (2.34) gives

$$-k\frac{\partial T(x,H,z,t)}{\partial y} = -q''_o.$$
(f)

(iii) Checking. *Dimensional check*: Each term in (f) has units of flux.

(5) Comments. (i) To write boundary conditions, origin and coordinate axes must be selected first.

(ii) Since the heat flux at the free surface points in the negative y-direction, a minus sign is introduced on the right hand side of equation (f).

2.10 Non-dimensional Form of the Governing Equations: Dynamic and Thermal Similarity Parameters

Useful information can be obtained without solving the governing equations by rewriting them in dimensionless form. This procedure is carried out to: (1) identify the governing parameters, (2) plan experiments, and (3) guide in the presentation of experimental results and theoretical solutions. To appreciate the importance of this process we consider an object of characteristic length L which is exchanging heat by convection with an ambient fluid. For simplicity we assume constant properties. In general the unknown variables are: u, v, w, p and T. These variables depend on the four independent variables x, y, z and t. In addition various quantities affect the solutions. They are: p_∞, T_∞, V_∞, T_s, L, g and fluid properties c_p, k, β, μ, and ρ. Furthermore, the geometry of the object is also a factor. To map the effect of these quantities experimentally or numerically for a single geometry requires extensive effort. However, in dimensionless formulation these quantities are consolidated into four dimensionless groups called *parameters*. This dramatically simplifies the mapping process.

2.10.1 Dimensionless Variables

To non-dimensionalize the dependent and independent variables, we use characteristic quantities that are constant throughout the flow and temperature fields. These quantities are g, L, T_s, T_∞, p_∞, ρ_∞, and V_∞. We consider Cartesian coordinates and define the following dimensionless dependent and independent variables:

$$\vec{V}^* = \frac{\vec{V}}{V_\infty}, \quad p^* = \frac{(p - p_\infty)}{\rho_\infty V_\infty^2}, \quad T^* = \frac{(T - T_\infty)}{(T_s - T_\infty)}, \quad \vec{g}^* = \frac{\vec{g}}{g},$$

$$x^* = \frac{x}{L}, \quad y^* = \frac{y}{L}, \quad z^* = \frac{z}{L}, \quad t^* = \frac{V_\infty}{L} t.$$

(2.35)

Note that in the above the subscript ∞ refers to the characteristic condition, say far away from the object. V_∞ is the magnitude of the velocity vector at ∞ and g is the magnitude of the gravitational acceleration vector. Equation (2.35) is first used to construct the dimensionless form of the operators ∇ and D/Dt

$$\nabla = \frac{\partial}{\partial x} + \frac{\partial}{\partial y} + \frac{\partial}{\partial z} = \frac{\partial}{L \partial x^*} + \frac{\partial}{L \partial y^*} + \frac{\partial}{L \partial z^*} = \frac{1}{L} \nabla^*, \quad (2.36a)$$

$$\nabla^2 = \frac{\partial^2}{\partial x^2} + \frac{\partial^2}{\partial y^2} + \frac{\partial^2}{\partial z^2} = \frac{\partial}{L^2 \partial x^{*2}} + \frac{\partial}{L^2 \partial y^{*2}} + \frac{\partial}{L^2 \partial z^{*2}} = \frac{1}{L^2} \nabla^{*2},$$

(2.36b)

$$\frac{D}{Dt} = \frac{D}{D(Lt^*/V_\infty)} = \frac{V_\infty}{L} \frac{D}{Dt^*}.$$

(2.36c)

2.10.2 Dimensionless Form of Continuity

Substituting (2.35) and (2.36) into continuity equation (2.2c) gives

$$\frac{D\rho}{Dt^*} + \rho \nabla \cdot \vec{V}^* = 0.$$

(2.37)

We note that the dimensionless form of continuity reveals no parameters.

2.10.3 Dimensionless Form of the Navier-Stokes Equations of Motion

Substituting (2.35) and (2.36) into (2.29) gives

$$\frac{D\vec{V}^*}{Dt^*} = -\frac{Gr}{Re^2}T^*\vec{g}^* - \nabla^*P^* + \frac{1}{Re}\nabla^{*2}\vec{V}^*, \tag{2.38}$$

where the parameters Re and Gr are the *Reynolds* and *Grashof* numbers, defined as

$$Re \equiv \frac{\rho V_\infty L}{\mu} = \frac{V_\infty L}{\nu}, \quad \textit{Reynolds number,} \tag{2.39}$$

$$Gr \equiv \frac{\beta g(T_s - T_\infty)L^3}{\nu^2}, \quad \textit{Grashof number.} \tag{2.40}$$

2.10.4 Dimensionless Form of the Energy Equation

We consider two special cases of the energy equation.

(i) Incompressible, constant conductivity

Substituting (2.35) and (2.36) into (2.19) gives

$$\frac{DT^*}{Dt^*} = \frac{1}{RePr}\nabla^{*2}T^* + \frac{Ec}{Re}\Phi^*, \tag{2.41a}$$

where the parameters Pr and Ec are the *Prandtl* and *Eckert* numbers, defined as

$$Pr = \frac{c_p\mu}{k} = \frac{\mu/\rho}{k/\rho c_p} = \frac{\nu}{\alpha}, \quad \textit{Prandtl number,} \tag{2.42}$$

$$Ec = \frac{V_\infty^2}{c_p(T_s - T_\infty)}, \quad \textit{Eckert number.} \tag{2.43}$$

The dimensionless dissipation function Φ^* is determined by substituting (2.35) and (2.36) into (2.17)

$$\Phi^* = 2\left[\left(\frac{\partial u^*}{\partial x^*}\right)^2 + \left(\frac{\partial v^*}{\partial y^*}\right)^2 + \cdots\right]. \tag{2.44}$$

(ii) Ideal gas, constant conductivity and viscosity

Substituting (2.35) and (2.36) into (2.22) yields

$$\frac{DT^*}{Dt^*} = \frac{1}{RePr}\nabla^{*2}T^* + Ec\frac{Dp^*}{Dt^*} + \frac{Ec}{Re}\Phi^*. \qquad (2.41b)$$

2.10.5 Significance of the Governing Parameters

The non-dimensional form of the governing equations (2.37), (2.38), and (2.41) are governed by four parameters: *Re, Pr, Gr,* and *Ec.* Thus the temperature solution for convection can be expressed as

$$T^* = f(x^*, y^*, z^*, t^*; Re, Pr, Gr, Ec). \qquad (2.45)$$

The following observations are made:

(1) The *Reynolds number* is associated with viscous flow while the *Prandtl number* is a heat transfer parameter which is a fluid property. The *Grashof number* represents buoyancy effect and the *Eckert number* is associated with viscous dissipation and is important in high speed flow and very viscous fluids.

(2) In dimensional formulation six quantities, p_∞, T_∞, T_s, V_∞, L, g and five properties c_p, k, β, μ, and ρ, affect the solution. In dimensionless formulation these factors are consolidated into four dimensionless parameters: *Re, Pr, Gr* and *Ec.*

(3) The number of parameters can be reduced in two special cases: (i) If fluid motion is dominated by forced convection (negligible free convection), the Grashof number can be eliminated. (ii) If viscous dissipation is negligible, the Eckert number can be dropped. Thus under these common conditions the solution is simplified to

$$T^* = f(x^*, y^*, z^*, t^*; Re, Pr). \qquad (2.46)$$

(4) The implication of (2.45) and (2.46) is that geometrically similar bodies have the same dimensionless velocity and temperature solutions if the similarity parameters are the same for all bodies.

(5) By identifying the important dimensionless parameters governing a given problem, experimental investigations can be planned accordingly. Instead of varying the relevant physical quantities, one can vary the

similarity parameters. This will vastly reduce the number of experiments needed. The same is true if numerical results are to be generated.

(6) Presentation of results such as heat transfer coefficient, pressure drop, and drag, whether experimental or numerical, is most efficiently done when expressed in terms of dimensionless parameters.

2.10.6 Heat Transfer Coefficient: The Nusselt Number

Having identified the important dimensionless parameters in convection heat transfer we now examine the dependency of the heat transfer coefficient h on these parameters. We begin with equation (1.10) which gives h

$$h = \frac{-k}{(T_s - T_\infty)} \frac{\partial T(x,0,z)}{\partial y} . \tag{1.10}$$

Using (2.30) to express temperature gradient and h in dimensionless form, (1.10) becomes

$$\frac{hx}{k} = -x^* \frac{\partial T^*(x^*,0,z^*)}{\partial y^*} , \tag{4.47}$$

where the dimensionless heat transfer coefficient hx / k is known as the *Nusselt number*. Since it depends on the location x^* it is referred to as the *local Nusselt number* and is given the symbol Nu_x. Thus we define

$$Nu_x = \frac{hx}{k} . \tag{2.48}$$

Similarly, the *average Nusselt number* $\overline{Nu_L}$ for a surface of length L is based on the average heat transfer coefficient $\overline{h}$ and is defined as

$$\overline{Nu_L} = \frac{\overline{h}L}{k} , \tag{2.49}$$

where $\overline{h}$ for the one-dimensional case is given by

$$\overline{h} = \frac{1}{L} \int_0^L h(x)dx . \tag{2.50}$$

Since T^* depends on four parameters, it follows from (2.45), (2.47) and (2.48) that the local Nusselt number also depends on the same four parameters and is expressed as

$$Nu_x = f(x^*; Re, Pr, Gr, Ec). \tag{2.51}$$

This is an important result since it suggests how experiments should be planned and provides an appropriate form for correlation equations for the Nusselt number. As was pointed out in Section 2.10.5, for the special case of negligible buoyancy and viscous dissipation, (2.51) is simplified to

$$Nu_x = f(x^*; Re, Pr). \tag{2.52}$$

Similarly, for free convection with negligible dissipation we obtain

$$Nu_x = f(x^*; Gr, Pr). \tag{2.53}$$

Equations (2.51)–(2.53) are for the local Nusselt number. For the average Nusselt number, which is based on the average heat transfer coefficient, the variable x^* is eliminated according to (2.50). Thus (2.51) takes the form

$$\overline{Nu}_L = \frac{\bar{h}L}{k} = f(Re, Pr, Gr, Ec). \tag{2.54}$$

Equations (2.52) and (2.53) are similarly modified.

It should be noted that much has been learned by expressing the governing equations in dimensionless form without solving them. However, although we now know what the Nusselt number depends on, the form of the functional relations given in (2.51)–(2.54) can only be determined by solving the governing equations or through experiments.

Example 2.5: Heat Transfer Coefficient for Flow over Cylinders

You carried out two experiments to determine the average heat transfer coefficient for flow normal to a cylinder. The diameter of one cylinder is $D_1 = 3$ cm and that of the other is $D_2 = 5$ cm. The free stream velocity over D_1 is $V_1 = 15$ m/s and the velocity over D_2 is $V_2 = 98$ m/s. Measurements showed that the average heat transfer *coefficient for D_1 is $\overline{h}_1 = 244\,W/m^2\text{-}°C$ and for D_2 is $\overline{h}_2 = 144\,W/m^2\text{-}°C$. In both experiments you used the same fluid. To check your results you decided to compare your data with the following correlation equation for flow normal to a cylinder:*

$$\overline{Nu}_D = \frac{hD}{k} = C\,Re_D^{0.6}\,Pr^n\,, \tag{a}$$

where C and n are constants. What do you conclude regarding the accuracy of your data?

(1) Observations. (i) Experimental results for $\overline{h}_1$ and $\overline{h}_2$ should be compared with those predicted by the correlation equation. (ii) The heat transfer coefficient appears in the definition of the Nusselt number $\overline{Nu}_D$. (iii) The correlation equation can not be used to determine $\overline{h}_1$ and $\overline{h}_2$ since the fluid and the constants C and n are not given. However, the equation can be used to determine the ratio $\overline{h}_1/\overline{h}_2$. (iv) The absence of the Grashof and Eckert numbers in the correlation equation implies that it is applicable to cases where buoyancy and viscous dissipation are negligible.

(2) Problem Definition. Determine $\overline{h}_1/\overline{h}_2$ using experimental data and the correlation equation.

(3) Solution Plan. Apply the correlation equation to determine $\overline{h}_1/\overline{h}_2$ and compare with the experimentally obtained ratio.

(4) Plan Execution.

   (i) Assumptions. (i) Correlation equation (a) is valid for both experiments. (ii) Fluid properties are constant.

   (ii) Analysis. Noting that $\overline{Nu}_D = \dfrac{\overline{h}D}{k}$ and $Re_D = \dfrac{VD}{v}$, equation (a) is rewritten as

$$\frac{\overline{h}D}{k} = C\left(\frac{VD}{v}\right)^{0.6} Pr^n\,, \tag{b}$$

where

   D = diameter, m
   $\overline{h}$ = heat transfer coefficient, W/m²-°C
   k = thermal conductivity, W/m-°C
   Pr = Prandtl number

V = free stream velocity, m/s

v = kinematic viscosity, m²/s

Solving equation (b) for $\overline{h}$

$$\overline{h} = \frac{CkV^{0.6}Pr^n}{v^{0.6}D^{0.4}}.$$ (c)

Applying (c) to the two experiments

$$\overline{h}_1 = \frac{CkV_1^{0.6}Pr^n}{v^{0.6}D_1^{0.4}},$$ (d)

and

$$\overline{h}_2 = \frac{CkV_2^{0.6}Pr^n}{v^{0.6}D_2^{0.4}}.$$ (e)

Taking the ratio of (d) and (e) gives

$$\frac{\overline{h}_1}{\overline{h}_2} = \left(\frac{V_1}{V_2}\right)^{0.6}\left(\frac{D_2}{D_1}\right)^{0.4}.$$ (f)

(iii) Computations. Substituting the experimental data for V_1, V_2, D_1 and D_2 into (f)

$$\frac{\overline{h}_1}{\overline{h}_2} = \left[\frac{15(\text{m/s})}{98(\text{m/s})}\right]^{0.6}\left[\frac{5(\text{cm})}{3(\text{cm})}\right]^{0.4} = 0.4$$

The experimentally obtained ratio $\overline{h}_1/\overline{h}_2$ is

$$\frac{\overline{h}_1}{\overline{h}_2} = \frac{244(\text{W/m}^2\text{ -}^\circ\text{ C})}{144(\text{W/m}^2\text{ -}^\circ\text{ C})} = 1.69$$

The two results differ by a factor of 4.2. This points to an error in the experimental data.

(iv) Checking. *Dimensional check*: Equation (f) is dimensionally consistent since each term is dimensionless.

Limiting check: If $V_1 = V_2$ and $D_1 = D_2$, then $\overline{h}_1 = \overline{h}_2$. Equation (f) confirms this.

Qualitative check: If V is increased, $\overline{h}$ should increase. This is substantiated by (c).

(5) Comments. (i) The assumption that the correlation equation is valid for both experiments is critical. If, for example, the effects of viscous dissipation and/or buoyancy are significant in the two experiments, equation (a) is not applicable.

(ii) The analysis suggests that there is an error in the experimental data. However, it is not possible to establish whether one experiment is wrong or both are wrong.

(iii) A more conclusive check can be made if C, n and the fluid are known.

2.11 Scale Analysis

Scale analysis, or scaling, is a procedure by which estimates of useful results are obtained without solving the governing equations. It should be emphasized that scaling gives order of magnitude answers, and thus the approximation is crude. Scaling is accomplished by assigning order of magnitude values to dependent and independent variables in an equation. Excellent applications of scaling in heat transfer is found in reference [2].

Example 2.6: Melting Time of Ice Sheet

An ice sheet of thickness L is at the freezing temperature T_f. One side is suddenly maintained at temperature T_o which is above the freezing temperature. The other side is insulated. Conservation of energy at the melting front gives

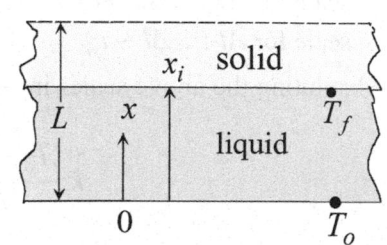

$$-k\frac{\partial T}{\partial x} = \rho \mathcal{L}\frac{dx_i}{dt}, \qquad \text{(a)}$$

where

    $k = $ *thermal conductivity*
    $T = $ *temperature distribution in the liquid phase*
    $t = $ *time*

$x = coordinate$

$x_i = interface\ location$

$\mathcal{L} = latent\ heat\ of\ fusion$

Use scale analysis to determine the time needed for the entire sheet to melt.

(1) Observations. (i) The entire sheet melts when $x_i = L$. (ii) The largest temperature difference is $T_o - T_f$. (iii) Scaling of equation (a) should be helpful in determining melt time.

(2) Problem Definition. Determine the time $t = t_o$ when $x_i(t) = L$.

(3) Solution Plan. Apply scale analysis to equation (a).

(4) Plan Execution.

   (i) Assumptions. (i) Sheet is perfectly insulated at $x = L$. (ii) Liquid phase is stationary.

   (ii) Analysis. Equation (a) is approximated by

$$-k\frac{\Delta T}{\Delta x} = \rho\mathcal{L}\frac{\Delta x_i}{\Delta t}. \tag{b}$$

We now select scales for the variables in (a).

   scale for ΔT : $\Delta T \sim (T_o - T_f)$
   scale for Δx : $\Delta x \sim L$
   scale for Δx_i : $\Delta x_i \sim L$
   scale for Δt : $\Delta t \sim t_o$

Substituting the above scales into (a)

$$k\frac{(T_o - T_f)}{L} \sim \rho\mathcal{L}\frac{L}{t_o}.$$

Solving for melt time t_o

$$t_o \sim \frac{\rho\mathcal{L}L^2}{k(T_o - T_f)}. \tag{c}$$

   (iii) Checking. *Dimensional check*: Each term in (c) should have units of time:

$$t_o = \frac{\rho\,(\text{kg/m}^3)\,\mathcal{L}(\text{J/kg})L^2\,(\text{m}^2)}{k\,(\text{W/m-}^\circ\text{C})(T_o - T_f)(^\circ\text{C})} = \text{s}.$$

Limiting check: (1) If the latent heat of fusion $\mathcal{L}$ is infinite, melt time should be infinite. Setting $\mathcal{L} = \infty$ in (c) gives $t_o = \infty$.

(2) If sheet thickness is zero, melt time should vanish. Setting $L = 0$ in (c) gives $t_o = 0$.

Qualitative check: Melt time should be directly proportional to mass, latent heat and thickness and inversely proportional to conductivity and temperature difference $(T_o - T_f)$. This is confirmed by solution (c).

(5) Comments.   (i) With little effort an estimate of the melt time is obtained without solving the governing equations for the two phase region.

(ii) An exact solution based on quasi-steady process gives the melt time t_o as

$$t_o = \frac{\rho\mathcal{L}L^2}{2k(T_o - T_f)}. \tag{d}$$

Thus scaling gives an approximate answer within a factor of 2.

REFERENCES

[1] Schlichting, H, *Boundary Layer Theory*, 7th ed., translated into English by J. Kestin, McGraw-Hill, 1979.

[2] Bejan, A., *Convection Heat Transfer*, 2nd ed., Wiley, 1995.

PROBLEMS

2.1   [a] Consider transient (unsteady), incompressible, three dimensional flow. Write the continuity equation in Cartesian coordinates for this flow.

[b] Repeat [a] for steady state.

2.2   Far away from the inlet of a tube, entrance effects diminish and streamlines become parallel and the flow is referred to as *fully developed*. Write the continuity equation in the fully developed region for incompressible fluid.

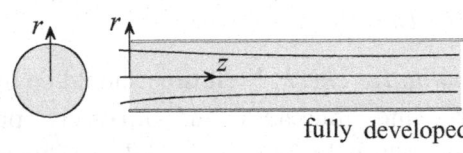

fully developed

2.3   Consider incompressible flow between parallel plates. Far away from the entrance the axial velocity component does not vary with the axial distance.

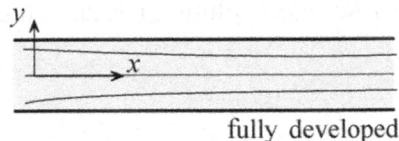

fully developed

[a] Determine the velocity component in the y-direction.

[b] Does your result in [a] hold for steady as well as unsteady flow? Explain.

2.4. The radial and tangential velocity components for incompressible flow through a tube are zero. Show that the axial velocity does not change in the flow direction. Is this valid for steady as well as transient flow?

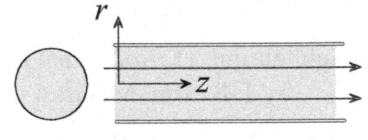

2.5   Show that $\tau_{xy} = \tau_{yx}$.

2.6   A fluid flows axially between parallel plates. Assume: Steady state, Newtonian fluid, constant density, constant viscosity, negligible gravity, and parallel streamlines. Write the three components of the momentum equations for this flow.

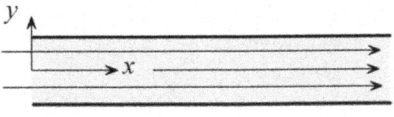

2.7   A fluid flows axially (z-direction) through a tube. Assume: Steady state, Newtonian fluid, constant density, constant viscosity, negligible gravity, and parallel streamlines. Write the three components of the momentum equations for this flow.

2.8   Consider two-dimensional flow, (x,y), between parallel plates. Assume: Newtonian fluid, constant density and viscosity. Write the two components of the momentum equations for this flow. How many unknowns do the equations have? Can they be solved for the unknowns? If not what other equation(s) is needed to obtain a solution?

2.9   Consider two-dimensional flow, (r,z), through a tube. Assume: Newtonian, constant density and viscosity. Write the two components of the momentum equations for this flow. How many unknowns do the equations have? Can the equations be solved for the unknowns? If not what other equation(s) is needed to obtain a solution?

2.10 In Chapter 1 it is stated that fluid motion and fluid nature play a role in convection heat transfer. Does the energy equation substantiate this observation? Explain.

2.11 A fluid flows axially (x-direction) between parallel plates. Assume: Newtonian fluid, steady state, constant density, constant viscosity, constant conductivity, negligible gravity, and parallel streamlines. Write the energy equation for this flow.

2.12 An ideal gas flows axially (x-direction) between parallel plates. Assume: Newtonian fluid, steady state, constant viscosity, constant conductivity, negligible gravity, and parallel stream- lines. Write the energy equation for this flow.

2.13 Consider two-dimensional free convection over a vertical plate. Assume: Newtonian fluid, steady state, constant viscosity, Boussinesq approximation, and negligible dissipation. Write the governing equations for this case. Can the flow field be determined independently of the temperature field?

2.14 Discuss the condition(s) under which the Navier-Stokes equations of motion can be solved independently of the energy equation.

2.15 Consider a thin film of liquid condensate which is falling over a flat surface by virtue of gravity. Neglecting variations in the z-direction and assuming Newtonian fluid, steady state, constant properties, and parallel streamlines

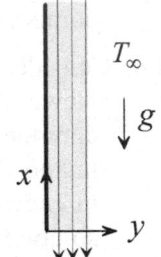

[a] Write the momentum equation(s) for this flow.
[b] Write the energy equation including dissipation effect.

2.16 A wedge is maintained at T_1 along one side and T_2 along the opposite side. A solution for the flow field is obtained based on Newtonian fluid and constant properties. The fluid approaches the wedge with uniform velocity and temperature. Examination of the solution shows that the velocity distribution is not symmetrical with respect to the x-axis. You are asked to support the argument that the solution is incorrect.

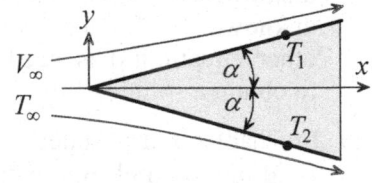

2.17 Starting with the equations of motion for constant properties and Boussinesq model

$$\frac{D\vec{V}}{Dt} = -\beta(T - T_\infty)\vec{g} - \frac{1}{\rho}\nabla(p - p_\infty) + \nu\nabla^2\vec{V},$$

and the energy equation for an ideal gas with constant k

$$\rho c_p \frac{DT}{Dt} = k\nabla^2 T + \frac{Dp}{Dt} + \mu\Phi,$$

show that the dimensionless form of these equations is

$$\frac{D\vec{V}^*}{Dt^*} = -\frac{Gr}{Re^2}T^*\vec{g}^* - \nabla^* P^* + \frac{1}{Re}\nabla^{*2}\vec{V}^*,$$

and

$$\frac{DT^*}{Dt^*} = \frac{1}{RePr}\nabla^{*2}T^* + E\frac{DP^*}{Dt^*} + \frac{Ec}{Re}\Phi^*.$$

2.18 Consider two-dimensional (x and y), steady, constant properties, parallel flow between two plates separated by a distance H. The lower plate is stationary while the upper plate moves axially with a velocity U_o. The upper plate is maintained at uniform temperature T_o and the lower plate is cooled with a flux q_o''. Taking into consideration dissipation, write the Navier-Stokes equations of motion, energy equation and boundary conditions at the two plates.

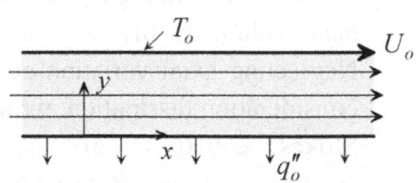

2.19 A shaft of radius r_1 rotates concentrically inside a sleeve of inner radius r_2. Lubrication oil fills the clearance between the shaft and the sleeve. The sleeve is maintained at uniform temperature T_o. Neglecting axial variation and taking into consideration dissipation, write the Navier-Stokes equations of motion, energy equation and boundary conditions for this flow. Assume constant properties.

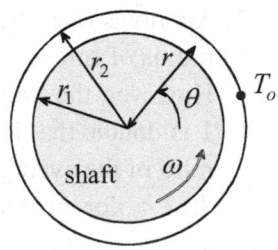

2.20 A rod of radius r_i moves axially with velocity U_o inside a concentric tube of radius r_o. A fluid having constant properties fills the space between the shaft and tube. The tube surface is maintained at uniform temperature T_o. Write the Navier-Stokes equations of motion, energy equation and surface boundary conditions taking into consideration dissipation. Assume that the streamlines are parallel to the surface.

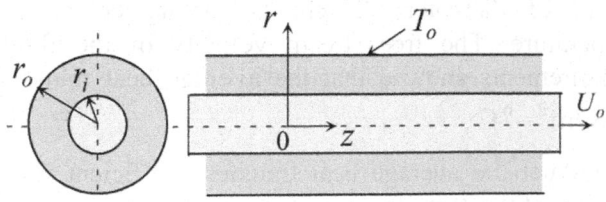

2.21   A rod of radius r_i rotates concentrically inside a tube of inner radius r_o. Lubrication oil fills the clearance between the shaft and the tube. Tube surface is maintained at uniform temperature T_o. The rod generates heat volumetrically at uniform rate q'''. Neglecting axial variation and taking into consideration dissipation, write the Navier-Stokes equations of motion, energy equation and boundary conditions for this flow. Assume constant properties.

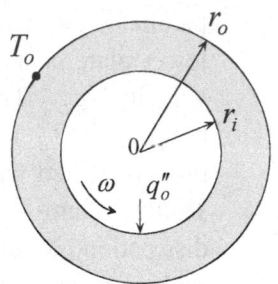

2.22   Air flows over the two spheres shown. The radius of sphere 2 is double that of sphere 1. However, the free stream velocity for sphere 1 is double that for sphere 2. Determine the ratio of the average heat transfer coefficients $\overline{h}_1 / \overline{h}_2$ for the two spheres.

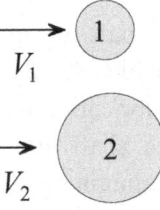

2.23   The average Nusselt number for laminar free convection over an isothermal vertical plate is determined analytically and is given by

$$\overline{Nu_L} = \frac{\overline{h}L}{k} = \frac{4}{3}\left[\frac{Gr_L}{4}\right]^{1/4} f(Pr),$$

where Gr_L is the Grashof number based on the length of the plate L and $f(Pr)$ is a function of the Prandtl number. Determine the percent change in the average heat transfer coefficient if the length of the plate is doubled.

2.24   An experiment was performed to determine the average heat transfer coefficient for forced convection over spheres. In the experiment a sphere of diameter 3.2 cm is maintained at uniform surface temperature. The free stream velocity of the fluid is 23.4 m/s. Measurements showed that the average heat transfer coefficient is $62 \, \text{W/m}^2 - ^\circ\text{C}$.

[a] Predict the average heat transfer coefficient for the same fluid which is at the same free stream temperature flowing over a sphere of diameter 6.4 cm which is maintained at the same surface temperature. The free stream velocity is 11.7 m/s.

[b] Which sphere transfers more heat?

2.25 Atmospheric air at $25\,^\circ\text{C}$ flows with a mean velocity of $10\,\text{m/s}$ between parallel plates. The plates are maintained at $115\,^\circ\text{C}$.

[a] Calculate the Eckert number. Can dissipation be neglected?

[b] Use scale analysis to compare the magnitude of the normal conduction, $k\,\partial^2 T / \partial y^2$, with dissipation, $\mu\,(\partial u / \partial y)^2$. Is dissipation negligible compared to conduction?

2.26 An infinitely large plate is immersed in an infinite fluid. The plate is suddenly moved along its plane with velocity U_o. Neglect gravity and assume constant properties.

[a] Show that the axial Navier-Stokes equation is given by

$$\rho\frac{\partial u}{\partial t} = \mu\frac{\partial^2 u}{\partial y^2}.$$

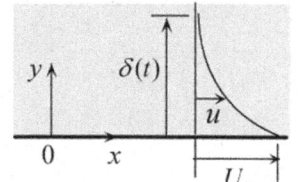

[b] Due to viscous forces, the effect of plate motion penetrates into the fluid. The penetration depth $\delta(t)$ increases with time. Use scaling to derive an expression for $\delta(t)$.

2.27 An infinitely large plate is immersed in an infinite fluid at uniform temperature T_i. The plate is suddenly maintained at temperature T_o. Assume constant properties and neglect gravity.

[a] Show that the energy equation is given by

$$\frac{\partial T}{\partial t} = \alpha\frac{\partial^2 T}{\partial y^2}.$$

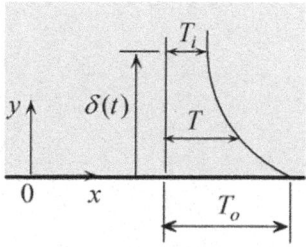

[b] Due to conduction, the effect of plate temperature propagates into the fluid. The penetration depth $\delta(t)$ increases with time. Use scaling to derive an expression for $\delta(t)$.

3

Exact One-Dimensional Solutions

3.1 Introduction

Although the energy equation for constant properties is linear, the Navier-Stokes equations of motion are non-linear. Thus, in general, convection problems are non-linear since the temperature field depends on the flow field. Nevertheless, exact solutions to certain simplified cases can easily be constructed. One of the objectives of this chapter is to develop an appreciation for the physical significance of each term in the equations of continuity, Navier-Stokes and energy and to identify the conditions under which certain terms can be neglected. Simplification of the governing equations is critical to constructing solutions. The general procedure in solving convection problems is, whenever possible, to first determine the flow field and then the temperature field.

3.2 Simplification of the Governing Equations

Simplified convection models are based on key assumptions that lead to tractable solutions. We will present these assumptions and study their application to the governing equations.

(1) Laminar flow. The assumption of laminar flow eliminates the effect of fluctuations. Mathematically this means that all time derivatives are set equal to zero at steady state.

(2) Constant properties. Returning to Table 2.1, we recall that for constant density (incompressible fluid), viscosity and conductivity, the velocity and temperature fields are *uncoupled*. This means that the velocity field can be determined first by solving the continuity and momentum equations without using the energy equation.

(3) Parallel streamlines. Consider the parallel flow of Fig. 3.1. This flow pattern is also referred to as *fully developed*. Since the velocity component normal to a streamline is zero, it follows that

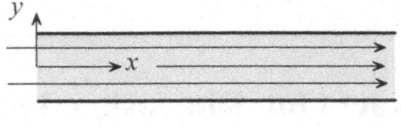

Fig. 3.1

$$v = 0. \tag{3.1}$$

Using this result into the continuity equation for two-dimensional constant density fluid, gives

$$\frac{\partial u}{\partial x} = 0. \tag{3.2}$$

Since (3.2) is valid everywhere in the flow field, it follows that

$$\frac{\partial^2 u}{\partial x^2} = 0. \tag{3.3}$$

Significant simplification is obtained when these results are substituted in the Navier-Stokes and energy equations. This is illustrated in Examples 3.1, 3.2 and 3.3. It should be emphasized that equations (3.1)-(3.3) are valid for constant density flow with parallel streamlines.

(4) Constant axial variation of temperature. For the case of axial flow in the x-direction, this condition leads to

$$\frac{\partial T}{\partial x} = \text{constant}. \tag{3.4}$$

Equation (3.4) is exact for certain channel flows and a reasonable approximation for others. The following are conditions that *may* lead to the validity of (3.4):

(a) Parallel streamlines.
(b) Uniform surface heat flux.
(c) Far away from the entrance of a channel (very long channels).

(5) Negligible axial conduction. For certain channel flows with high product of Reynolds and Prandtl numbers ($RePr > 100$), axial conduction may be neglected. Thus

$$\frac{\partial^2 T}{\partial x^2} = 0 . \tag{3.5}$$

(6) Concentric streamlines. In Fig. 3.2 a shaft rotates concentrically inside a sleeve. The streamlines are concentric circles. For axisymmetric conditions and no axial variations, we have

$$\frac{\partial v_\theta}{\partial \theta} = \frac{\partial T}{\partial \theta} = 0 . \tag{3.6}$$

It follows that

$$\frac{\partial^2 v_\theta}{\partial \theta^2} = \frac{\partial^2 T}{\partial \theta^2} = 0 . \tag{3.7}$$

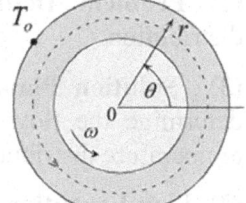

Fig. 3.2

3.3 Exact Solutions

The simplifications described in the previous section will be invoked to obtain exact solutions. We will consider various cases to show how these assumptions lead to tractable differential equations and solutions to the temperature distribution [1-4].

3.3.1 Couette Flow

In this class of flow configuration a fluid between parallel plates is set in motion by moving one or both plates in their own plane. Fluid motion can also be driven by axial pressure gradient. A general case of Couette flow includes the effects of both plate motion and pressure gradient. The plates are assumed infinite and thus there are no end effects. It will be shown that the streamlines in this flow are parallel to the plates.

Example 3.1: Couette Flow with Dissipation

Two infinitely large parallel plates form a channel of width H. An incompressible fluid fills the channel. The lower plate is stationary while the upper plate moves with constant velocity U_o. The lower plate is insulated and the upper plate is maintained at uniform temperature T_o. Taking into consideration dissipation, determine the temperature distribution in the fluid and the rate of heat transfer at the moving plate.

Assume laminar flow and neglect gravity, end effects and pressure variation in the channel.

(1) Observations. (i) The moving plate sets fluid in motion in the x-direction. (ii) Since the plates are infinite, the flow field does not vary in the axial direction x. (iii) Pressure gradient is zero. (iv) The fluid is incompressible (constant density). (v) Use Cartesian coordinates.

(2) Problem Definition.   Determine the velocity and temperature distribution.

(3) Solution Plan. Apply continuity and Navier-Stokes equations to determine the flow field.   Apply the energy equation to determine the temperature distribution.

(4) Plan Execution.

   (i) Assumptions. (1) Continuum, (2) Newtonian fluid, (3) steady state, (4) laminar flow, (5) constant properties (density, viscosity, conductivity and specific heat), (6) infinite plates, (7) no end effects, (8) parallel plates, and (9) negligible gravitational effect.

   (ii) Analysis. Since the objective is the determination of temperature distribution and heat transfer rate, it is logical to begin the analysis with the energy equation. The energy equation for constant properties is given by (2.19b)

$$\rho c_p \left(\frac{\partial T}{\partial t} + u \frac{\partial T}{\partial x} + v \frac{\partial T}{\partial y} + w \frac{\partial T}{\partial z} \right) = k \left(\frac{\partial^2 T}{\partial x^2} + \frac{\partial^2 T}{\partial y^2} + \frac{\partial^2 T}{\partial z^2} \right) + \mu\Phi \, , \, (2.19b)$$

where the dissipation function Φ is given by (2.17)

$$\Phi = 2 \left[\left(\frac{\partial u}{\partial x} \right)^2 + \left(\frac{\partial v}{\partial y} \right)^2 + \left(\frac{\partial w}{\partial z} \right)^2 \right]$$
$$+ \left[\left(\frac{\partial u}{\partial y} + \frac{\partial v}{\partial x} \right)^2 + \left(\frac{\partial v}{\partial z} + \frac{\partial w}{\partial y} \right)^2 + \left(\frac{\partial w}{\partial x} + \frac{\partial u}{\partial z} \right)^2 \right] - \frac{2}{3} \left(\frac{\partial u}{\partial x} + \frac{\partial v}{\partial y} + \frac{\partial w}{\partial z} \right)^2 .$$

$$(2.17)$$

Thus it is clear from (2.19b) and (2.17) that the determination of temperature distribution requires the determination of the velocity components u, v and w. This is accomplished by applying continuity and the Navier-Stokes equations. We begin with the continuity equation in Cartesian coordinates

$$\frac{\partial \rho}{\partial t} + u\frac{\partial \rho}{\partial x} + v\frac{\partial \rho}{\partial y} + w\frac{\partial \rho}{\partial z} + \rho\left[\frac{\partial u}{\partial x} + \frac{\partial v}{\partial y} + \frac{\partial w}{\partial z}\right] = 0. \qquad (2.2b)$$

For constant density

$$\frac{\partial \rho}{\partial t} = \frac{\partial \rho}{\partial x} = \frac{\partial \rho}{\partial y} = \frac{\partial \rho}{\partial z} = 0. \qquad (a)$$

Since the plates are infinite

$$\frac{\partial}{\partial x} = \frac{\partial}{\partial z} = w = 0. \qquad (b)$$

Substituting (a) and (b) into (2.2b), gives

$$\frac{\partial v}{\partial y} = 0. \qquad (c)$$

Integrating (c)

$$v(x,y) = f(x). \qquad (d)$$

To determine the "constant" of integration $f(x)$ we apply the no-slip boundary condition at the lower plate

$$v(x,0) = 0. \qquad (e)$$

Equations (d) and (e) give

$$f(x) = 0.$$

Substituting into (d)

$$v(x,y) = 0. \qquad (f)$$

Since the vertical component v vanishes everywhere, it follows that the streamlines are parallel. To determine the horizontal component u we apply the Navier-Stokes equation in the x-direction, (2.10x)

$$\rho\left(\frac{\partial u}{\partial t} + u\frac{\partial u}{\partial x} + v\frac{\partial u}{\partial y} + w\frac{\partial u}{\partial z}\right) = \rho g_x - \frac{\partial p}{\partial x} + \mu\left(\frac{\partial^2 u}{\partial x^2} + \frac{\partial^2 u}{\partial y^2} + \frac{\partial^2 u}{\partial z^2}\right).$$

(2.10x)

This equation is simplified as follows:

Steady state: $\qquad\qquad\qquad \dfrac{\partial u}{\partial t} = 0.$ $\qquad\qquad\qquad$ (g)

Negligible gravity effect: $\qquad g_x = 0.$ $\qquad\qquad\qquad$ (h)

No axial pressure variation: $\qquad \dfrac{\partial p}{\partial x} = 0.$ $\qquad\qquad\qquad$ (i)

Substituting (b) and (f)-(i) into (2.10x) gives

$$\frac{d^2 u}{dy^2} = 0.$$

(j)

The solution to (j) is

$$u = C_1 y + C_2,$$

(k)

where C_1 and C_2 are constants of integration. The two boundary conditions on u are:

$$u(0) = 0 \text{ and } u(H) = U_o.$$

(l)

These conditions give

$$C_1 = \frac{U_o}{H} \text{ and } C_2 = 0,$$

(m)

Substituting (m) into (k)

$$\frac{u}{U_o} = \frac{y}{H}.$$

(3.8)

With the velocity distribution determined, we return to the dissipation function and energy equation. Substituting (b) and (f) into (2.17) gives

$$\Phi = \left(\frac{\partial u}{\partial y}\right)^2 . \tag{n}$$

Using solution (3.8) into (n) gives

$$\Phi = \frac{U_o^2}{H^2} . \tag{o}$$

Noting that for steady state $\partial T / \partial t = 0$ and using (b), (f) and (o), the energy equation (2.19b) simplifies to

$$k\frac{d^2 T}{dy^2} + \mu \frac{U_o^2}{H^2} = 0 . \tag{p}$$

In arriving at (p), axial temperature variation was neglected. This is valid for infinite plates at uniform surface temperature and no variation of velocity and temperature normal to the flow direction. Equation (p) is solved by direct integration to obtain

$$T = -\frac{\mu U_o^2}{2kH^2} y^2 + C_3 y + C_4 , \tag{q}$$

where C_3 and C_4 are constants of integration. The boundary conditions on (q) are

$$-k\frac{dT(0)}{dy} = 0 \text{ and } T(H) = T_o . \tag{r}$$

These boundary conditions and solution (q) give

$$C_3 = 0 \text{ and } C_4 = T_o + \frac{\mu U_o^2}{2k} . \tag{s}$$

Substituting (s) into (q) and rearranging the result in dimensionless form, give

$$\frac{T - T_o}{\dfrac{\mu U_o^2}{k}} = \frac{1}{2}\left(1 - \frac{y^2}{H^2}\right). \tag{3.9}$$

The heat flux at the moving plate is determined by applying Fourier's law at $y = H$

$$q''(H) = -k\frac{dT(H)}{dy}.$$

Substituting (3.9) into the above, gives

$$q''(H) = \frac{\mu U_o^2}{H}.$$ (3.10)

(iii) **Checking.** *Dimensional check*: Each term in (3.8) and (3.9) is dimensionless. Units of (3.10) should be W/m^2

$$q''(H) = \frac{\mu(\text{kg/m} - \text{s})U_o^2(\text{m}^2/\text{s}^2)}{H(\text{m})} = \frac{\text{kg}}{\text{s}^3} = \frac{W}{\text{m}^2}.$$

Differential equation check: Velocity solution (3.8) satisfies equation (j) and temperature solution (3.9) satisfies (p).

Boundary conditions check: Velocity solution (3.8) satisfies boundary conditions (l) and temperature solution (3.9) satisfies boundary conditions (r).

Limiting check: (i) If the upper plate is stationary the fluid will also be stationary. Setting $U_o = 0$ in (3.8) gives $u(y) = 0$.

(ii) If the upper plate is stationary the dissipation will vanish, temperature distribution will be uniform equal to T_o and surface flux at the upper plate should be zero. Setting $U_o = 0$ in (o), (3.9) and (3.10) give $\Phi = 0$, $T(y) = T_o$ and $q''(H) = 0$.

(iii) If the fluid is inviscid the dissipation term will vanish and the temperature should be uniform equal to T_o. Setting $\mu = 0$ in (3.9) gives $T(y) = T_o$.

(iv) Global conservation of energy. All energy dissipation due to friction is conducted in the y-direction. Energy dissipation is equal to the rate of work done by the plate to overcome frictional resistance. Thus

$$W = \tau(H)U_o,$$ (t)

where

$\quad W$ = work done by the plate.

$\tau(H)$ = shearing stress at the moving plate, given by,

$$\tau(H) = \mu \frac{du(H)}{dy}.$$ (u)

Substituting (3.8) into (u)

$$\tau(H) = \mu \frac{U_o}{H}.$$ (v)

Combining (v) and (t), gives

$$W = \frac{\mu U_o^2}{H}.$$ (w)

This result is identical to surface heat flux given in (3.10).

(5) Comments. (i) Treating the plate as infinite is one of the key simplifying assumptions. This eliminates the x-coordinate as a variable and results in governing equations that are ordinary. Alternatively, one could state that the streamline are parallel. This means that $v = \partial v / \partial y = 0$. Substituting this into the continuity equation for two-dimensional incompressible flow gives $\partial u / \partial x = 0$. This is identical to equation (b) which is based on assuming infinite plate.

(ii) Maximum temperature occurs at the insulated surface $y = 0$. Setting $y = 0$ in (3.9) gives

$$T(0) - T_o = \frac{\mu U_o^2}{2k}.$$

3.3.2 Poiseuille Flow

This class of problems deals with axial flow in long channels or tubes. Fluid motion is driven by axial pressure gradient. The channel or tube is assumed infinite and thus end effects are neglected. The flow is characterized by parallel streamlines.

Example 3.2: Flow in a Tube at Uniform Surface Temperature

Incompressible fluid flows in a long tube of radius r_o. The fluid is set in motion due to an axial pressure gradient $\partial p / \partial z$. The

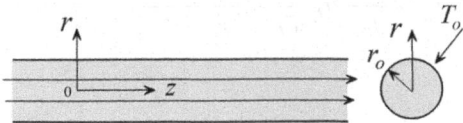

surface of the tube is maintained at uniform temperature T_o. Taking into consideration dissipation, assuming axisymmetric laminar flow and neglecting gravity, axial temperature variation, and end effects, determine

[a] *Fluid temperature distribution.*

[b] Surface *heat flux.*

[c] *Nusselt number based on* $T(0) - T_o$.

(1) Observations. (i) Fluid motion is driven by axial pressure drop. (ii) For a very long tube the flow field does not vary in the axial direction z. (iii) The fluid is incompressible (constant density). (iv) Heat is generated due to viscous dissipation. It is removed from the fluid by conduction at the surface. (v) The Nusselt number is a dimensionless heat transfer coefficient. (vi) To determine surface heat flux and heat transfer coefficient requires the determination of temperature distribution. (vii) Temperature distribution depends on the velocity distribution. (viii) Use cylindrical coordinates.

(2) Problem Definition.   Determine the velocity and temperature distribution.

(3) Solution Plan. Apply continuity and Navier-Stokes equations in cylindrical coordinates to determine the flow field. Apply the energy equation to determine temperature distribution. Fourier's law gives surface heat flux. Equation (1.10) gives the heat transfer coefficient.

(4) Plan Execution.

   (i) Assumptions. (1) Continuum, (2) Newtonian fluid, (3) steady state, (4) laminar flow, (5) axisymmetric flow, (6) constant properties (density, viscosity and conductivity), (7) no end effects, (8) uniform surface temperature, and (9) negligible gravitational effect.

   (ii) Analysis. [a] Since temperature distribution is obtained by solving the energy equation, we begin the analysis with the energy equation. The energy equation in cylindrical coordinates for constant properties is given by (2.24)

$$\rho c_p \left(\frac{\partial T}{\partial t} + v_r \frac{\partial T}{\partial r} + \frac{v_\theta}{r} \frac{\partial T}{\partial \theta} + v_z \frac{\partial T}{\partial z} \right) = k \left[\frac{1}{r} \frac{\partial}{\partial r} \left(r \frac{\partial T}{\partial r} \right) + \frac{1}{r^2} \frac{\partial^2 T}{\partial \theta^2} + \frac{\partial^2 T}{\partial z^2} \right] + \mu \Phi,$$

$$(2.24)$$

where the dissipation function Φ is given by (2.25)

$$\Phi = 2\left(\frac{\partial v_r}{\partial r}\right)^2 + 2\left(\frac{1}{r}\frac{\partial v_\theta}{\partial \theta} + \frac{v_r}{r}\right)^2 + 2\left(\frac{\partial v_z}{\partial z}\right)^2 + \left(\frac{\partial v_\theta}{\partial r} - \frac{v_\theta}{r} + \frac{1}{r}\frac{\partial v_r}{\partial \theta}\right)^2$$

$$+ \left(\frac{1}{r}\frac{\partial v_z}{\partial \theta} + \frac{\partial v_\theta}{\partial z}\right)^2 + \left(\frac{\partial v_r}{\partial z} + \frac{\partial v_z}{\partial r}\right)^2. \tag{2.25}$$

Equations (2.24) and (2.25) show that the determination of temperature distribution requires the determination of the velocity components v_r, v_θ and v_z. The flow field is determined by solving the continuity and the Navier-Stokes equations. We begin with the continuity equation in cylindrical coordinates

$$\frac{\partial \rho}{\partial t} + \frac{1}{r}\frac{\partial}{\partial r}(\rho r v_r) + \frac{1}{r}\frac{\partial}{\partial \theta}(\rho v_\theta) + \frac{\partial}{\partial z}(\rho v_z) = 0. \tag{2.4}$$

For constant density

$$\frac{\partial \rho}{\partial t} = \frac{\partial \rho}{\partial r} = \frac{\partial \rho}{\partial \theta} = \frac{\partial \rho}{\partial z} = 0. \tag{a}$$

For axisymmetric flow

$$v_\theta = \frac{\partial}{\partial \theta} = 0. \tag{b}$$

For a long tube with no end effects, axial changes in velocity vanish

$$\frac{\partial v_z}{\partial z} = 0. \tag{c}$$

Substituting (a)-(c) into (2.4)

$$\frac{d}{dr}(r v_r) = 0. \tag{d}$$

Integrating (d)

$$r v_r = f(z). \tag{e}$$

To determine the "constant" of integration $f(z)$ we apply the no-slip boundary condition at the surface

$$v_r(r_o, z) = 0.$$ (f)

Equations (e) and (f) give

$$f(z) = 0.$$

Substituting into (e)

$$v_r(r, z) = 0.$$ (g)

Since the radial component v_r vanishes everywhere, it follows that the streamlines are parallel to the surface. To determine the axial component v_z we apply the Navier-Stokes equation in the z-direction, (2.11z)

$$\rho\left(v_r \frac{\partial v_z}{\partial r} + \frac{v_\theta}{r} \frac{\partial v_z}{\partial \theta} + v_z \frac{\partial v_z}{\partial z} + \frac{\partial v_z}{\partial t}\right) =$$

$$\rho g_z - \frac{\partial p}{\partial z} + \mu\left[\frac{1}{r} \frac{\partial}{\partial r}\left(r \frac{\partial v_z}{\partial r}\right) + \frac{1}{r^2} \frac{\partial^2 v_z}{\partial \theta^2} + \frac{\partial^2 v_z}{\partial z^2}\right].$$ (2.11z)

This equation is simplified as follows:
Steady state

$$\frac{\partial}{\partial t} = 0.$$ (h)

Negligible gravity effect

$$g_r = g_z = 0.$$ (i)

Substituting (b), (c) and (g)-(i) into (2.11z) gives

$$-\frac{\partial p}{\partial z} + \mu \frac{1}{r} \frac{d}{dr}\left(r \frac{dv_z}{dr}\right) = 0.$$ (3.11)

Since v_z depends on r only, equation (3.11) can be written as

$$\frac{\partial p}{\partial z} = \mu \frac{1}{r} \frac{d}{dr}\left(r \frac{dv_z}{dr}\right) = g(r).$$ (j)

Integrating (j) with respect to z

$$p = g(r)z + C_o,$$ (k)

where C_o is constant of integration. We turn our attention now to the radial component of Navier-Stokes equation, (2.11r)

$$\rho v \left(v_r \frac{\partial v_r}{\partial r} + \frac{v_\theta}{r} \frac{\partial v_r}{\partial \theta} - \frac{v_\theta^2}{r} + v_z \frac{\partial v_r}{\partial z} + \frac{\partial v_r}{\partial t} \right) =$$

$$\rho g_r - \frac{\partial p}{\partial r} + \mu \left[\frac{\partial}{\partial r} \left(\frac{1}{r} \frac{\partial}{\partial r} (r v_r) \right) + \frac{1}{r^2} \frac{\partial^2 v_r}{\partial \theta^2} - \frac{2}{r^2} \frac{\partial v_\theta}{\partial \theta} + \frac{\partial^2 v_r}{\partial z^2} \right].$$

(2.11r)

Substituting (b), (g) and (i) into (2.11r), gives

$$\frac{\partial p}{\partial r} = 0.$$ (l)

Integrating (l)

$$p = f(z),$$ (m)

where $f(z)$ is "constant" of integration. We now have two solutions for the pressure p: (k) and (m). Equating the two, gives

$$p = g(r)z + C_o = f(z).$$ (n)

One side of (n) shows that the pressure depends on z only while the other side shows that it depends on r and z. This, of course, is a contradiction. The only possibility for reconciling this is by requiring that

$$g(r) = C,$$ (o)

where C is a constant. Substituting (o) into (j)

$$\frac{\partial p}{\partial z} = \mu \frac{1}{r} \frac{d}{dr} \left(r \frac{dv_z}{dr} \right) = C.$$ (p)

Thus the axial pressure gradient in the tube is constant. Equation (p) can now be integrated to give the axial velocity distribution. Integrating once

$$r\frac{dv_z}{dr} = \frac{1}{2\mu}\frac{dp}{dz}r^2 + C_1.$$

Separating variables and integrating again

$$v_z = \frac{1}{4\mu}\frac{dp}{dz}r^2 + C_1 \ln r + C_2, \qquad (q)$$

where C_1 and C_2 are constants of integration. The two boundary conditions on v_z are

$$\frac{dv_z(0)}{dr} = 0, \quad v_z(r_o) = 0. \qquad (r)$$

Equations (q) and (r) give C_1 and C_2

$$C_1 = 0, \quad C_2 = -\frac{1}{4\mu}\frac{dp}{dz}r_o^2.$$

Substituting into (q)

$$v_z = -\frac{1}{4\mu}\frac{dp}{dz}(r^2 - r_o^2). \qquad (3.12)$$

With the velocity distribution determined we return to the energy equation (2.24) and the dissipation function (2.25). To simplify the problem, we will assume that axial temperature variation is negligible. Thus

$$\frac{\partial T}{\partial z} = \frac{\partial^2 T}{\partial z^2} = 0. \qquad (s)$$

It should be emphasized that this is an approximation and not an exact condition. Substituting (b), (c), (g), (h) and (s) into (2.24)

$$k\frac{1}{r}\frac{d}{dr}\left(r\frac{dT}{dr}\right) + \mu\Phi = 0. \qquad (t)$$

Using (b), (c) and (g) into (2.25) gives the dissipation function for this flow

$$\Phi = \left(\frac{dv_z}{dr} \right)^2 .$$

Substituting the velocity solution (3.12) into the above, gives

$$\Phi = \left(\frac{1}{2\mu} \frac{dp}{dz} \right)^2 r^2 . \qquad (u)$$

Using (u) to eliminate Φ in (t) and rearranging, we obtain

$$\frac{d}{dr} \left(r \frac{dT}{dr} \right) = -\frac{1}{4k\mu} \left(\frac{dp}{dz} \right)^2 r^3 . \qquad (3.13)$$

Integrating the above twice

$$T = -\frac{1}{64k\mu} \left(\frac{dp}{dz} \right)^2 r^4 + C_3 \ln r + C_4 . \qquad (v)$$

Two boundary conditions are needed to evaluate the constants of integration C_3 and C_4. They are:

$$\frac{dT(0)}{dr} = 0 \text{ and } T(r_o) = T_o . \qquad (w)$$

Equations (v) and (w) give the two constants

$$C_3 = 0, \quad C_4 = T_o + \frac{1}{64k\mu} \left(\frac{dp}{dz} \right)^2 r_o^4 .$$

Substituting the above into (v) gives

$$T = T_o + \frac{r_o^4}{64k\mu} \left(\frac{dp}{dz} \right)^2 \left(1 - \frac{r^4}{r_o^4} \right) . \qquad (3.14a)$$

This solution can be expressed in dimensionless form as

$$\frac{T - T_o}{\dfrac{r_o^4}{64 k \mu}\left(\dfrac{dp}{dz}\right)^2} = \left(1 - \frac{r^4}{r_o^4}\right). \qquad (3.14b)$$

[b] Surface heat flux $q''(r_o)$ is obtained by applying Fourier's law

$$q''(r_o) = -k\frac{dT(r_o)}{dr}.$$

Using (3.14) into the above

$$q''(r_o) = \frac{r_o^3}{16\mu}\left(\frac{dp}{dz}\right)^2. \qquad (3.15)$$

[c] The Nusselt number is defined as

$$Nu = \frac{hD}{k} = \frac{2hr_o}{k}, \qquad (x)$$

where D is tube diameter. The heat transfer coefficient h is determined using equation (1.10)

$$h = -\frac{k}{[T(0) - T_o]}\frac{dT(r_o)}{dr}. \qquad (y)$$

Substituting (3.14a) into (y)

$$h = \frac{4k}{r_o}. \qquad (z)$$

Substituting (z) into (x)

$$Nu = 8. \qquad (3.16)$$

(iii) **Checking.** *Dimensional check*: Each term in (3.12) has units of velocity. Each term in (3.14a) has units of temperature. Each term in (3.15) has units of W/m^2.

Differential equation check: Velocity solution (3.12) satisfies equation (p) and temperature solution (3.14) satisfies (3.13).

Boundary conditions check: Velocity solution (3.12) satisfies boundary conditions (r) and temperature solution (3.14) satisfies boundary conditions (w).

Limiting check: (i) If pressure is uniform ($dp/dz = 0$) the fluid will be stationary. Setting $dp/dz = 0$ in (3.12) gives $v_z = 0$.

(ii) If pressure is uniform ($dp/dz = 0$) the fluid will be stationary and no dissipation takes place and thus surface heat transfer should vanish. Setting $dp/dz = 0$ in (3.15) gives $q''(r_o) = 0$.

(iii) Global conservation of energy. Heat transfer rate leaving the tube must be equal to the rate of work required to pump the fluid. Pump work for a tube section of length L is

$$W = (p_1 - p_2)Q, \qquad (\text{z-1})$$

where

p_1 = upstream pressure

p_2 = downstream pressure

Q = volumetric flow rate, given by

$$Q = 2\pi \int_0^{r_o} v_z r \, dr .$$

Substituting (3.12) into the above and integrating

$$Q = -\frac{\pi}{8\mu} \frac{dp}{dz} r_o^4 . \qquad (\text{z-2})$$

Combining (z-1) and (z-2)

$$W = -\frac{\pi r_o^4}{8\mu} \frac{dp}{dz} (p_1 - p_2) . \qquad (\text{z-3})$$

Work per unit area W'' is

$$W'' = \frac{W}{2\pi r_o L} .$$

Substituting (z-3) into the above

$$W'' = -\frac{r_o^3}{16\mu}\frac{dp}{dz}\frac{(p_1 - p_2)}{L}.$$ (z-4)

However

$$\frac{(p_1 - p_2)}{L} = -\frac{dp}{dz}.$$

Combining this result with (z-4) gives

$$W'' = \frac{r_o^3}{16\mu}\left(\frac{dp}{dz}\right)^2.$$

This result is identical to surface heat transfer rate given in (3.15).

(5) Comments. (i) Neglecting axial variation of temperature is a key factor in simplifying the problem. This assumption eliminates the z-coordinate as a variable and results in governing equations that are ordinary.

(ii) Solution (3.14) shows that the maximum temperature occurs at the center, $r = 0$.

(iii) The Nusselt number is constant, independent of Reynolds and Prandtl numbers.

(iv) A more appropriate definition of the heat transfer coefficient is based on the mean temperature, T_m, rather than the centerline temperature. Thus, (y) is modified to

$$h = \frac{-k}{T_m - T_o}\frac{dT(r_o)}{dr}.$$

3.3.3 Rotating Flow

Angular fluid motion can be generated by rotating a cylinder. A common example is fluid motion in the clearance space between a journal and its bearing. Under certain conditions the streamlines for such flows are concentric circles.

Example 3.3: Lubrication Oil Temperature in Rotating Shaft

Lubrication oil fills the clearance between a shaft and its housing. The radius of the shaft is r_i and its angular velocity is ω. The housing radius

is r_o and its temperature is T_o. Assuming laminar
flow and taking into consideration dissipation,
determine the maximum temperature rise in the oil
and the heat generated due to dissipation.

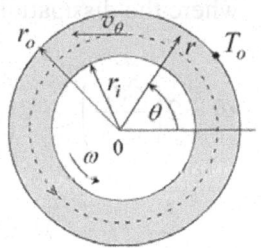

Solution

(1) Observations. (i) Fluid motion is driven by
shaft rotation (ii) The housing is stationary. (iii) Axial variation in velocity
and temperature are negligible for a very long shaft. (iv) Velocity and
temperature do not vary with angular position. (v) Heat generated by
viscous dissipation is removed from the oil at the housing. (vi) The fluid is
incompressible (constant density). (vii) No heat is conducted through the
shaft. (viii) The maximum temperature occurs at the shaft. (ix) Use
cylindrical coordinates.

(2) Problem Definition.   Determine the velocity and temperature
distribution in the oil.

(3) Solution Plan.   Apply continuity and Navier-Stokes equations in
cylindrical coordinates to determine the flow field.   Use the energy
equation to determine temperature distribution. Apply Fourier's law at the
housing to determine the rate of energy generated by dissipation.

(4) Plan Execution.

   (i) Assumptions. (1) Continuum, (2) Newtonian fluid, (3) steady state,
(4) laminar flow, (5) axisymmetric flow, (6) constant properties (density,
conductivity, specific heat, and viscosity), (7) no end effects, (8) uniform
surface temperature and (9) negligible gravitational effect.

   (ii) Analysis.   Temperature distribution is obtained by solving the
energy equation. Thus we begin the analysis with the energy equation. The
energy equation in cylindrical coordinates for constant properties is given
by (2.24)

$$\rho c_p \left(\frac{\partial T}{\partial t} + v_r \frac{\partial T}{\partial r} + \frac{v_\theta}{r} \frac{\partial T}{\partial \theta} + v_z \frac{\partial T}{\partial z} \right) =$$

$$k \left[\frac{1}{r} \frac{\partial}{\partial r} \left(r \frac{\partial T}{\partial r} \right) + \frac{1}{r^2} \frac{\partial^2 T}{\partial \theta^2} + \frac{\partial^2 T}{\partial z^2} \right] + \mu \Phi, \qquad (2.24)$$

where the dissipation function Φ is given by (2.25)

$$\Phi = 2\left(\frac{\partial v_r}{\partial r}\right)^2 + 2\left(\frac{1}{r}\frac{\partial v_\theta}{\partial \theta} + \frac{v_r}{r}\right)^2 + 2\left(\frac{\partial v_z}{\partial z}\right)^2 + \left(\frac{\partial v_\theta}{\partial r} - \frac{v_\theta}{r} + \frac{1}{r}\frac{\partial v_r}{\partial \theta}\right)^2$$

$$+ \left(\frac{1}{r}\frac{\partial v_z}{\partial \theta} + \frac{\partial v_\theta}{\partial z}\right)^2 + \left(\frac{\partial v_r}{\partial z} + \frac{\partial v_z}{\partial r}\right)^2. \tag{2.25}$$

The solution to (2.24) requires the determination of the velocity components v_r, v_θ and v_z. These are obtained by solving the continuity and the Navier-Stokes equations in cylindrical coordinates. The continuity equation is given by equation (2.4)

$$\frac{\partial \rho}{\partial t} + \frac{1}{r}\frac{\partial}{\partial r}(\rho r v_r) + \frac{1}{r}\frac{\partial}{\partial \theta}(\rho v_\theta) + \frac{\partial}{\partial z}(\rho v_z) = 0. \tag{2.4}$$

For constant density

$$\frac{\partial \rho}{\partial t} = \frac{\partial \rho}{\partial r} = \frac{\partial \rho}{\partial \theta} = \frac{\partial \rho}{\partial z} = 0. \tag{a}$$

For axisymmetric flow

$$\frac{\partial}{\partial \theta} = 0. \tag{b}$$

For a long shaft with no end effects axial changes are negligible

$$\frac{\partial v_z}{\partial z} = 0. \tag{c}$$

Substituting (a)-(c) into (2.4)

$$\frac{d}{dr}(r v_r) = 0. \tag{d}$$

Integrating (d)

$$r v_r = C. \tag{e}$$

To determine the constant of integration C we apply the no-slip boundary condition at the housing surface

$$v_r(r_o) = 0.$$ (f)

Equations (e) and (f) give

$$C = 0.$$

Substituting into (e)

$$v_r(r,z) = 0.$$ (g)

Since the radial component v_r vanishes everywhere, it follows that the streamlines are concentric circles. To determine the tangential velocity v_θ we apply the Navier-Stokes equation in the θ-direction, equation (2.11θ)

$$\rho\left(v_r \frac{\partial v_\theta}{\partial r} + \frac{v_\theta}{r}\frac{\partial v_\theta}{\partial \theta} + \frac{v_r v_\theta}{r} + v_z \frac{\partial v_\theta}{\partial z} + \frac{\partial v_\theta}{\partial t}\right) =$$

$$\rho g_\theta - \frac{1}{r}\frac{\partial p}{\partial \theta} + \mu\left[\frac{\partial}{\partial r}\left(\frac{1}{r}\frac{\partial}{\partial r}(rv_\theta)\right) + \frac{1}{r^2}\frac{\partial^2 v_\theta}{\partial \theta^2} + \frac{2}{r^2}\frac{\partial v_r}{\partial \theta} + \frac{\partial^2 v_\theta}{\partial z^2}\right].$$

$$(2.11\theta)$$

For steady state

$$\frac{\partial}{\partial t} = 0.$$ (h)

Neglecting gravity and applying (b), (c), (g) and (h), equation (2.11θ) simplifies to

$$\frac{d}{dr}\left(\frac{1}{r}\frac{d}{dr}(rv_\theta)\right) = 0.$$ (3.17)

Integrating (3.17) twice

$$v_\theta = \frac{C_1}{2}r + \frac{C_2}{r},$$ (i)

where C_1 and C_2 are constants of integration. The two boundary conditions on v_θ are

$$v_\theta(r_i) = \omega r_i, \quad v_\theta(r_o) = 0. \tag{j}$$

Boundary conditions (j) give C_1 and C_2

$$C_1 = -\frac{2\omega r_i^2}{r_o^2 - r_i^2}, \quad C_2 = \frac{\omega r_i^2 r_o^2}{r_o^2 - r_i^2}. \tag{k}$$

Substituting (k) into (i) and rearranging in dimensionless form, gives

$$\frac{v_\theta(r)}{\omega r_i} = \frac{(r_o/r_i)^2 (r_i/r) - (r/r_i)}{(r_o/r_i)^2 - 1}. \tag{3.18}$$

We now return to the energy equation (2.24) and the dissipation function (2.25). Using (b), (c), (g) and (h), equation (2.24) simplifies to

$$k\frac{1}{r}\frac{d}{dr}\left(r\frac{dT}{dr}\right) + \mu\Phi = 0. \tag{l}$$

The dissipation function (2.25) is simplified using (b), (c) and (g)

$$\Phi = \left(\frac{dv_\theta}{dr} - \frac{v_\theta}{r}\right)^2. $$

Substituting the velocity solution (3.18) into the above, gives

$$\Phi = \left[\frac{2\omega r_i^2}{1 - (r_i/r_o)^2}\right]^2 \frac{1}{r^4}. \tag{m}$$

Combining (m) and (l) and rearranging, we obtain

$$\frac{d}{dr}\left(r\frac{dT}{dr}\right) = -\frac{\mu}{k}\left[\frac{2\omega r_i^2}{1 - (r_i/r_o)^2}\right]^2 \frac{1}{r^3}. \tag{3.19}$$

Integrating (3.19) twice

$$T(r) = -\frac{\mu}{4k}\left[\frac{2\omega r_i^2}{1-(r_i/r_o)^2}\right]^2\frac{1}{r^2} + C_3\ln r + C_4, \tag{n}$$

where C_3 and C_4 are the integration constants. Two boundary conditions are needed to determine C_3 and C_4. They are:

$$\frac{dT(r_i)}{dr} = 0 \text{ and } T(r_o) = T_o. \tag{o}$$

Equations (n) and (o) give the two constants

$$C_3 = -\frac{\mu}{2k}\left[\frac{2\omega r_i^2}{1-(r_i/r_o)^2}\right]^2\frac{1}{r_i^2},$$

and

$$C_4 = T_o + \frac{\mu}{4k}\left[\frac{2\omega r_i^2}{1-(r_i/r_o)^2}\right]^2\left[\frac{1}{r_o^2}+\frac{2}{r_i^2}\ln r_o\right].$$

Substituting the above into (n)

$$T(r) = T_o + \frac{\mu}{4k}\left[\frac{2\omega r_i}{1-(r_i/r_o)^2}\right]^2\left[(r_i/r_o)^2 - (r_i/r)^2 + 2\ln(r_o/r)\right]. \tag{3.20a}$$

This solution can be expressed in dimensionless form as

$$\frac{T(r)-T_o}{\dfrac{\mu}{4k}\left[\dfrac{2\omega r_i}{1-(r_i/r_o)^2}\right]^2} = (r_i/r_o)^2 - (r_i/r)^2 + 2\ln(r_o/r). \tag{3.20b}$$

The maximum temperature is at the shaft's surface. Setting $r = r_i$ in (3.20a) gives

$$T(r_i) - T_o = \frac{\mu}{4k}\left[\frac{2\omega r_i}{1-(r_i/r_o)^2}\right]^2\left[(r_i/r_o)^2 + 2\ln(r_o/r_i) - 1\right]. \tag{3.21}$$

Energy generated due to dissipation per unit shaft length, $q'(r_o)$, is determined by applying Fourier's law at the housing. Thus

$$q'(r_o) = -2\pi r_o k \frac{dT(r_o)}{dr}.$$

Using (3.20a), the above gives

$$q'(r_o) = 4\pi \mu \frac{(\omega r_i)^2}{1 - (r_i / r_o)^2}. \tag{3.22}$$

(iii) Checking. *Dimensional check*: each term in solutions (3.18) and (3.20b) is dimensionless. Equation (3.22) has the correct units of W/m.

Differential equation check: Velocity solution (3.18) satisfies equation (3.17) and temperature solution (3.20) satisfies (3.19).

Boundary conditions check: Velocity solution (3.18) satisfies boundary conditions (j) and temperature solution (3.20) satisfies boundary conditions (o).

Limiting check: (i) If the shaft does not rotate the fluid will be stationary. Setting $\omega = 0$ in (3.18) gives $v_\theta = 0$.

(ii) If the shaft does not rotate no dissipation takes place and thus surface heat transfer should vanish. Setting $\omega = 0$ in (3.22) gives $q'(r_o) = 0$.

Global conservation of energy: Heat transfer rate from the housing must equal to work required to overcome friction at the shaft's surface. The rate of shaft work per unit length is given by

$$W' = 2\pi r_i \tau(r_i) \omega r_i, \tag{p}$$

where

W' = work done on the fluid per unit shaft length.
$\tau(r_i)$ = shearing stress at the shaft's surface, given by

$$\tau(r_i) = \mu \left[\frac{dv_\theta}{dr} - \frac{v_\theta}{r} \right]_{r=r_i}. \tag{q}$$

Substituting (3.18) into the above

$$\tau(r_i) = 2\mu \frac{\omega}{1-(r_i/r_o)^2}. \tag{r}$$

Combining (p) and (r) and rearranging, gives

$$W' = 4\pi\mu \frac{(\omega r_i)^2}{1-(r_i/r_o)^2}. \tag{s}$$

This result is identical to surface heat transfer rate given in (3.22)

(5) Comments. (i) The key simplifying assumption is axisymmetry. This results in concentric streamlines with vanishing normal velocity and angular changes.

(ii) Temperature rise of the lubricating oil and energy dissipation increase as the clearance between the shaft and the housing is decreased. This is evident from equations (3.22) and (s) which show that in the limit as $(r_i/r_o) \to 1$, $W' = q' \to \infty$.

(iii) Velocity and temperature distributions are governed by a single parameter (r_i/r_o).

REFERENCES

[1] Pai, Shih-I, *Viscous flow Theory*, D. Van Nostrand, 1956.

[2] Bird, R.B., W.E. Stewart and E.N. Lightfoot, *Transport Phenomena*, Wiley, 1960.

[3] Schlichting, H, *Boundary Layer Theory*, 7th ed., translated into English by J. Kestin, McGraw-Hill, 1979.

[4] Burmiester, L.C., *Convective Heat Transfer*, Wiley, 1083.

PROBLEMS

3.1   A large plate moves with constant velocity U_o parallel to a stationary plate at a distance H. An incompressible fluid fills the channel formed by the plates. The stationary plate is at temperature T_1 and the moving plate is at temperature T_o. Taking into consideration dissipation, determine the maximum temperature and the heat flux at the moving plate. Assume laminar flow and neglect gravity effect and pressure variation in the channel.

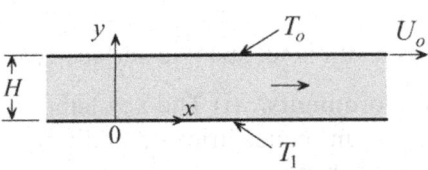

3.2   A large plate moves with constant velocity U_o parallel to a stationary plate separated by a distance H. An incompressible fluid fills the channel formed by the plates. The upper plate is maintained at uniform temperature T_o and the stationary plate is insulated. A pressure gradient dp/dx is applied to the fluid. Taking into consideration dissipation, determine the temperature of the insulated plate and the heat flux at the upper plate. Assume laminar flow and neglect gravity effect.

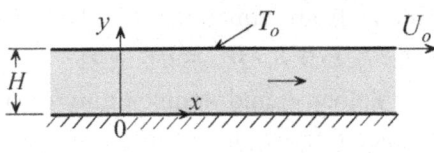

3.3   Incompressible fluid is set in motion between two large parallel plates by moving the upper plate with constant velocity U_o and holding the lower plate stationary. The clearance between the plates is H. The lower plate is insulated while the upper plate exchanges heat with the ambient by convection. The heat transfer coefficient is h and the ambient temperature is T_∞. Taking into consideration dissipation determine the temperature of the insulated plate and the heat flux at the moving plate. Assume laminar flow and neglect gravity effect.

3.4   Two parallel plates are separated by a distance $2H$. The plates are moved in opposite directions with constant velocity U_o. Each plate is maintained at uniform temperature T_o. Taking into consideration dissipation, determine the heat flux at the plates. Assume laminar flow and neglect gravity effect.

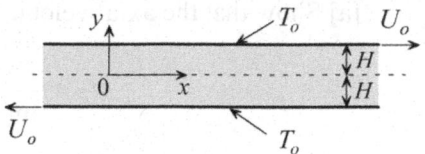

3.5 Incompressible fluid flows in a long tube of radius r_o. Fluid motion is driven by an axial pressure gradient dp/dz. The tube exchanges heat by convection with an ambient fluid. The heat transfer coefficient is h and the ambient temperature is T_∞. Taking into consideration dissipation, assuming laminar incompressible axisymmetric flow, and neglecting gravity, axial temperature variation and end effects, determine:

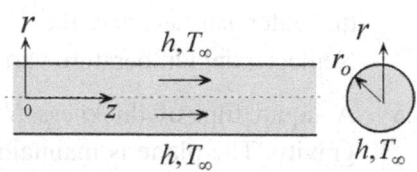

[a] Surface temperature.
[b] Surface heat flux.
[c] Nusselt number based on $T(0) - T(r_o)$.

3.6 Fluid flows axially in the annular space between a cylinder and a concentric rod. The radius of the rod is r_i and that of the cylinder is r_o. Fluid motion in the annular space is driven by an axial pressure gradient dp/dz. The cylinder is maintained at uniform temperature T_o. Assume incompressible laminar axisymmetric flow and neglect gravity and end effects. Show that the axial velocity is given by

$$v_z = \frac{r_o^2}{4\mu}\frac{dp}{dz}\left[(r/r_o)^2 - \frac{1-(r_i/r_o)^2}{\ln(r_o/r_i)}\ln(r/r_o)-1\right].$$

3.7 A rod of radius r_i is placed concentrically inside a cylinder of radius r_o. The rod moves axially with constant velocity U_o and sets the fluid in the annular space in motion. The cylinder is maintained at uniform temperature T_o. Neglect gravity and end effects, and assume incompressible laminar axisymmetric flow.

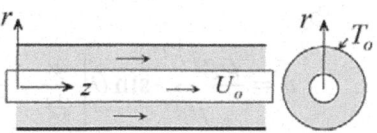

[a] Show that the axial velocity is given by

$$v_z = \frac{U_o}{\ln(r_i / r_o)} \ln(r / r_o).$$

[b] Taking into consideration dissipation, determine the heat flux at the outer surface and the Nusselt number based on $[T(r_i) - T_o]$. Neglect axial temperature variation.

3.8   A liquid film of thickness H flows down an inclined plane due to gravity. The plane is maintained at uniform temperature T_o and the free film surface is insulated. Assume incompressible laminar flow and neglect axial variation of velocity and temperature and end effects.

[a] Show that the axial velocity is given by

$$u = \frac{\rho g H^2}{\mu} \sin\theta \left[\frac{y}{H} - \frac{1}{2} \frac{y^2}{H^2} \right].$$

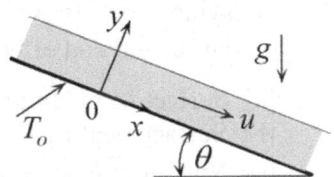

[b] Taking into consideration dissipation, determine the heat flux at the inclined plane.

3.9   A liquid film of thickness H flows down an inclined plane due to gravity. The plane exchanges heat by convection with an ambient fluid. The heat transfer coefficient is h and the ambient temperature is T_∞. The inclined surface is insulated. Assume incompressible laminar flow and neglect axial variation of velocity and temperature and end effects.

[a] Show that the axial velocity is given by

$$u = \frac{\rho g H^2}{\mu} \sin\theta \left[\frac{y}{H} - \frac{1}{2} \frac{y^2}{H^2} \right].$$

[b] Taking into consideration dissipation, determine the heat flux at the free surface.

3.10 Lubricating oil fills the clearance space between a rotating shaft and its housing. The shaft radius is $r_i = 6$ cm and housing radius is

$r_i = 6.1\,\mathrm{cm}$. The angular velocity of the shaft is $\omega = 3000\,\mathrm{RPM}$ and the housing temperature is $T_o = 40^\circ\mathrm{C}$. Taking into consideration dissipation, determine the maximum oil temperature and the heat flux at the housing. Neglect end effects and assume incompressible laminar flow. Properties of lubricating oil are: $k = 0.138\,\mathrm{W/m\text{–}^\circ C}$ and $\mu = 0.0356\,\mathrm{kg/m\text{–}s}$.

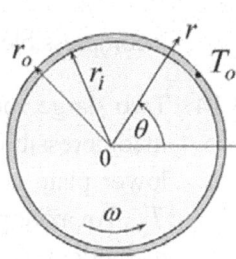

3.11 Consider lubrication oil in the clearance between a shaft and its housing. The radius of the shaft is r_i and that of the housing is r_o. The shaft rotates with an angular velocity ω and the housing exchanges heat by convection with the ambient fluid. The heat transfer coefficient is h and the ambient temperature is T_∞. Taking into consideration dissipation, determine the maximum temperature of the oil and surface heat flux at the housing. Assume incompressible laminar flow and neglect end effects.

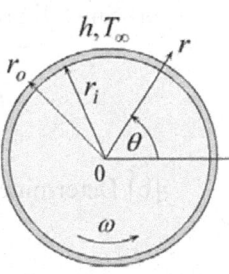

3.12 A rod of radius r_i is placed concentrically inside a sleeve of radius r_o. Incompressible fluid fills the clearance between the rod and the sleeve. The sleeve is maintained at uniform temperature T_o while rotating with constant angular velocity ω. Taking into consideration dissipation, determine the maximum fluid temperature and surface heat flux at the sleeve. Assume incompressible laminar flow and neglect end effects.

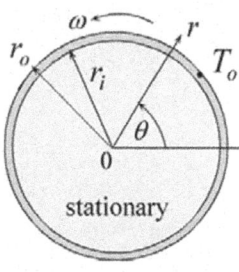

3.13 A hollow shaft of outer radius r_o rotates with constant angular velocity ω while immersed in an infinite fluid at uniform temperature T_∞. Taking into consideration dissipation, determine surface heat flux. Assume

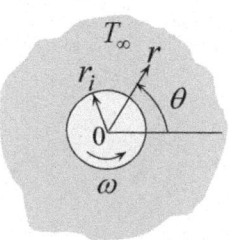

incompressible laminar flow and neglect end effects.

3.14 Two large porous plates are separated by a distance H. An incompressible fluid fills the channel formed by the plates. The lower plate is maintained at temperature T_1 and the upper plate at T_2. An axial pressure gradient dp / dx is applied to the fluid to set it in motion. A fluid at temperature T_1 is injected through the lower plate with a normal velocity v_o. Fluid is removed along the upper plate at velocity v_o. The injected fluid is identical to the channel fluid. Neglect gravity, dissipation and axial variation of temperature.

[a] Show that the axial velocity is given by

$$u = -\frac{Hv}{v_o}\frac{1}{\mu}\frac{dp}{dx}\left[\frac{y}{H} - \frac{1-\exp(v_o y / v)}{1-\exp(v_o H / v)}\right].$$

[b] Determine surface heat flux at each plate.

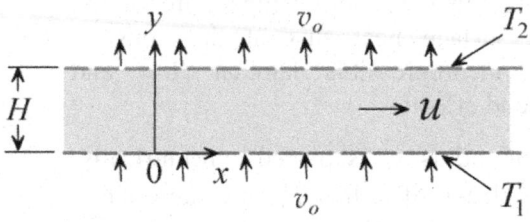

BOUNDARY LAYER FLOW: APPLICATION TO EXTERNAL FLOW

4.1 Introduction

The mathematical complexity of convection heat transfer is traced to the non-linearity of the Navier-Stokes equations of motion and the coupling of flow and thermal fields. The *boundary layer* concept, first introduced by Prandtl [1] in 1904, provides major simplifications. This concept is based on the notion that under special conditions certain terms in the governing equations are much smaller than others and therefore can be neglected without significantly affecting the accuracy of the solution. This raises two questions:

(1) *What are the conditions under which terms in the governing equations can be dropped?*

(2) *What terms can be dropped?*

These questions will be answered first by using intuitive arguments and then by scale analysis.

4.2 The Boundary Layer Concept: Simplification of the Governing Equations

4.2.1 Qualitative Description

Consider fluid flow over the semi-infinite heated surface shown in Fig. 4.1. The foundation of the boundary layer concept is the following observation: under certain conditions the effect of

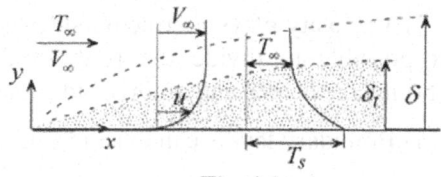

Fig. 4.1

viscosity is confined to a thin region near the surface. This region, whose edge is defined by δ, is referred to as the *velocity* or *viscous boundary layer*. Similarly, under certain conditions the effect of thermal interaction between the surface and the moving fluid is confined to a thin region near the surface defined by δ_t. This region is referred to as the *thermal boundary layer*. It should be noted that the boundaries of these regions are not sharply or uniquely defined.

We consider the conditions for the formation of the two boundary layers. The conditions for the velocity boundary layer model are: (1) slender body without flow separation and (2) high Reynolds number $(Re > 100)$. The conditions for the thermal boundary layer model are: (1) slender body without flow separation and (2) high product of Reynolds and Prandtl numbers $(Re\, Pr > 100)$. This product is called the *Peclet number*, given by

$$\text{Peclet Number} = Pe = RePr = \frac{\rho V_\infty L}{\mu} \frac{c_p \mu}{k} = \frac{\rho c_p V_\infty L}{k}. \qquad (4.1)$$

Before examining the mathematical implication of the boundary layer concept we make the following observations:

(1) Fluid velocity at the surface vanishes. This is the no-slip condition due to the viscous nature of the fluid. However, fluid velocity changes rapidly as the thickness of the boundary layer is traversed, as shown in Fig. 4.1. At the edge of the boundary layer the velocity approaches its free stream value V_∞. Similarly, fluid temperature changes within the thickness of the thermal boundary layer from surface temperature T_s to free stream value T_∞ at the edge of the thermal boundary layer.

(2) In general, at high Re and Pr both velocity and thermal boundary layers are thin. For example, air flowing at 10 m/s parallel to a 1.0 m long plate will have a viscous boundary layer thickness of 6 mm at the downstream end.

(3) Viscosity plays no role outside the viscous boundary layer. This makes it possible to divide the flow field into a viscosity dominated region (boundary layer), and an inviscid region (outside the boundary layer).

(4) Boundary layers can exist in both forced and free convection flows.

4.2.2 The Governing Equations

To examine the mathematical consequences of the boundary layer concept we begin with the governing equations for a simplified case. We make the following assumptions: (1) steady state, (2) two-dimensional, (3) laminar, (4) uniform properties, (5) no dissipation, and (6) no gravity. Based on these assumptions the continuity, momentum, and energy equations are

$$\frac{\partial u}{\partial x} + \frac{\partial v}{\partial y} = 0 , \tag{2.3}$$

$$u\frac{\partial u}{\partial x} + v\frac{\partial u}{\partial y} = -\frac{1}{\rho}\frac{\partial p}{\partial x} + v\left(\frac{\partial^2 u}{\partial x^2} + \frac{\partial^2 u}{\partial y^2}\right) , \tag{2.10x}$$

$$u\frac{\partial v}{\partial x} + v\frac{\partial v}{\partial y} = -\frac{1}{\rho}\frac{\partial p}{\partial y} + v\left(\frac{\partial^2 v}{\partial x^2} + \frac{\partial^2 v}{\partial y^2}\right) , \tag{2.10y}$$

$$\rho c_p\left(u\frac{\partial T}{\partial x} + v\frac{\partial T}{\partial y}\right) = k\left(\frac{\partial^2 T}{\partial x^2} + \frac{\partial^2 T}{\partial y^2}\right) . \tag{2.19}$$

These equations will be simplified using the boundary layer concept.

4.2.3 Mathematical Simplification

It is legitimate to ask whether all terms in the governing equations are equally significant. We will argue that certain terms play a minor role in the solution and thus can be neglected to simplify the equations. Indeed, this is what the boundary layer concept enables us to do. We will first use intuitive arguments to simplify the equations. A more rigorous approach will then be presented using scaling to arrive at the same simplifications.

4.2.4 Simplification of the Momentum Equations

(i) Intuitive Arguments

Starting with the x-momentum equation (2.10x), we wish to establish if one of the two viscous terms on the right-hand-side, $\partial^2 u/\partial x^2 + \partial^2 u/\partial y^2$, is small compared to the other. Imagine that a very small insect, so small that it does not disturb the flow, is placed inside the viscous boundary layer at position 0, as shown in Fig. 4.2. The insect finds the fluid velocity at this location too high and wishes to move to a position of lower velocity. If the insect is allowed to take one short step to any of the four locations 1, 2, 3, or 4, where should it go? If your answer is to

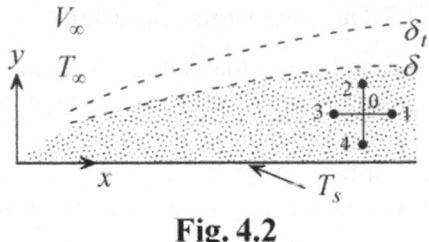

Fig. 4.2

position 4, you are correct. This decision is based on the intuitive sense that changes in axial velocity u with respect to the y are much more pronounced compared to changes with respect to x. After all, the axial velocity u changes from zero at the wall to V_∞ across a short distance δ in the y-direction. However, this does not give a clue regarding the relative magnitudes of $\partial^2 u / \partial x^2$ and $\partial^2 u / \partial y^2$. Additional communication with the insect is required. Imagine that the insect is initially a step away from the surface. Taking a step towards the surface is the ultimate choice since the velocity drops to zero. On the other hand, if the insect is at the edge of the boundary layer, taking a step towards the surface will hardly bring about a change in u. Mathematically, this means that there is a significant change in the axial velocity gradient with respect to y. However, taking one, two, or three steps in the x-direction will bring no significant change in u. This means that changes in the axial gradient of u with respect to x are negligible. We conclude that

$$\frac{\partial^2 u}{\partial x^2} << \frac{\partial^2 u}{\partial y^2} . \tag{4.2}$$

Thus, the term $\partial^2 u / \partial x^2$ can be dropped from equation (2.10x).

We now examine the pressure term in (2.10x) and (2.10y). We argue that for a slender body the streamlines are nearly parallel having small vertical velocity component. Thus

$$\frac{\partial p}{\partial y} \approx 0 . \tag{4.3}$$

It follows that p depends on x only, i.e. $p \approx p(x)$. Therefore

$$\frac{\partial p}{\partial x} \approx \frac{dp}{dx} \approx \frac{dp_\infty}{dx} . \tag{4.4}$$

Here dp_∞ / dx is the pressure gradient at the edge of the boundary layer, $y = \delta$, where the fluid can be assumed inviscid. Substituting (4.2) and (4.4) into (2.10x) gives the *x-momentum boundary layer equation*

$$u\frac{\partial u}{\partial x} + v\frac{\partial u}{\partial y} = -\frac{1}{\rho}\frac{dp_\infty}{dx} + v\frac{\partial^2 u}{\partial y^2}, \qquad (4.5)$$

where $v = \mu/\rho$. On the other hand, equation (2.10y) simplifies to (4.3). Continuity equation (2.3) and the x-momentum boundary layer equation (4.5) contain three unknowns: u, v, and p_∞. However, since p_∞ is the pressure at the edge of the boundary layer $y = \delta$, it can be independently obtained from the solution to the governing equations for inviscid flow outside the boundary layer.

(ii) Scale Analysis

Boundary layer approximations (4.2)-(4.4) will now be arrived at using scaling. Here we follow the procedure detailed in reference [2]. Scaling is used to estimate the order of magnitude of each term in Navier-Stokes equations and drop terms of higher order. In this procedure a scale (measure) is assigned to each variable in an equation.

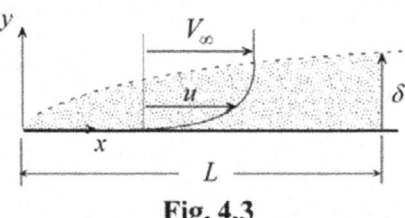

Fig. 4.3

Consider the flow over a slender body such as the flat plate shown in Fig. 4.3. The free stream velocity is V_∞, characteristic length is L, and the boundary layer thickness is δ. We postulate that δ is small compared to the characteristic length L, that is

$$\frac{\delta}{L} \ll 1. \qquad (4.6)$$

Assuming that equation (4.6) is valid, we pose three questions:

(1) *What terms in the governing equations can be dropped?*
(2) *Is normal pressure gradient negligible compared to axial pressure gradient?*
(3) *Under what conditions is (4.6) valid?*

To answer these questions the dependent variables u and v and independent variables x and y are assigned the following measures or scales:

$$u \sim V_\infty, \tag{4.7a}$$

$$x \sim L, \tag{4.7b}$$

$$y \sim \delta. \tag{4.7c}$$

Equation (4.7) is applied to continuity (2.3) to develop a scale for v. Rewriting (2.3) as

$$\frac{\partial v}{\partial y} = -\frac{\partial u}{\partial x}.$$

Using (4.7), the above gives

$$\frac{v}{\delta} \sim \frac{V_\infty}{L}.$$

Solving for v

$$v \sim V_\infty \frac{\delta}{L}. \tag{4.7d}$$

Based on assumption (4.6) it follows that $v \ll V_\infty$. Equation (4.7) is now used to determine the order of magnitude of the inertia and viscous terms of the x-momentum equation (2.10x).

First inertia term:

$$u \frac{\partial u}{\partial x} \sim V_\infty \frac{V_\infty}{L}. \tag{a}$$

Second inertia term:

$$v \frac{\partial u}{\partial y} \sim v \frac{V_\infty}{\delta}.$$

Using (4.7d) to eliminate v in the above, gives

$$v \frac{\partial u}{\partial y} \sim V_\infty \frac{V_\infty}{L}. \tag{b}$$

We conclude that the two inertia terms are of the same magnitude. Following the same procedure, the order of magnitude of the two viscous terms in (2.10x) are determined.

First viscous term:

$$\frac{\partial^2 u}{\partial x^2} \sim \frac{V_\infty}{L^2}. \tag{c}$$

Second viscous term:

$$\frac{\partial^2 u}{\partial y^2} \sim \frac{V_\infty}{\delta^2}. \tag{d}$$

Since, according to (4.6), $\delta \ll L$, comparing (c) with (d) shows that

$$\frac{\partial^2 u}{\partial x^2} \ll \frac{\partial^2 u}{\partial y^2}. \tag{4.2}$$

Thus $\partial^2 u / \partial x^2$ in equation (2.10x) can be neglected. This conclusion is identical to the result obtained using the intuitive approach of the insect model.

Scaling of the two viscous terms in the y-component of the momentum equation, (2.10y), shows that

$$\frac{\partial^2 v}{\partial x^2} \ll \frac{\partial^2 v}{\partial y^2}. \tag{4.8}$$

Using (4.2) and (4.8), equations (2.10x) and (2.10y) simplify to

$$u \frac{\partial u}{\partial x} + v \frac{\partial u}{\partial y} = -\frac{1}{\rho} \frac{\partial p}{\partial x} + v \frac{\partial^2 u}{\partial y^2}, \tag{4.9x}$$

$$u \frac{\partial v}{\partial x} + v \frac{\partial v}{\partial y} = -\frac{1}{\rho} \frac{\partial p}{\partial y} + v \frac{\partial^2 v}{\partial y^2}. \tag{4.9y}$$

Having answered the first question we turn our attention to the second question regarding pressure gradient. The order of magnitude of $\partial p / \partial x$

and $\partial p / \partial y$ is determined using scaling. A balance between axial pressure and inertia in (4.9x) gives

$$\frac{\partial p}{\partial x} \sim \rho u \frac{\partial u}{\partial x}.$$

The above is scaled using (4.7)

$$\frac{\partial p}{\partial x} \sim \rho \frac{V_\infty^2}{L}. \tag{e}$$

Similarly, a balance between pressure and inertia in (4.9y) and scaling, gives

$$\frac{\partial p}{\partial y} \sim \rho \frac{V_\infty^2}{L} \frac{\delta}{L}. \tag{f}$$

Comparison between (e) and (f) using assumption (4.6) shows that

$$\frac{\partial p}{\partial y} << \frac{\partial p}{\partial x}. \tag{4.10}$$

Note that the same result is obtained by balancing pressure gradient against viscous forces instead of inertia in (2.10x) and (2.10y). Equation (4.10) has important consequences on the determination of boundary layer pressure. For two-dimensional flow, pressure depends on the variables x and y. That is

$$p = p(x, y),$$

and

$$dp = \frac{\partial p}{\partial x} dx + \frac{\partial p}{\partial y} dy.$$

Dividing through by dx and rearranging

$$\frac{dp}{dx} = \frac{\partial p}{\partial x} \left[1 + \frac{(\partial p / \partial y)}{(\partial p / \partial x)} \frac{dy}{dx} \right]. \tag{4.11}$$

The gradient dy / dx is scaled as

$$\frac{dy}{dx} \sim \frac{\delta}{L}. \tag{g}$$

Substituting (e)-(g) into (4.11)

$$\frac{dp}{dx} = \frac{\partial p}{\partial x} \left[1 + (\delta / L)^2 \right]. \tag{h}$$

Invoking (4.6), the above simplifies to

$$\frac{dp}{dx} \approx \frac{\partial p}{\partial x}. \tag{i}$$

We thus conclude that boundary layer pressure depends on the axial direction x and that variation in the y-direction is negligible. That is, at a given location x the pressure $p(x)$ inside the boundary layer is the same as the pressure $p_\infty(x)$ at the edge of the boundary layer $y = \delta$. Thus

$$p(x, y) \approx p_\infty(x). \tag{j}$$

We now examine the consequences of this result on the x-momentum equation. Differentiating (j) and using (i)

$$\frac{\partial p}{\partial x} \approx \frac{dp_\infty}{dx}. \tag{4.12}$$

Substituting (4.12) into (4.9x)

$$u \frac{\partial u}{\partial x} + v \frac{\partial u}{\partial y} = -\frac{1}{\rho} \frac{dp_\infty}{dx} + v \frac{\partial^2 u}{\partial y^2}. \tag{4.13}$$

This is the x-momentum equation for boundary layer flow. It is the same as that obtained using our intuitive approach. Note that (4.13) is arrived at using scaling of the y-momentum equation. It is important to recall that this result is based on the key assumption that $\delta / L \ll 1$. It remains to answer the third question regarding the condition under which this assumption is valid.

The first two terms in (4.13) representing inertia are of the same order as shown in (a) and (b). The last term in (4.13) represents viscous force. According to (d) this term is scaled as

$$v\frac{\partial^2 u}{\partial y^2} \sim v\frac{V_\infty}{\delta^2}.$$

(k)

A balance between inertia (a) and viscous force (k) gives

$$\frac{V_\infty^2}{L} \sim v\frac{V_\infty}{\delta^2}.$$

Rearranging the above

$$\frac{\delta}{L} \sim \sqrt{\frac{v}{V_\infty L}},$$

(4.14a)

This result is rewritten as

$$\frac{\delta}{L} \sim \frac{1}{\sqrt{Re_L}}.$$

(4.14b)

where Re_L is the Reynolds number defined as

$$Re_L = \frac{V_\infty L}{v}.$$

(4.15)

Thus $\delta / L \ll 1$ is valid when $\sqrt{Re_L} \gg 1$. For example, if $Re_L = 100$, $\delta / L = 0.1$. Equation (4.14) is generalized to provide a measure of the variation of boundary layer thickness along a surface. Setting $x = L$ in (4.14) and (4.15), gives

$$\frac{\delta}{x} \sim \frac{1}{\sqrt{Re_x}}.$$

(4.16)

We return to the y-component of the momentum equation, (4.9y), and note that each term is of order δ. Thus all terms in this equation are neglected, leading to the important boundary layer simplification of negligible pressure gradient in the y-direction, as described in (4.3).

4.2.5 Simplification of the Energy Equation

We now simplify the energy equation for two-dimensional constant properties flow by neglecting higher order terms in equation (2.19)

$$\rho c_p \left(u \frac{\partial T}{\partial x} + v \frac{\partial T}{\partial y} \right) = k \left(\frac{\partial^2 T}{\partial x^2} + \frac{\partial^2 T}{\partial y^2} \right). \tag{2.19}$$

(i) Intuitive Argument

We wish to determine if one of the conduction terms in (2.19), $\partial^2 T / \partial x^2 + \partial^2 T / \partial y^2$, is small compared to the other. Returning to the small insect of Fig. 4.2., we pretend that the surface is hot and the fluid is at a lower temperature. The insect is placed at location 0 inside the thermal boundary layer shown in Fig. 4.2. It finds the environment too hot and wishes to move to a cooler location. It is allowed to take a small step to location 1, 2, 3, or 4. Where would you advise the insect to go? If your answer is to location 2, you are correct. This is interpreted as recognizing that temperature changes in the y-direction are much more pronounced than changes in the x-direction. To evaluate the relative magnitudes of $\partial^2 T / \partial x^2$ and $\partial^2 T / \partial y^2$, additional observations are required. Imagine that the insect is initially near the surface. Taking a step away from the surface brings significant relief. However, taking a step in the x-direction brings minor change in temperature. Suppose instead the insect is near the edge of the boundary layer. Taking a step away from the surface will essentially result in a minor change in temperature as would moving axially. From this we conclude that changes in the axial temperature gradient with respect to x are small compared to changes in the normal temperature gradient with respect to y. That is

$$\frac{\partial^2 T}{\partial x^2} << \frac{\partial^2 T}{\partial y^2}. \tag{4.17}$$

Thus, axial conduction, $\partial^2 T / \partial x^2$, can be dropped from the energy equation to obtain

$$u \frac{\partial T}{\partial x} + v \frac{\partial T}{\partial y} = \alpha \frac{\partial^2 T}{\partial y^2}, \tag{4.18}$$

where α is thermal diffusivity. This equation is known as the *boundary layer energy equation*.

(ii) Scale Analysis

Scaling will now be used to examine the order of magnitude of each term in energy equation (2.19). Again we consider the flow over a slender body of characteristic length L. The free stream velocity and temperature are V_∞ and T_∞. The thermal boundary layer thickness is δ_t. We postulate that δ_t is small compared to the characteristic length L, that is

$$\frac{\delta_t}{L} \ll 1. \tag{4.19}$$

Assuming that equation (4.19) is valid, we pose two questions:

(1) *What terms in energy equation* (2.19) *can be dropped?*
(2) *Under what condition is equation* (4.19) *valid?*

To answer these questions we assign scales to the variables in the energy equation. The scale for x is given by equation (4.7b)

$$x \sim L. \tag{4.7b}$$

Scales for y and ΔT are

$$y \sim \delta_t, \tag{4.20}$$

$$\Delta T \sim T_s - T_\infty. \tag{4.21}$$

Scales for u and v depend on whether δ_t is larger or smaller than δ. Thus two cases are identified as illustrated in Fig. 4.4:

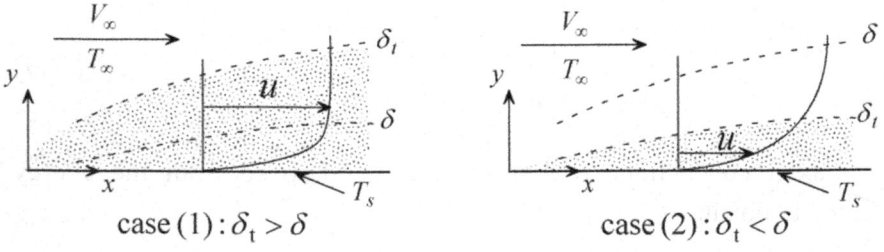

case $(1): \delta_t > \delta$ case $(2): \delta_t < \delta$

Fig. 4.4

Case (1): $\delta_t > \delta$. For this case the axial velocity u within the thermal boundary layer is of the order of the free stream velocity. Thus the scale for u is

$$u \sim V_\infty . \tag{4.22}$$

Following the formulation of (4.7d), scaling of the continuity equation gives

$$v \sim V_\infty \frac{\delta_t}{L} . \tag{4.23}$$

Using (4.7b) and (4.20-4.23), the two convection terms in equation (2.19) scale as

$$u \frac{\partial T}{\partial x} \sim V_\infty \frac{\Delta T}{L} , \tag{a}$$

and

$$v \frac{\partial T}{\partial y} \sim V_\infty \frac{\Delta T}{L} . \tag{b}$$

Thus the two convection terms are of the same order. We now examine the order of magnitude of the two conduction terms in (2.19). They scale as

$$\frac{\partial^2 T}{\partial x^2} \sim \frac{\Delta T}{L^2} , \tag{c}$$

and

$$\frac{\partial^2 T}{\partial y^2} \sim \frac{\Delta T}{\delta_t^2} . \tag{d}$$

Comparing (c) with (d) and using (4.19), we conclude that

$$\frac{\partial^2 T}{\partial x^2} << \frac{\partial^2 T}{\partial y^2} . \tag{e}$$

This is identical to the intuitive result of (4.17). Thus the boundary layer energy equation simplifies to

$$u \frac{\partial T}{\partial x} + v \frac{\partial T}{\partial y} = \alpha \frac{\partial^2 T}{\partial y^2}. \tag{4.18}$$

To answer the second question we note that each term in (4.18) is equally important. A balance between convection and conduction gives

$$u \frac{\partial T}{\partial x} \sim \alpha \frac{\partial^2 T}{\partial y^2}.$$

Scaling each term in the above

$$V_\infty \frac{\Delta T}{L} \sim \alpha \frac{\Delta T}{\delta_t^2}.$$

Rearranging

$$\frac{\delta_t}{L} \sim \sqrt{\frac{\alpha}{V_\infty L}}.$$

Using the definition of α, the above gives

$$\frac{\delta_t}{L} \sim \sqrt{\frac{k}{\rho c_p V_\infty L}}.$$

Using the definitions of Prandtl and Reynolds numbers, the above is rewritten as

$$\frac{\delta_t}{L} \sim \frac{1}{\sqrt{Pr Re_L}}. \tag{4.24}$$

Thus

$$\frac{\delta_t}{L} \ll 1 \text{ when } \sqrt{Pr Re_L} \gg 1. \tag{4.25}$$

The product of the Prandtl and Reynolds numbers appears in various convection problems and is called the *Peclet number*, *Pe*, defined as

$$Pe = Pr Re_L. \tag{4.26}$$

As an example, a Peclet number of 100 gives $\delta_t / L \sim 0.1$. It should be noted that the above result applies to the case of $\delta_t > \delta$. It remains to establish the condition under which $\delta_t > \delta$. Taking the ratio of (4.24) to (4.14b) gives

$$\frac{\delta_t}{\delta} \sim \frac{1}{\sqrt{Pr}}.$$ (4.27)

Thus the criterion for $\delta_t > \delta$ is

$$\delta_t > \delta \quad \text{when} \quad \sqrt{Pr} \ll 1.$$ (4.28)

Case (2): $\delta_t < \delta$. Examination of Fig. 4.4 shows that for this case the axial velocity u within the thermal boundary layer is smaller than the free stream velocity. Pretending that the velocity profile is linear, similarity of triangles gives a scale for u as

$$u \sim V_\infty \frac{\delta_t}{\delta}.$$ (4.29)

Following the formulation of (4.7d), scaling of the continuity equation gives

$$v \sim V_\infty \frac{\delta_t^2}{L\delta}.$$ (4.30)

Using (4.29) and (4.30) and following the procedure used in Case (1), we arrive at the conclusion that the two convection terms are of the same order and that axial conduction is negligible compared to normal conduction.

To answer the second question we again perform a balance between convection and conduction

$$u \frac{\partial T}{\partial x} \sim \alpha \frac{\partial^2 T}{\partial y^2}.$$

Using (4.29) for u, scaling each term in the above gives

$$V_\infty \frac{\delta_t}{\delta} \frac{\Delta T}{L} \sim \alpha \frac{\Delta T}{\delta_t^2}.$$

Using the definition of α and rearranging

$$\left(\delta_t / L\right)^3 \sim \frac{k}{\rho c_p V_\infty L} \frac{\delta}{L}.$$

Applying (4.14b) to eliminate δ / L in the above, we obtain

$$\frac{\delta_t}{L} \sim \frac{1}{Pr^{1/3} \sqrt{Re_L}}. \qquad (4.31)$$

Thus the condition for the assumption in (4.19) that $\delta_t / L \ll 1$ is

$$\frac{\delta_t}{L} \ll 1 \quad \text{when} \quad Pr^{1/3} \sqrt{Re_L} \gg 1. \qquad (4.32)$$

We next establish the condition under which $\delta_t < \delta$. Taking the ratio of (4.31) to (4.14b)

$$\frac{\delta_t}{\delta} \sim \frac{1}{Pr^{1/3}}. \qquad (4.33)$$

Thus the criterion for $\delta_t < \delta$ is

$$\delta_t < \delta \quad \text{when} \quad Pr^{1/3} \gg 1. \qquad (4.34)$$

4.3 Summary of Boundary Layer Equations for Steady Laminar Flow

In formulating the governing equations for convection heat transfer we have made several simplifying assumptions in order to limit the mathematical complexity. These assumptions are: (1) Continuum, (2) Newtonian fluid, (3) two-dimensional process, (5) negligible changes in kinetic and potential energy and (4) constant properties. The additional assumptions leading to boundary layer simplifications are: (6) slender surface, (7) high Reynolds number ($Re > 100$), and (8) high Peclet number ($Pe > 100$). Finally, we introduce the following additional simplifications: (9) steady state, (10) laminar flow, (11) no dissipation ($\Phi = 0$), (12) no gravity, and (13) no energy generation ($q''' = 0$). The governing boundary layer equations for these conditions are:

Continuity:

$$\frac{\partial u}{\partial x} + \frac{\partial v}{\partial y} = 0 . \tag{2.3}$$

x-Momentum:

$$u\frac{\partial u}{\partial x} + v\frac{\partial u}{\partial y} = -\frac{1}{\rho}\frac{dp_\infty}{dx} + v\frac{\partial^2 u}{\partial y^2} . \tag{4.13}$$

Energy

$$u\frac{\partial T}{\partial x} + v\frac{\partial T}{\partial y} = \alpha\frac{\partial^2 T}{\partial y^2} . \tag{4.18}$$

Note the following: (1) The continuity equation is not simplified for boundary layer flow because both terms in (2.3) are of the same order of magnitude. (2) The pressure term in (4.13) is obtained from the solution to inviscid flow outside the boundary layer. Thus (2.3) and (4.13) have two unknowns: u and v. (3) To include the effect of buoyancy, the term $[-\rho\beta g(T-T_\infty)]$ should be added to the right-hand-side of (4.13). This assumes that gravity points in the positive x-direction. (4) In applying these equations one must keep in mind all the assumptions and restrictions leading to their formulation.

4.4 Solutions: External Flow

We consider external flow over a surface in which the fluid is infinite in extent. Of interest is thermal interaction between a surface and the external fluid. Thermal interaction is fully characterized once fluid temperature distribution is determined. However, temperature distribution depends on velocity distribution. For the special case of constant properties, velocity distribution is independent of temperature. Since this assumption will be made throughout, in each case the solution to the velocity distribution will be determined first and used to obtain the corresponding temperature solution. The exceptions are problems involving free convection where velocity and temperature must be solved simultaneously.

4.4.1 Laminar Boundary Layer Flow over Semi-infinite Flat Plate: Uniform Surface Temperature

Consider the classic problem of flow over a semi-infinite flat plate shown in Fig. 4.5. The plate is maintained at uniform temperature T_s and the upstream fluid temperature is T_∞. The upstream velocity is uniform and parallel to the plate. Invoking all

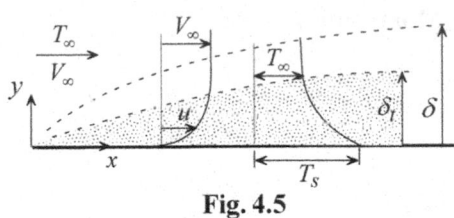

Fig. 4.5

the assumptions summarized in Section 4.3, the three governing equations (continuity, momentum, and energy) are given in (2.3), (4.13), and (4.18). It should be recalled that for uniform flow over a semi-infinite flat plate transition from laminar to turbulent flow takes place at $Re_t = V_\infty x_t / v \approx 500,000$ (see Section 2.3).

(i) Velocity Distribution. In addition to the velocity distribution, of interest is the determination of the boundary layer thickness $\delta(x)$ and wall shearing stress $\tau_o(x)$. These important flow characteristics are easily determined once the velocity solution is obtained. Before an analytic solution is presented, scaling will be used to estimate $\delta(x)$ and $\tau_o(x)$.

(a) Governing equations and boundary conditions

The continuity and x-momentum equations for this flow are:

$$\frac{\partial u}{\partial x} + \frac{\partial v}{\partial y} = 0, \tag{2.3}$$

$$u\frac{\partial u}{\partial x} + v\frac{\partial u}{\partial y} = -\frac{1}{\rho}\frac{dp_\infty}{dx} + v\frac{\partial^2 u}{\partial y^2}. \tag{4.13}$$

The velocity boundary conditions are:

$$u(x,0) = 0, \tag{4.35a}$$

$$v(x,0) = 0, \tag{4.35b}$$

$$u(x,\infty) = V_\infty, \tag{4.35c}$$

$$u(0,y) = V_\infty. \tag{4.35d}$$

(b) Scale analysis: boundary layer thickness, wall shear and friction coefficient

In Section 4.2.4 we used scale analysis to obtain an order of magnitude solution to the boundary layer thickness $\delta(x)$, given in (4.16)

$$\frac{\delta}{x} \sim \frac{1}{\sqrt{Re_x}} . \tag{4.16}$$

Wall shearing stress τ_o is determined using (2.7a)

$$\tau_{xy} = \tau_{yx} = \mu \left(\frac{\partial v}{\partial x} + \frac{\partial u}{\partial y} \right) . \tag{2.7a}$$

Applying (2.7a) at the wall $y = 0$ where $v(x,0) = 0$, gives τ_o

$$\tau_o = \mu \frac{\partial u(x,0)}{\partial y} . \tag{4.36}$$

To determine the order of magnitude of τ_o the following scales are assigned to u and y:

$$u \sim V_\infty , \tag{4.7a}$$

$$y \sim \delta . \tag{4.7c}$$

Equation (4.36) is scaled using (4.7)

$$\tau_o \sim \mu \frac{V_\infty}{\delta} . \tag{a}$$

Using (4.16) to eliminate δ in (a)

$$\tau_o \sim \mu \frac{V_\infty}{x} \sqrt{Re_x} . \tag{b}$$

Introducing the definition of the *Darcy friction coefficient* C_f

$$C_f = \frac{\tau_o}{(1/2)\rho V_\infty^2}.$$ (4.37a)

Using (b) to eliminate τ_o in (4.37), yields

$$C_f \sim \frac{1}{\sqrt{Re_x}}.$$ (4.37b)

(c) Blasius solution: similarity method

Equations (2.3) and (4.13) are solved analytically for the velocity components u and v. These two equations contain three unknowns: u, v, and p_∞. However, as was previously pointed out, pressure in boundary layer problems is independently obtained from the inviscid flow solution outside the boundary layer. Focusing on this inviscid region we note that it can be modeled as uniform inviscid flow over the slightly curved edge of the viscous boundary layer δ. However, since this layer is thin, we make the assumption that the edge coincides with the plate. We do this as a first approximation since the edge of the boundary layer is not yet known. Thus, the inviscid problem becomes that of uniform flow over a flat plate of zero thickness. Since the fluid is assumed inviscid, the plate does not disturb the flow and the velocity remains uniform. Therefore the solution to the inviscid flow outside the boundary layer is

$$u = V_\infty, \quad v = 0, \quad p = p_\infty = \text{constant}.$$ (4.38)

Thus the pressure gradient is

$$\frac{dp_\infty}{dx} = 0.$$ (4.39)

Substituting (4.39) into (4.13) we obtain the boundary layer momentum equation for this problem

$$u\frac{\partial u}{\partial x} + v\frac{\partial u}{\partial y} = v\frac{\partial^2 u}{\partial y^2}.$$ (4.40)

Equation (4.40) is nonlinear and must be solved simultaneously with the continuity equation (2.3). The solution, which is briefly outlined here, was obtained by Blasius [1] in 1908. He used *similarity transformation* to combine the two independent variables x and y into a single variable $\eta(x, y)$ and postulated that u/V_∞ depends on η only. For this problem the correct form of the transformation variable η is

$$\eta(x, y) = y\sqrt{\frac{V_\infty}{vx}}. \tag{4.41}$$

The velocity $u(x, y)$ is assumed to depend on η according to

$$\frac{u}{V_\infty} = \frac{df}{d\eta}, \tag{4.42}$$

where $f = f(\eta)$ is a function which is yet to be determined. Two observations are made regarding similarity transformation (4.41): (1) The inclusion of $(V_\infty/v)^{1/2}$ in the definition of η, as will be seen later, is for convenience only. (2) Formal procedures are available for identifying the appropriate transformation variable for a given problem [2].

   With u defined in (4.42), continuity equation (2.3) is integrated to give the vertical velocity component v. Rewriting (2.3)

$$\frac{\partial v}{\partial y} = -\frac{\partial u}{\partial x}.$$

Multiplying both sides by dy and integrating gives v

$$v = -\int \frac{\partial u}{\partial x} dy. \tag{a}$$

To evaluate the integral in (a) we use (4.41) and (4.42) to express dy and $\partial u/\partial x$ in terms of the variable η. Differentiating (4.41) with respect to y and rearranging, yields

$$dy = \sqrt{\frac{vx}{V_\infty}} d\eta. \tag{b}$$

Using the chain rule, the derivative $\partial u/\partial x$ is expressed in terms of η

$$\frac{\partial u}{\partial x} = \frac{du}{d\eta}\frac{\partial \eta}{\partial x}.$$

Using (4.41) and (4.42) into the above

$$\frac{\partial u}{\partial x} = -\frac{V_\infty}{2x}\frac{d^2 f}{d\eta^2}\eta. \tag{c}$$

Substituting (b) and (c) into (a) and rearranging

$$\frac{v}{V_\infty} = \frac{1}{2}\sqrt{\frac{v}{V_\infty x}}\int \frac{d^2 f}{d\eta^2}\eta\, d\eta.$$

Integration by parts gives

$$\frac{v}{V_\infty} = \frac{1}{2}\sqrt{\frac{v}{V_\infty x}}\left(\eta\frac{df}{d\eta} - f\right). \tag{4.43}$$

With continuity satisfied, the momentum equation will be transformed and the function $f(\eta)$ determined. In addition to u, v and $\partial u / \partial x$, the derivatives $\partial u / \partial y$ and $\partial^2 u / \partial y^2$ must be expressed in terms of η. Using the chain rule and equations (4.41) and (4.42), we obtain

$$\frac{\partial u}{\partial y} = \frac{du}{d\eta}\frac{\partial \eta}{\partial y} = V_\infty \frac{d^2 f}{d\eta^2}\sqrt{\frac{V_\infty}{v x}}, \tag{d}$$

$$\frac{\partial^2 u}{\partial y^2} = V_\infty \frac{d^3 f}{d\eta^3}\frac{V_\infty}{v x}. \tag{e}$$

Substituting (4.42), (4.43), and (c)-(e) into (4.40)

$$2\frac{d^3 f}{d\eta^3} + f(\eta)\frac{d^2 f}{d\eta^2} = 0. \tag{4.44}$$

Thus, the governing partial differential equations are successfully transformed into an ordinary differential equation. Note that the original independent variables x and y are eliminated in this transformation. Note further that (4.44) is independent of characteristic velocity V_∞ and property ν. This is a direct consequence of including the factor $(V_\infty / \nu)^{1/2}$ in the definition of η in (4.41).

To complete the transformation, the boundary conditions must also be transformed in terms of the new variable η. Using (4.41)–(4.43), boundary conditions (4.35a-4.35d) transform to

$$\frac{df(0)}{d\eta} = 0, \quad (4.45a)$$

$$f(0) = 0, \quad (4.45b)$$

$$\frac{df(\infty)}{d\eta} = 1, \quad (4.45c)$$

$$\frac{df(\infty)}{d\eta} = 1. \quad (4.45d)$$

Note that transformed equation (4.44) is third order requiring three boundary conditions. Boundary conditions (4.35c) and (4.35d) coalesce into a single condition, as shown in (4.45c) and (4.45d). Although the mathematical problem is reduced to solving a third order ordinary differential equation (4.44), the difficulty is that this equation is nonlinear. Blasius obtained a power series solution for the function $f(\eta)$. Since the solution is not expressed in terms of simple functions that are convenient to use, tabulated values for f and its derivatives are available for the determination of u and v. Table 4.1 gives a summary of these functions. Beside giving the velocity distribution, Blasius solution also gives the boundary layer thickness $\delta(x)$ and the wall shearing stress $\tau_o(x)$. Defining δ as the distance y from the plate where the velocity ratio $u/V_\infty = 0.994$, Table 4.1 gives

Table 4.1 Blasius solution [1]			
$\eta = y\sqrt{\dfrac{V_\infty}{\nu x}}$	f	$\dfrac{df}{d\eta} = \dfrac{u}{V_\infty}$	$\dfrac{d^2 f}{d\eta^2}$
0.0	0.0	0.0	0.33206
0.4	0.02656	0.13277	0.33147
0.8	0.10611	0.26471	0.32739
1.2	0.23795	0.39378	0.31659
1.6	0.42032	0.51676	0.29667
2.0	0.65003	0.62977	0.26675
2.4	0.92230	0.72899	0.22809
2.8	1.23099	0.81152	0.18401
3.2	1.56911	0.87609	0.13913
3.6	1.92954	0.92333	0.09809
4.0	2.30576	0.95552	0.06424
4.4	2.69238	0.97587	0.03897
4.8	3.08534	0.98779	0.02187
5.0	3.28329	0.99155	0.01591
5.2	3.48189	0.99425	0.01134
5.4	3.68094	0.99616	0.00793
5.6	3.88031	0.99748	0.00543
6.0	4.27964	0.99898	0.00240
7.0	5.27926	0.99992	0.00022
8.0	6.27923	1.00000	0.00001

$$\delta = 5.2 \sqrt{\frac{v x}{V_\infty}} .$$

This can be expressed as

$$\frac{\delta}{x} = \frac{5.2}{\sqrt{Re_x}} . \tag{4.46}$$

where Re_x is the local Reynolds number. We are now in a position to compare this result with scaling prediction of (4.16)

$$\frac{\delta}{x} \sim \frac{1}{\sqrt{Re_x}} . \tag{4.16}$$

The comparison shows that scaling predicts the correct dependency on the local Reynolds number with the constant 5.2 in Blasius solution approximated by unity.

Wall shearing stress τ_o is obtained using (4.36)

$$\tau_o = \mu \frac{\partial u(x,0)}{\partial y} . \tag{4.36}$$

Substituting (d) into (4.36) and using Table 4.1

$$\tau_0 = \mu V_\infty \sqrt{\frac{V_\infty}{v x}} \frac{d^2 f(0)}{d\eta^2} = 0.33206 \, \mu V_\infty \sqrt{\frac{V_\infty}{v x}} . \tag{4.47}$$

Substituting (4.47) into (4.37a) gives the friction coefficient C_f

$$C_f = \frac{0.664}{\sqrt{Re_x}} . \tag{4.48}$$

Note that scaling prediction of C_f is given by equation (4.37b)

$$C_f \sim \frac{1}{\sqrt{Re_x}} . \tag{4.37b}$$

Again scaling predicts the correct dependency on the local Reynolds number with the constant 0.664 in Blasius solution approximated by unity.

(ii) Temperature Distribution.

We return to the problem shown in Fig. 4.5 for uniform flow over an isothermal semi-infinite plate. The determination of the thermal boundary layer thickness δ_t, surface heat flux, heat transfer coefficient, and Nusselt

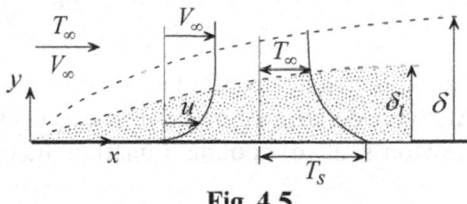

Fig. 4.5

number, hinges on the determination of the temperature distribution in the fluid.

(a) Governing equation and boundary conditions

Based on all the assumptions summarized in Section 4.3, the temperature distribution is governed by energy equation (4.18).

$$u\frac{\partial T}{\partial x} + v\frac{\partial T}{\partial y} = \alpha\frac{\partial^2 T}{\partial y^2}. \qquad (4.18)$$

The boundary conditions for this problem are:

$$T(x,0) = T_s, \qquad (4.49a)$$

$$T(x,\infty) = T_\infty, \qquad (4.49b)$$

$$T(0,y) = T_\infty. \qquad (4.49c)$$

(b) Scale analysis: Thermal boundary layer thickness, heat transfer coefficient and Nusselt number

In Section 4.2.5 we used scale analysis to obtain an order of magnitude estimate for the thermal boundary layer thickness δ_t. We generalize the results by setting $L = x$ in equations (4.24) and (4.31):

Case (1): $\delta_t > \delta$ $(Pr \ll 1)$

$$\frac{\delta_t}{x} \sim \frac{1}{\sqrt{PrRe_x}}. \qquad (4.50)$$

Case (2): $\delta_t < \delta$ $(Pr \gg 1)$

$$\frac{\delta_t}{x} \sim \frac{1}{Pr^{1/3}\sqrt{Re_x}}.$$

(4.51)

The heat transfer coefficient h was introduced in Section 1.6 of Chapter 1. Analytic determination of h is based on Fourier's law of conduction and Newton's law of cooling. Equating the two laws gives

$$h = -k\frac{\dfrac{\partial T(x,0)}{\partial y}}{T_s - T_\infty}.$$

(1.10)

Using the scales of (4.20) and (4.21), the above gives

$$h \sim \frac{k}{\delta_t},$$

(4.52)

where δ_t is given by (4.50) and (4.51).

Case (1): $\delta_t > \delta$ $(Pr \ll 1)$. Substituting (4.50) into (4.52)

$$h \sim \frac{k}{x}\sqrt{PrRe_x}, \quad \text{for } Pr \ll 1.$$

(4.53)

Defining the local Nusselt number Nu_x as

$$Nu_x = \frac{hx}{k}.$$

(4.54)

Substituting (4.53) into (4.54)

$$Nu_x \sim Pr^{1/2}\sqrt{Re_x}, \quad \text{for } Pr \ll 1.$$

(4.55)

Case (2): $\delta_t \ll \delta$ $(Pr \gg 1)$. Substituting (4.51) into (4.52)

$$h \sim \frac{k}{x}Pr^{1/3}\sqrt{Re_x}, \quad \text{for } Pr \gg 1.$$

(4.56)

The corresponding Nusselt number is

$$Nu_x \sim Pr^{1/3}\sqrt{Re_x}, \quad \text{for } Pr \gg 1. \tag{4.57}$$

(c) Pohlhausen's solution: Temperature distribution, thermal boundary layer thickness, heat transfer coefficient, and Nusselt number

Boundary layer energy equation (4.18) is solved analytically for the temperature distribution $T(x,y)$. The solution was obtained in 1921 by Pohlhausen [1] using similarity transformation. For convenience, equation (4.18) is expressed in terms of dimensionless temperature θ defined as

$$\theta = \frac{T - T_s}{T_\infty - T_s}. \tag{4.58}$$

Substituting (4.58) into (4.18)

$$u\frac{\partial \theta}{\partial x} + v\frac{\partial \theta}{\partial y} = \alpha\frac{\partial^2 \theta}{\partial y^2}. \tag{4.59}$$

Boundary conditions (4.49) become

$$\theta(x,0) = 0, \tag{4.60a}$$

$$\theta(x,\infty) = 1, \tag{4.60b}$$

$$\theta(0,y) = 1. \tag{4.60c}$$

To solve (4.59) and (4.60) using similarity method, the two independent variables x and y are combined into a single variable $\eta(x, y)$. For this problem the correct form of the transformation variable η is the same as that used in Blasius solution

$$\eta(x, y) = y\sqrt{\frac{V_\infty}{\nu x}}. \tag{4.41}$$

The solution $\theta(x, y)$ is assumed to depend on η as

$$\theta(x, y) = \theta(\eta).$$

Velocity components u and v in (4.59) are given by Blasius solution

$$\frac{u}{V_\infty} = \frac{df}{d\eta},$$

(4.42)

$$\frac{v}{V_\infty} = \frac{1}{2}\sqrt{\frac{v}{V_\infty x}}\left(\eta\frac{df}{d\eta} - f\right).$$

(4.43)

Substituting (4.41)-(4.43) into (4.59) and noting that

$$\frac{\partial\theta}{\partial x} = \frac{d\theta}{d\eta}\frac{\partial\eta}{\partial x} = -\frac{\eta}{2x}\frac{d\theta}{d\eta},$$

$$\frac{\partial\theta}{\partial y} = \frac{d\theta}{d\eta}\frac{\partial\eta}{\partial y} = \sqrt{\frac{V_\infty}{vx}}\frac{d\theta}{d\eta},$$

$$\frac{\partial^2\theta}{\partial y^2} = \frac{V_\infty}{vx}\frac{d^2\theta}{d\eta^2},$$

gives

$$\frac{d^2\theta}{d\eta^2} + \frac{Pr}{2}f(\eta)\frac{d\theta}{d\eta} = 0.$$

(4.61)

Thus, the governing partial differential equation is successfully transformed into an ordinary differential equation. The following observations are made regarding (4.61):

(1) The Prandtl number Pr is the single parameter characterizing the equation.

(2) This is a linear second order ordinary differential equation requiring two boundary conditions.

(3) The function $f(\eta)$ appearing in (4.61) represents the effect of fluid motion on temperature distribution. It is obtained from Blasius solution.

To complete the transformation, boundary conditions (4.60) must also be transformed in terms of the new variable η. Using (4.41), the three boundary conditions (4.60a-4.60c) transform to

$$\theta(0) = 0,$$

(4.62a)

$$\theta(\infty) = 1,$$

(4.62b)

$$\theta(\infty) = 1.$$

(4.62c)

Note that boundary conditions (4.60b) and (4.60c) coalesce into a single condition, as shown in (4.62b) and (4.62c). Equation (4.61) is solved by separating the variables, integrating and using boundary conditions (4.62). Integration details are found in Appendix B. The temperature solution is

$$\theta(\eta) = 1 - \frac{\int_{\eta}^{\infty} \left[\frac{d^2 f}{d\eta^2}\right]^{Pr} d\eta}{\int_{0}^{\infty} \left[\frac{d^2 f}{d\eta^2}\right]^{Pr} d\eta}. \tag{4.63}$$

Differentiating (4.63), evaluating the result at $\eta = 0$ and using Table 4.1 gives the temperature gradient at the surface

$$\frac{d\theta(0)}{d\eta} = \frac{[0.332]^{Pr}}{\int_{0}^{\infty} \left[\frac{d^2 f}{d\eta^2}\right]^{Pr} d\eta}. \tag{4.64}$$

The integrals in (4.63) and (4.64) are evaluated numerically. The integrand $d^2 f / d\eta^2$ is obtained from Blasius solution and is given in Table 4.1. The integration result is presented graphically in Fig. 4.6 for several values of the Prandtl number.

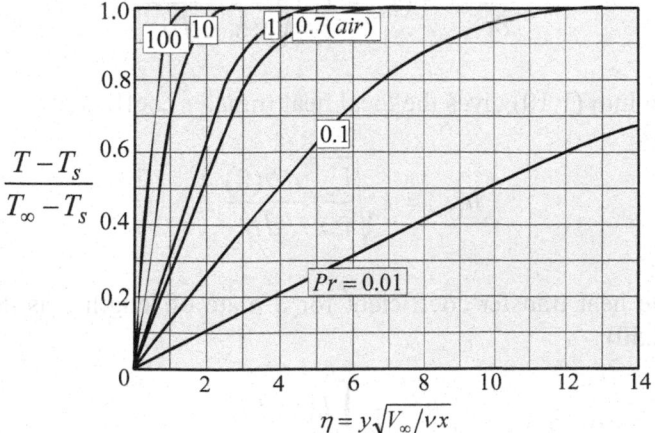

Fig. 4.6 Pohlhausen's solution for temperature distribution for laminar flow over a semi-infinite isotheral flat plate

With the temperature distribution determined, attention is focused on the thermal boundary layer thickness, heat transfer coefficient, and Nusselt number. The thermal boundary layer thickness δ_t is determined from Fig. 4.6. The edge of the thermal layer is defined as the distance y where $T \approx T_\infty$. This corresponds to

$$\theta = \frac{T - T_s}{T_\infty - T_s} \approx 1, \quad \text{at } y = \delta_t . \tag{4.65}$$

Using this definition of δ_t, Fig. 4.6 shows that $\delta_t(x)$ depends on the Prandtl number and that it decreases as the Prandtl number is increased.

The heat transfer coefficient h is determined using equation (1.10)

$$h = -k \frac{\dfrac{\partial T(x,0)}{\partial y}}{T_s - T_\infty}, \tag{1.10}$$

where

$$\frac{\partial T(x,0)}{\partial y} = \frac{dT}{d\theta} \frac{d\theta(0)}{d\eta} \frac{\partial \eta}{\partial y}.$$

Using (4.41) and (4.58) into the above

$$\frac{\partial T(x,0)}{\partial y} = (T_\infty - T_s) \sqrt{\frac{V_\infty}{\nu x}} \frac{d\theta(0)}{d\eta}.$$

Substituting into (1.10) gives the local heat transfer coefficient

$$h(x) = k \sqrt{\frac{V_\infty}{\nu x}} \frac{d\theta(0)}{d\eta}. \tag{4.66}$$

The average heat transfer coefficient for a plate of length L is defined in equation (2.50)

$$\bar{h} = \frac{1}{L} \int_0^L h(x)\,dx . \tag{2.50}$$

Substituting (4.66) into (2.50) and integrating

$$\overline{h} = 2\frac{k}{L}\sqrt{Re_L}\,\frac{d\theta(0)}{d\eta}.$$ (4.67)

The local Nusselt number is obtained by substituting (4.66) into (4.54)

$$Nu_x = \frac{d\theta(0)}{d\eta}\sqrt{Re_x}.$$ (4.68)

The corresponding average Nusselt number is

$$\overline{Nu_L} = 2\frac{d\theta(0)}{d\eta}\sqrt{Re_L}.$$ (4.69)

Total heat transfer rate q_T from a plate of length L and width W is obtained by applying Newton's law of cooling

$$q_T = \int_0^L h(x)(T_s - T_\infty)W\,dx = (T_s - T_\infty)W\int_0^L h(x)dx = (T_s - T_\infty)WL\overline{h}.$$

Noting that WL is the surface area A, the above becomes

$$q_T = (T_s - T_\infty)A\overline{h}.$$ (4.70)

Examination of equations (4.66)-(4.69) shows that the heat transfer coefficient and Nusselt number depend on the temperature gradient at the surface, $d\theta(0)/d\eta$. This key factor depends on the Prandtl number and is determined from (4.64). The integral in (4.64) is evaluated numerically using Blasius data in Table 4.1. Values of $d\theta(0)/d\eta$ corresponding to various Prandtl numbers are given in Table 4.2 [3]. The following equations give good approximation of $d\theta(0)/d\eta$

$$\frac{d\theta(0)}{d\eta} = 0.500\,Pr^{1/2}, \quad 0.005< Pr <0.05.\ (4.71a)$$

Table 4.2	
Pr	$\dfrac{d\theta(0)}{d\eta}$
0.001	0.0173
0.01	0.0516
0.1	0.140
0.5	0.259
0.7	0.292
1.0	0.332
7.0	0.645
10.0	0.730
15.0	0.835
50	1.247
100	1.572
1000	3.387

$$\frac{d\theta(0)}{d\eta} = 0.332 \, Pr^{1/3}, \quad 0.6 < Pr < 10, \tag{4.71b}$$

$$\frac{d\theta(0)}{d\eta} = 0.339 \, Pr^{1/3}, \quad Pr > 10. \tag{4.71c}$$

To evaluate scaling prediction of the Nusselt number, we consider two cases corresponding to $Pr \ll 1$ and $Pr \gg 1$. Combining (4.71) with (4.68) gives

$$Nu_x = 0.564 \, Pr^{1/2} \sqrt{Re_x}, \quad \text{for } Pr < 0.05, \tag{4.72a}$$

$$Nu_x = 0.332 \, Pr^{1/3} \sqrt{Re_x}, \quad \text{for } 0.6 < Pr < 10, \tag{4.72b}$$

$$Nu_x = 0.339 \, Pr^{1/3} \sqrt{Re_x}, \quad \text{for } Pr > 10, \tag{4.72c}$$

The corresponding scaling results are given in (4.55) and (4.57)

$$Nu_x \sim Pr^{1/2} \sqrt{Re_x}, \quad \text{for } Pr \ll 1, \tag{4.55}$$

$$Nu_x \sim Pr^{1/3} \sqrt{Re_x}, \quad \text{for } Pr \gg 1. \tag{4.57}$$

Comparing (4.72a) with (4.55) and (4.72c) with (4.57) shows that scaling predicts the correct dependency on the local Reynolds number and the Prandtl number. However, scaling approximates the coefficients 0.564 and 0.339 of the analytic solution with unity.

The use of Pohlhausen's solution to determine heat transfer characteristics requires the determination of fluid properties such as kinematic viscosity, thermal conductivity, and Prandtl number. All fluid properties in Pohlhausen's solution are assumed constant. In fact they are temperature dependent. When carrying out computations using Pohlhausen's solution, properties are evaluated at the film temperature T_f, defined as

$$T_f = (T_s + T_\infty)/2. \tag{4.73}$$

4.4.2 Applications: Blasius Solution, Pohlhausen's Solution, and Scaling

Three examples will be presented in this section to illustrate the application of Blasius solution, Pohlhausen's solution, and scaling to the solution of convection problems.

Example 4.1: Insect in Search of Advice

Air at 30°C flows with uniform velocity $V_\infty = 4$ m/s over a flat plate. A tiny insect finds itself at location 0 near the surface of the plate. Air velocity u at this location is too high for the insect. It wants to take a one 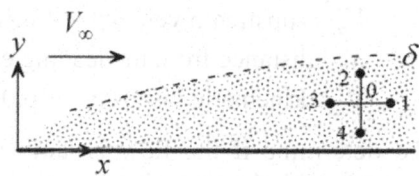 *millimeter step to any of the locations 1, 2, 3, or 4. What will the velocity u be at these locations if the insect starts at x = 150 mm and y = 2 mm? Is the insect inside the viscous boundary layer?*

(1) Observations. (i) This is an external forced convection boundary layer problem. (ii) Changes in velocity between locations 1 and 3 should be small compared to those between 2 and 4. (iii) Location 4 should have the lowest velocity. (iv) If the flow is laminar, Blasius solution can be used to determine the velocity distribution and boundary layer thickness. (v) The flow is laminar if Reynolds number is less than 500,000.

(2) Problem Definition. Determine the axial velocity at the five given locations.

(3) Solution Plan. Check the Reynolds number to determine if boundary layer approximations can be made and if the flow is laminar. If it is, use Blasius solution, Table 4.1, to determine the axial velocity and the boundary layer thickness at the five locations.

(4) Plan Execution.

(i) **Assumptions.** All assumptions leading to Blasius solution are applicable. These are: (1) Continuum, (2) Newtonian fluid, (3) steady state, (4) constant properties, (5) two-dimensional, (6) laminar flow ($Re_x < 5\times10^5$), (7) viscous boundary layer flow ($Re_x > 100$), (8) uniform upstream velocity, (9) flat plate, (10) negligible changes in kinetic and potential energy and (11) no buoyancy ($\beta = 0$ or $g = 0$).

(ii) **Analysis.** The Reynolds number is computed to establish if the flow is laminar and if boundary layer approximations can be made. The Reynolds number is defined as

$$Re_x = \frac{V_\infty x}{\nu},$$ (a)

where

Re_x = Reynolds number
V_∞ = upstream velocity = 4 m/s
x = distance from the leading edge of the plate, m
ν = kinematic viscosity = 16.01×10^{-6} m^2/s

To determine if the flow is laminar or turbulent, compare the Reynolds number with the transition Reynolds number. For flow over a flat plate the transition Reynolds number Re_{x_t} is

$$Re_{x_t} = 5 \times 10^5.$$ (b)

The flow is laminar if $Re_x < Re_{x_t}$. Viscous boundary layer approximations are valid for

$$Re_x > 100.$$ (c)

Evaluating the Reynolds number at $x = 151$ mm, equation (a) gives

$$Re_x = \frac{4(\text{m/s})0.151(\text{m})}{16.01 \times 10^{-6}(\text{m}^2/\text{s})} = 37,726$$

Therefore, boundary layer approximations can be made and the flow is laminar. Use Blasius solution to determine the velocity component u at any location and boundary layer thickness δ. At each location, the variable η is computed and used in Table 4.1 to determine the corresponding velocity ratio u/V_∞. This variable is defined as

$$\eta = y\sqrt{\frac{V_\infty}{\nu x}},$$ (d)

where

y = normal distance from surface, m
η = dimensionless variable

Blasius solution also gives the boundary layer thickness as

$$\frac{\delta}{x} = \frac{5.2}{\sqrt{Re_x}} . \qquad (4.46)$$

(iii) **Computations.** At each location (x, y), equation (d) is used to compute η. The computed η is used in Table 4.1 to determine u/V_∞. Sample computation is shown for location 0. The results for the five locations 0, 1, 2, 3 and 4 are tabulated below.

At location 0 where $x = 150$ mm and $y = 2$ mm. Equation (d) gives

$$\eta = 0.002(m) \sqrt{\frac{4(m/s)}{16.01 \times 10^{-6}(m^2/s)0.15(m)}} = 2.581$$

At this value of η, Table 4.1 gives

$$u/V_\infty = = 0.766, \quad u = 0.766 \times 4(m/s) = 3.064 \text{ m/s}$$

Location	x (m)	y (m)	η	u/V_∞	u(m/s)
0	0.150	0.002	2.581	0.766	3.064
1	0.151	0.002	2.573	0.765	3.06
2	0.150	0.003	3.872	0.945	3.78
3	0.149	0.002	2.590	0.768	3.072
4	0.150	0.001	1.291	0.422	1.688

The boundary layer thickness at the location of the insect is determined using (4.46) where $x = 0.15$ m and $Re_x = 37{,}726$

$$\delta = \frac{5.2}{\sqrt{Re_x}} x = \frac{5.2}{\sqrt{37{,}726}} 0.151(m) = 0.004 \text{ m} = 4 \text{ mm}$$

Thus the insect is within the boundary layer.

(iv) **Checking.** *Dimensional check*: Computations showed that equations (a) and (d) are dimensionally correct.

Qualitative check: The velocity at the five locations follows an expected behavior; minor changes in velocity in the x-direction and significant changes in the y-direction.

(5) **Comments.** (i) The insect should move to location 4 where the axial velocity is lowest.

(ii) Changes in axial velocity with respect to x, at the same distance y from the plate, are minor.

(iii) Changes in axial velocity with respect to y, at the same distance x, are significant.

(iv) The tabulated values of u are approximate since they are determined by interpolations of Table 4.1.

(v) What is important for the insect is the magnitude of the velocity vector $V = (u^2 + v^2)^{1/2}$ and not the axial component u. However, since $v \ll u$ in boundary layer flow, using u as a measure of total velocity is reasonable.

Example 4.2: Laminar Convection over a Flat Plate

Water flows with a velocity of 0.25 m/s over a 75 cm long plate. Free stream temperature is 35°C and surface temperature is 85°C. [a] Derive an equation for the thermal boundary layer thickness δ_t in terms of the Reynolds number. [b] Determine the heat transfer coefficient at $x = 7.5$ cm and 75 cm. [c] Determine the heat transfer rate for a plate 50 cm wide. [d] Can Pohlhausen's solution be used to determine the heat flux at the trailing end of the plate if its length is doubled?

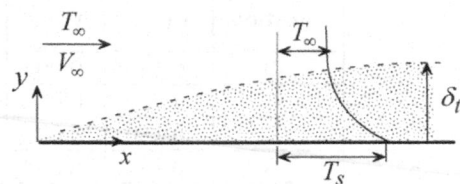

(1) Observations. (i) This is an external forced convection over a flat plate. (ii) The thermal boundary layer thickness increases with distance along the plate. (iii) Newton's law of cooling gives surface heat flux and heat transfer rate from the plate. (iv) The heat transfer coefficient changes with distance along the plate. (v) Pohlhausen's solution is applicable only if the flow is laminar and all other assumptions leading to this solution are valid. (vi) Doubling the length doubles the Reynolds number.

(2) Problem Definition. Determine the water temperature distribution.

(3) Solution Plan. Compute the Reynolds and Peclet numbers to establish if this is a laminar boundary layer problem. If it is, use Pohlhausen's solution to determine the thermal boundary layer thickness, heat transfer coefficient, heat transfer rate, and surface heat flux.

(4) Plan Execution.

(i) Assumptions. The assumptions listed in Section 4.3, which lead to Pohlhausen's solution, are made: (1) Continuum, (2) Newtonian fluid, (3) two-dimensional process, (4) negligible changes in kinetic and potential energy, (5) constant properties, (6) boundary layer flow, (7) steady state, (8) laminar flow, (9) no dissipation, (10) no gravity, (11) no energy generation, (12) flat plate, (13) negligible plate thickness, (14) uniform upstream velocity V_∞, (15) uniform upstream temperature T_∞, (16) uniform surface temperature T_s, and (16) no radiation.

(ii) Analysis and Computations. Calculate the Reynolds and Peclet numbers to determine if boundary layer approximations can be made and if the flow is laminar or turbulent. Boundary layer approximations are valid if the body is streamlined and if

$$Re_x > 100 \ \text{ and } \ Pe = Re_x \, Pr > 100 , \tag{a}$$

where

$$Re_x = V_\infty x / v$$
Pe = Peclet number
Pr = Prandtl number
V_∞ = free stream velocity = 0.25 m/s
x = distance along plate, m
v = kinematic viscosity, m²/s

The transition Reynolds number Re_t for flow over a semi-infinite plate is

$$Re_t = 5 \times 10^5 . \tag{b}$$

Properties of water are evaluated at the film temperature, T_f, defined in (4.73)

$$T_f = (T_s + T_\infty)/2 , \tag{c}$$

where

T_s = surface temperature = 85°C

T_∞ = free stream temperature = 35°C

Substituting into (c) gives

$T_f = (85 + 35)(°C)/2 = 60°C$

Water properties at this temperature are:

k = thermal conductivity = 0.6507 W/m-°C

$Pr = 3.0$

$v = 0.4748 \times 10^{-6} \text{ m}^2/\text{s}.$

Thus at $x = 7.5$ cm Re_x and Pe are

$$Re_x = \frac{V_\infty x}{v} = \frac{0.25(\text{m/s})0.075(\text{m})}{0.4748 \times 10^{-6}(\text{m}^2/\text{s})} = 3.949 \times 10^4$$

and

$$Pe = Re_x Pr = 3.949 \times 10^4 \times 3 = 11.85 \times 10^4$$

Comparison with equations (a) and (b) shows that boundary layer approximations can be made and the flow is laminar at $x = 7.5$ m. At the trailing edge, $x = L = 75$ cm, the Reynolds number $Re_L = 3.949 \times 10^5$. Since this is less than the transition number it follows that the flow is laminar over the entire plate. Thus, Pohlhausen's solution is applicable.

[a] Determination of δ_t. At the edge of the thermal boundary layer $y = \delta_t$ and $T \approx T_\infty$. Thus, $\theta(\eta_t) = (T_\infty - T_s)/(T_\infty - T_s) \approx 1$. From Fig. 4.6 the value of η_t corresponding to $\theta(\eta_t) = 1$ and $Pr = 3$ is approximately 3.2. Therefore

or

$$\eta_t \approx 3.2 = \delta_t \sqrt{V_\infty/vx} \quad,$$

$$\frac{\delta_t}{x} = \frac{3.2}{\sqrt{V_\infty/vx}} = \frac{3.2}{\sqrt{Re_x}}. \tag{d}$$

[b] Heat transfer coefficient. The local heat transfer coefficient is given in (4.66)

$$h(x) = k\sqrt{\frac{V_\infty}{vx}} \frac{d\theta(0)}{d\eta}, \tag{4.66}$$

where $d\theta(0)/d\eta$ for $Pr = 3$ is given in (4.71b)

$$\frac{d\theta(0)}{d\eta} = 0.332 Pr^{1/3}, \quad 0.6 < Pr < 10. \tag{4.71b}$$

For $Pr = 3$, this gives

$$\frac{d\theta(0)}{d\eta} = 0.332\,(3)^{1/3} = 0.4788$$

Substituting into (4.66) for $x = 0.075$ m

$$h = 0.4788(0.6507)(\text{W/m}-^\circ\text{C})\sqrt{\frac{0.25(\text{m/s})}{0.4748\times10^{-6}\,(\text{m}^2/\text{s})0.075(\text{m})}}$$

$$= 825.5\,\text{W/m}^2-^\circ\text{C}$$

Similarly, at $x = 0.75$ m

$$h = 0.4788(0.6507)(\text{W/m}-^\circ\text{C})\sqrt{\frac{0.25(\text{m/s})}{0.4748\times10^{-6}\,(\text{m}^2/\text{s})0.75(\text{m})}}$$

$$= 261\,\text{W/m}^2-^\circ\text{C}$$

[c] Heat transfer rate. Equation (4.70) gives the total heat transfer rate from the plate

$$q_T = (T_s - T_\infty)\,A\overline{h}, \tag{4.70}$$

where

A = surface area = LW, m^2
$\overline{h}$ = average heat transfer coefficient, W/m^2-$^\circ$C
L = length of plate = 75 cm =0.75 m
q_T = total heat transfer rate from plate, W
W = width of plate = 50 cm = 0.5 m

The average heat transfer coefficient is given in (4.67)

$$\overline{h} = 2\frac{k}{L}\sqrt{Re_L}\,\frac{d\theta(0)}{d\eta}. \tag{4.67}$$

The Reynolds number at the trailing edge is $Re_L = 3.949\times10^5$. At this Reynolds number (4.67) gives

$$\overline{h} = 2\frac{0.6507(\text{W/m}^2-^\circ\text{C})}{0.75(\text{m})}\sqrt{3.949\times10^5}\;0.4788 = 522.1\;\text{W/m}^2-^\circ\text{C}$$

Substituting into (4.70)

$$q_T = 522.1\,(\text{W/m}^2 - °\text{C})(85 - 35)(°\text{C})0.75(\text{m})0.5(\text{m}) = 9789 \text{ W}$$

[d] Doubling the length of plate doubles the corresponding Reynolds number at the trailing end. There is a possibility that transition to turbulent flow may take place. For a plate of length $2L$, the Reynolds number is

$$Re_{2L} = 2\,(3.949 \times 10^5) = 7.898 \times 10^5$$

Since this Reynolds number is greater than $Re_t = 5 \times 10^5$, the flow at the trailing end is turbulent and consequently Pohlhausen's solution is not applicable.

(iii) **Checking.** *Dimensional check*: Computations showed that the Reynolds number is dimensionless and units of h and $\overline{h}$ are correct.

Qualitative check: As x is increased h decreases. Computation of the local heat transfer coefficient at $x = 0.075$ m and $x = 0.75$ m confirms this.

Quantitative check: The computed values of the heat transfer coefficients are within the range given in Table 1.1 for forced convection of liquids.

(5) Comments. (i) It is important to check the Reynolds number before applying Pohlhausen's solution.

(ii) The velocity boundary layer thickness δ is given by

$$\frac{\delta}{x} = \frac{5.2}{\sqrt{Re_x}}. \qquad (4.46)$$

Comparing (d) with equation (4.46) indicates that the thermal boundary layer thickness for water is smaller than the velocity boundary layer.

Example 4.3: Scaling Estimate of Heat Transfer Rate

Use scaling to determine the total heat transfer rate for the conditions described in Example 4.2

(1) Observation. (i) Heat transfer rate is determined using Newton's law of cooling. (ii) The heat transfer coefficient can be estimated using scaling.

(2) Problem Definition. Determine the heat transfer coefficient h.

(1) Solution Plan. Apply Newton's law of cooling and use scaling to determine h.

(2) Plan Execution.

(i) **Assumptions.** (1) Continuum, (2) Newtonian fluid, (3) two-dimensional process, (4) negligible changes in kinetic and potential energy, (5) constant properties, (6) boundary layer flow, (7) steady state, (8) no dissipation, (9) no gravity, (10) no energy generation and (11) no radiation.

(ii) **Analysis.** Application of Newton's law of cooling gives

$$q_T = (T_s - T_\infty) A \overline{h}, \qquad (4.70)$$

where

A = surface area = LW, m^2

$\overline{h}$ = average heat transfer coefficient, W/m^2-°C

L = length of plate = 75 cm =0.75 m

q_T = total heat transfer rate from plate, W

T_s = surface temperature = 85°C

T_∞ = free stream temperature = 35°C

W = width of plate = 50 cm = 0.5 m

The heat transfer coefficient is given by (1.10)

$$h = -k \frac{\dfrac{\partial T(x,0)}{\partial y}}{T_s - T_\infty}, \qquad (1.10)$$

where

k = thermal conductivity = 0.6507 W/m-°C

Following the analysis of Section 4.41, scaling of h for $Pr \gg 1$ gives

$$h \sim \frac{k}{x} Pr^{1/3} \sqrt{Re_x}, \quad \text{for } Pr \gg 1, \qquad (4.56)$$

where $Re_x = V_\infty x / v$ and Pr = 3. Setting $h \sim \overline{h}, x = L$, $A = WL$ and substituting (4.56) into (4.70)

$$q_T \sim (T_s - T_\infty) W \, k Pr^{1/3} \sqrt{Re_L}. \qquad (a)$$

(iii) **Computations.** The Reynolds number at the trailing end is $Re_L = 3.949 \times 10^5$. Substituting numerical values into (a)

$$q_T \sim (85 - 35)(^\circ \text{C})\, 0.5(\text{m})\, 0.6507 (\text{W/m}-^\circ \text{C})\, 3^{1/3} \sqrt{394900}$$

$$q_T \sim 14740 \ \text{W}$$

Using Pohlhausen's solution gives $q_T = 9789 \ \text{W}$.

(iv) Checking. *Dimensional Check*: Solution (a) is dimensionally correct.

(5) Comments. Scaling gives an order of magnitude estimate of the heat transfer coefficient. In this example, the error in scaling estimate of the heat transfer rate is 50%.

4.4.3 Laminar Boundary Layer Flow over Semi-infinite Flat Plate: Variable Surface Temperature [4]

Consider uniform flow over a semi-infinite flat plate shown in Fig. 4.7. Surface temperature varies with axial distance x according to

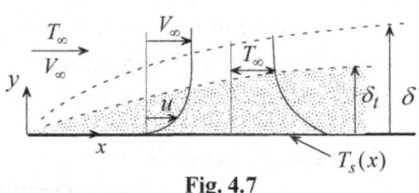

Fig. 4.7

$$T_s(x) - T_\infty = Cx^n, \quad (4.74)$$

where C and n are constants and T_∞ is free stream temperature. We wish to determine the temperature distribution, heat transfer coefficient, Nusselt number, and heat transfer rate. To solve this problem we invoke all the assumptions summarized in Section 4.3.

(i) Velocity Distribution.   Since properties are assumed constant, the velocity distribution is independent of the temperature distribution. Thus Blasius solution is applicable to this case and the velocity components are given by

$$\frac{u}{V_\infty} = \frac{df}{d\eta}, \quad (4.42)$$

$$\frac{v}{V_\infty} = \frac{1}{2}\sqrt{\frac{v}{V_\infty x}}\left(\eta \frac{df}{d\eta} - f\right), \quad (4.43)$$

where the similarity variable η is defined as

$$\eta(x,y) = y\sqrt{\frac{V_\infty}{\nu x}}. \tag{4.41}$$

(ii) Governing Equations for Temperature Distribution. Based on the assumptions listed in Section 4.3, the temperature distribution is governed by energy equation (4.18)

$$u\frac{\partial T}{\partial x} + v\frac{\partial T}{\partial y} = \alpha\frac{\partial^2 T}{\partial y^2}. \tag{4.18}$$

The boundary conditions for this problem are:

$$T(x,0) = T_s = T_\infty + Cx^n, \tag{a}$$

$$T(x,\infty) = T_\infty, \tag{b}$$

$$T(0,y) = T_\infty. \tag{c}$$

(iii) Solution. The solution to (4.18) is obtained by the method of similarity transformation. We define a dimensionless temperature θ as

$$\theta = \frac{T - T_s}{T_\infty - T_s}. \tag{4.58}$$

We assume

$$\theta(x,y) = \theta(\eta). \tag{4.75}$$

Using (4.41)-(4.43), (4.58), (4.74) and (4.75), energy equation (4.18) transforms to (see Appendix C for details)

$$\frac{d^2\theta}{d\eta^2} + nPr\frac{df}{d\eta}(1-\theta) + \frac{Pr}{2}f(\eta)\frac{d\theta}{d\eta} = 0. \tag{4.76}$$

Boundary conditions (a)-(c) become

$$\theta(0) = 0, \tag{4.77a}$$

$$\theta(\infty) = 1, \tag{4.77b}$$

$$\theta(\infty) = 1. \tag{4.77c}$$

Note that boundary conditions (b) and (c) coalesce into a single condition, as shown in (4.76b) and (4.76c). The local heat transfer coefficient and Nusselt number are determined using (1.10)

$$h = -k \frac{\dfrac{\partial T(x,0)}{\partial y}}{T_s - T_\infty} \, , \tag{1.10}$$

where

$$\frac{\partial T(x,0)}{\partial y} = \frac{dT}{d\theta} \frac{d\theta(0)}{d\eta} \frac{\partial \eta}{\partial y} \, .$$

Using (4.41), (4.58) and (a) into the above

$$\frac{\partial T(x,0)}{\partial y} = -Cx^n \sqrt{\frac{V_\infty}{vx}} \frac{d\theta(0)}{d\eta} \, .$$

Substituting into (1.10) gives the local heat transfer coefficient

$$h(x) = k \sqrt{\frac{V_\infty}{vx}} \frac{d\theta(0)}{d\eta} \, . \tag{4.78}$$

The average heat transfer coefficient for a plate of length L is defined in equation (2.50)

$$\overline{h} = \frac{1}{L} \int_0^L h(x) dx \, . \tag{2.50}$$

Substituting (4.78) into (2.50) and integrating

$$\overline{h} = 2 \frac{k}{L} \sqrt{Re_L} \frac{d\theta(0)}{d\eta} \, . \tag{4.79}$$

The local Nusselt number is obtained by substituting (4.78) into (4.54)

$$Nu_x = \frac{d\theta(0)}{d\eta} \sqrt{Re_x} \, . \tag{4.80}$$

The corresponding average Nusselt number is

$$\overline{Nu_L} = 2\frac{d\theta(0)}{d\eta}\sqrt{Re_L}\,. \qquad (4.81)$$

Thus the key factor in the determination of the heat transfer coefficient and Nusselt number is surface temperature gradient $d\theta(0)/d\eta$.

(iii) Results. The solution to (4.76) subject to boundary conditions (4.77) is obtained by numerical integration [4]. The solution depends on two parameters: the Prandtl number Pr and the exponent n in (4.74) which characterizes surface temperature variation. Temperature gradient at the surface, $d\theta(0)/d\eta$, is presented in Fig. 4.8 for three Prandtl numbers.

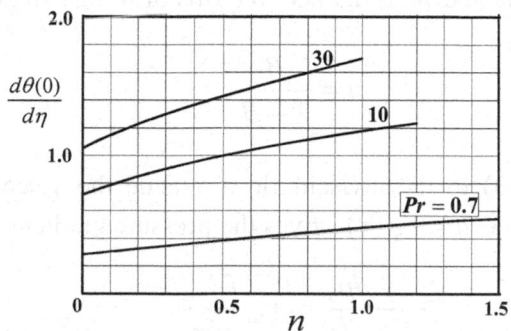

Fig. 4.8 Surface temperature gradient for plate with varying surface temperatue, $T_s - T_\infty = Cx^n$ [4]

4.4.4 Laminar Boundary Layer Flow over a Wedge: Uniform Surface Temperature

Consider symmetrical flow over a wedge of angle $\beta\pi$ shown in Fig. 4.9. The wedge is maintained at uniform surface temperature. Fluid velocity, temperature, and pressure upstream of the wedge are uniform. However, pressure and velocity outside the viscous boundary layer vary with distance x along the wedge. A summary of key features of this problem follows.

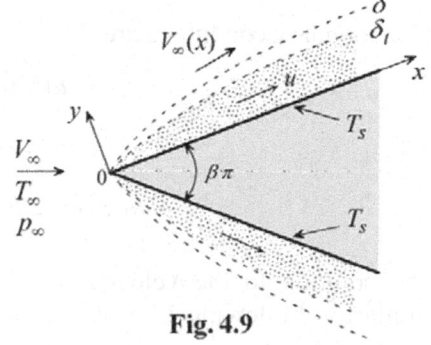

Fig. 4.9

Based on the assumptions listed in Section 4.3, the x-momentum equation for this case is:

$$u\frac{\partial u}{\partial x} + v\frac{\partial u}{\partial y} = -\frac{1}{\rho}\frac{dp_\infty}{dx} + v\frac{\partial^2 u}{\partial y^2}.$$

(4.13)

The solution to inviscid flow over the wedge gives the velocity outside the viscous boundary layer $V_\infty(x)$ as

$$V_\infty(x) = Cx^m,$$

(4.82)

where C is a constant and m is defined in terms of wedge angle as

$$m = \frac{\beta}{2-\beta}.$$

(4.83)

Application of (4.13) to the inviscid flow outside the viscous boundary layer where $v = 0$ and $u = V_\infty(x)$, gives the pressure gradient dp_∞ / dx

$$-\frac{1}{\rho}\frac{dp_\infty}{dx} = V_\infty\frac{\partial V_\infty}{\partial x}.$$

Substituting into (4.13)

$$u\frac{\partial u}{\partial x} + v\frac{\partial u}{\partial y} = V_\infty\frac{\partial V_\infty}{\partial x} + v\frac{\partial^2 u}{\partial y^2}.$$

(4.84)

The boundary conditions are

$$u(x,0) = 0,$$

(4.85a)

$$v(x,0) = 0,$$

(4.85b)

$$u(x,\infty) = V_\infty(x) = Cx^m.$$

(4.85c)

The solution to the velocity distribution is obtained by the method of similarity. Following Blasius approach, a similarity variable η is defined as

$$\eta(x,y) = y\sqrt{\frac{V_\infty(x)}{\nu x}} = y\sqrt{\frac{C}{\nu}} \, x^{(m-1)/2}. \qquad (4.86)$$

The velocity $u(x, y)$ is assumed to depend on η according to

$$\frac{u}{V_\infty(x)} = \frac{dF}{d\eta}. \qquad (4.87)$$

Continuity equation (2.3), (4.86), and (4.87) give the vertical velocity component v

$$v = -V_\infty(x)\sqrt{\frac{\nu}{xV_\infty(x)}}\frac{m+1}{2}\left[F - \frac{1-m}{1+m}\eta\frac{dF}{d\eta}\right]. \qquad (4.88)$$

Substituting (4.82) and (4.86)-(4.88) into (4.84)

$$\frac{d^3F}{d\eta^3} + \frac{m+1}{2}F\frac{d^2F}{d\eta^2} - m\left[\frac{dF}{d\eta}\right]^2 + m = 0. \qquad (4.89)$$

This is the transformed momentum equation. Boundary conditions (4.85) transform to

$$\frac{dF(0)}{d\eta} = 0, \qquad (4.90a)$$

$$F(0) = 0, \qquad (4.90b)$$

$$\frac{dF(\infty)}{d\eta} = 1. \qquad (4.90c)$$

Note the following regarding (4.89) and (4.90):

(1) The original variables x and y do not appear explicitly in these equations.

(2) Momentum equation (4.89) is a third order non-linear ordinary differential equation.

(3) The special case of $m = \beta = 0$ corresponds to a flat plate. Setting $m = 0$ in (4.89)   and (4.90) reduces to Blasius problem (4.44) and (4.45) with $F(\eta) = f(\eta)$.

Equation (4.89) is integrated numerically [5, 6]. The solution gives the function $F(\eta)$ and its derivative $dF / d\eta$. These in turn give the velocity components u and v.

To determine the temperature distribution we begin with the energy equation and thermal boundary conditions. The applicable equations for the wedge are the same as those of the semi-infinite flat plate, given by

$$u\frac{\partial \theta}{\partial x} + v\frac{\partial \theta}{\partial y} = \alpha \frac{\partial^2 \theta}{\partial y^2} \; , \tag{4.59}$$

$$\theta(x,0) = 0 \, , \tag{4.60a}$$

$$\theta(x,\infty) = 1 \, , \tag{4.60b}$$

$$\theta(0,y) = 1 \, . \tag{4.60c}$$

where the dimensionless temperature θ is defined as

$$\theta = \frac{T - T_s}{T_\infty - T_s} \, . \tag{4.58}$$

The difference between the flat plate and wedge problem is the velocity distribution. In the flat plate case the velocity is given by Blasius solution while in the wedge the solution to (4.89) gives the velocity distribution. Energy equation (4.59) is solved by the method of similarity transformation. We assume

$$\theta = \theta(\eta) \, , \tag{4.75}$$

where the similarity variable η is defined in (4.86). Substituting (4.86)-(4.88) and (4.75) into (4.59) and (4.60)

$$\frac{d^2\theta}{d\eta^2} + \frac{Pr}{2}(m+1)F(\eta)\frac{d\theta}{d\eta} = 0 \, , \tag{4.91}$$

$$\theta(0) = 0 , \tag{4.92a}$$

$$\theta(\infty) = 1, \tag{4.92b}$$

$$\theta(\infty) = 1. \tag{4.92c}$$

Thus, the governing partial differential equation is successfully transformed into an ordinary differential equation. The following observations are made regarding (4.91) and (4.92):

(1) Two parameters, Prandtl number Pr and the wedge size m, characterize the equation.

(2) This is a linear second order ordinary differential equation requiring two boundary conditions.

(3) The function $F(\eta)$ appearing in (4.91) represents the effect of fluid motion on temperature distribution. It is obtained from the solution to (4.89).

(4) Boundary conditions (4.60b) and (4.60c) coalesce into a single condition, as shown in (4.92b) and (4.92c).

(5) The special case of $m = \beta = 0$ corresponds to a flat plate. Setting $m = 0$ in (4.91) reduces to Pohlhausen's problem (4.61).

Following the procedure used in Appendix B, separating variables in equation (4.91), integrating twice and applying boundary conditions (4.92), gives the temperature solution as

$$\theta(\eta) = 1 - \frac{\displaystyle\int_{\eta}^{\infty} \exp\left[-\frac{(m+1)Pr}{2} \int_{0}^{\eta} F(\eta)d\eta \right] d\eta}{\displaystyle\int_{0}^{\infty} \exp\left[-\frac{(m+1)Pr}{2} \int_{0}^{\eta} F(\eta)d\eta \right] d\eta} . \tag{4.93}$$

The temperature gradient at the surface is obtained by differentiating (4.93) and evaluating the derivative at the surface, $\eta = 0$ to obtain

$$\frac{d\theta(0)}{d\eta} = \left\{ \int_{0}^{\infty} \exp\left[-\frac{(m+1)Pr}{2} \int_{0}^{\eta} F(\eta)d\eta \right] d\eta \right\}^{-1} . \tag{4.94}$$

The function $F(\eta)$ appearing in (4.93) and (4.94) is obtained from the numerical solution to flow field equation (4.89). The integrals in (4.93) and (4.94) are evaluated numerically. Results for the temperature gradient at the surface, $d\theta(0)/d\eta$, are given in Table 4.3 for four wedge angles at five Prandtl numbers [7]. Also shown in Table 4.3 is $F''(0)$ [5].

Surface temperature gradient $\dfrac{d\theta(0)}{d\eta}$ and surface velocity gradient $F''(0)$ for flow over an isothermal wedge							
m	wedge angle $\pi\beta$	$F''(0)$	$d\theta(0)/d\eta$ at five values of Pr				
			0.7	0.8	1.0	5.0	10.0
0	0	0.3206	0.292	0.307	0.332	0.585	0.730
0.111	$\pi/5$ (36°)	0.5120	0.331	0.348	0.378	0.669	0.851
0.333	$\pi/2$ (90°)	0.7575	0.384	0.403	0.440	0.792	1.013
1.0	π (180°)	1.2326	0.496	0.523	0.570	1.043	1.344

Table 4.3 is used to determine the heat transfer coefficient h and Nusselt number Nu. Equation (1.10) gives h

$$h = -k\frac{\dfrac{\partial T(x,0)}{\partial y}}{T_s - T_\infty}, \qquad (1.10)$$

where

$$\frac{\partial T(x,0)}{\partial y} = \frac{dT}{d\theta}\frac{d\theta(0)}{d\eta}\frac{\partial \eta}{\partial y}.$$

Using (4.58), (4.75) and (4.86) into the above

$$\frac{\partial T(x,0)}{\partial y} = (T_\infty - T_s)\sqrt{\frac{V_\infty(x)}{v x}}\frac{d\theta(0)}{d\eta}.$$

Substituting into (1.10) gives the local heat transfer coefficient

$$h(x) = k \sqrt{\frac{V_\infty(x)}{\nu x}} \frac{d\theta(0)}{d\eta} . \tag{4.95}$$

The local Nusselt number is obtained by substituting (4.95) into (4.54)

$$Nu_x = \frac{d\theta(0)}{d\eta} \sqrt{Re_x} , \tag{4.96}$$

where Re_x is the local Reynolds number defined as

$$Re_x = \frac{x V_\infty(x)}{\nu} . \tag{4.97}$$

Examination of (4.95) and (4.96) shows that the key factor in the determination of the heat transfer coefficient and Nusselt number is surface temperature gradient $d\theta(0)/d\eta$ listed in Table 4.3.

REFERENCES

[1] Schlichting, H, *Boundary Layer Theory*, 7th edition, translated into English by J. Kestin, McGraw-Hill, 1979.

[2] Hansen, A.G., *Similarity Analysis of Boundary Value Problems in Engineering*, Prentice-Hall, 1976.

[3] Kays, W.M. and M.E. Crawford, *Convection Heat Transfer, 3rd edition*, McGraw-Hill, 1993.

[4] Oosthuizen, P.H. and D. Naylor, *Introduction to Convection Heat Transfer Analysis*, McGraw-Hill, 1999.

[5] Falkner, W.M. and S.W. Skan, "Solution of the Boundary-Layer Equations," Phil. Mag., Vol. 12, 1931, pp. 865-896.

[6] Hartree, A.G., "On an Equation Occurring in Falkner and Skan's Approximate Treatment of the Equations of the Boundary Layer," Proc. Cambridge Phil. Soc., Vol. 33, Part II, 1937, pp. 223-239.

[7] Evens, H.L., "Mass Transfer through Laminar Boundary Layers. Further Similar Solutions to the *B*-equation for the case of *B* = 0," Int. J. Heat Mass Transfer, Vol. 5, 1962, pp. 35-57.

PROBLEMS

4.1   Put a check mark in the appropriate column for each of the following statements.

Statement	true	false	may be
(a) $(\partial u / \partial x) + (\partial v / \partial y) = 0$ is valid for transient flow.			
(b) The y-momentum equation is neglected in boundary layer flow.			
(c) Boundary layer equations are valid for all Reynolds numbers.			
(d) Pressure gradient is zero outside the boundary layer.			
(e) $\dfrac{\partial^2 u}{\partial x^2} << \dfrac{\partial^2 u}{\partial y^2}$ is for a streamlined body.			
(f) In boundary layer flow fluid velocity upstream of an object is undisturbed.			
(g) Axial pressure gradient is neglected in boundary layer flow.			
(i) Axial conduction is neglected in boundary layer flow.			

4.2   Examine the three governing equations, (2.3), (4.13) and (4.18) for two-dimensional, constant properties, laminar boundary layer flow.

[a] How many dependent variables do these equations have?

[b] How is the pressure p_∞ determined?

[c] If streamlines are parallel in the boundary layer, what terms will vanish?

[d] Can (2.3) and (4.13) be solved for the velocity field u and v independently of the energy equation (4.18)?

4.3   Air flows over a semi-infinite plate with a free stream velocity $V_\infty = 0.4$ m/s and a free stream temperature $T_\infty = 20^\circ$C. The plate is maintained at $T_s = 60^\circ$C. Can boundary layer approximations for the flow and temperature fields be applied at:

[a] location $x = 1.5$ mm?
[b] location $x = 15$ mm?

Note: Evaluate air properties at the average film temperature $T_f = (T_s + T_\infty)/2$.

4.4 Water at 25°C flows with uniform velocity $V_\infty = 2$ m/s over a streamlined object. The object is 8 cm long and its surface is maintained at $T_s = 85^\circ$C. Use scaling to:

[a] show that $\delta / L \ll 1$,
[b] evaluate the inertia terms $u\partial u / \partial x$ and $v\partial u / \partial y$,
[c] evaluate the viscous terms $\nu\partial^2 u / \partial x^2$ and $\nu\partial^2 u / \partial y^2$.

4.5 Water at 25°C flows with uniform velocity $V_\infty = 2$ m/s over a streamlined object. The object is 8 cm long and its surface is maintained at $T_s = 85^\circ$C. Use scaling to:

[a] show that $\delta / L \ll 1$,
[b] evaluate the convection terms $u\partial T / \partial x$ and $v\partial T / \partial y$,
[c] evaluate the conduction terms $\alpha\partial^2 T / \partial x^2$ and $\alpha\partial^2 T / \partial y^2$.

4.6 Atmospheric air at 25°C flows over a surface at 115°C. The free stream velocity is 10 m/s.

[a] Calculate the Eckert number.
[b] Use scale analysis to show that the dissipation term $\mu (\partial u / \partial y)^2$ is small compared to the conduction term $k (\partial^2 T / \partial y^2)$.

4.7 Air at 20°C flows over a streamlined surface with a free stream velocity of 10 m/s. Use scale analysis to determine the boundary layer thickness at a distance of 80 cm from the leading edge.

4.8 In boundary layer flow, pressure gradient normal to the flow direction is assumed zero. That is $\partial p / \partial y \approx 0$. If this is correct, how do you explain lift on the wing of an airplane in flight?

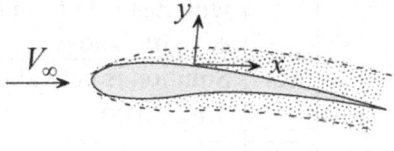

4.9 Derive an equation describing the vertical velocity component v at the edge of the boundary layer for two-dimensional incompressible flow over a semi-infinite flat plate. Assume laminar flow. Compare your result with scaling estimate.

4.10 Sketch the streamlines in boundary layer flow over a semi-infinite flat plate.

4.11 Define the thickness of the velocity boundary layer δ in Blasius solution as the distance y where the velocity $u = 0.988\ V_\infty$. Derive an expression for δ/x.

4.12 Water flows over a semi-infinite plate with an upstream velocity of 0.2 m/s. Blasius solution is used to calculate δ at three locations along the plate. Results are tabulated. Are these results valid? Explain.

x(cm)	δ(cm)
300	1.441
40	0.526
0.01	0.0083

4.13 Consider laminar boundary layer flow over a semi-infinite flat plate. Evaluate the wall shearing stress at the leading edge. Comment on your answer. Is it valid? If not explain why.

4.14 Water at $20\,^\circ\mathrm{C}$ flows over a $2\,\mathrm{m} \times 2\,\mathrm{m}$ plate with a free stream velocity of 0.18 m/s. Determine the force needed to hold the plate in place. Assume laminar boundary layer flow.

4.15 Consider Blasius solution for uniform flow over a semi-infinite plate. Put a check mark in the appropriate column for each of the following statements.

	Statement	true	false	may be
(a)	$dp_\infty/dx = 0$ because the flow is laminar.			
(b)	Wall shearing stress increases with distance from the leading edge of plate.			
(c)	Solution is not valid for $Re_x < 100$.			
(d)	Solution is not valid for $Re_x > 5 \times 10^5$.			
(e)	Solution is valid for $Re_x > 100$.			

(f)	Boundary layer thickness is uniquely defined.			
(g)	Solution is not valid for a curved plate.			
(h)	Solution for the wall shear at the leading edge $(x = 0)$ is not valid.			
(i)	The plate does not disturb upstream flow.			
(j)	Solution is not valid for $Re_x < 5 \times 10^5$.			

4.16 Imagine a cold fluid flowing over a thin hot plate. Using your intuition, would you expect the fluid just upstream of the plate to experience a temperature rise due to conduction from the hot plate? How do you explain the assumption in Pohlhausen's solution that fluid temperature is unaffected by the plate and therefore $T(0, y) = T_\infty$?

4.17 Consider laminar boundary layer flow over a semi-infinite flat plate. The plate is maintained at uniform temperature T_s. Assume constant properties and take into consideration dissipation.

[a] Does Blasius solution apply to this case? Explain.
[b] Does Pohlhausen's solution apply to this case? Explain.

4.18 A fluid with Prandtl number 9.8 flows over a semi-infinite flat plate. The plate is maintained at uniform surface temperature. Derive an expression for the variation of the thermal boundary layer thickness with distance along the plate. Assume steady state laminar boundary layer flow with constant properties and neglect dissipation. Express your result in dimensionless form.

4.19 Use Pohlhausen's solution to determine the heat flux at the leading edge of a plate. Comment on your answer. Is it valid? If not explain why.

4.20 Consider laminar boundary layer flow over a semi-infinite flat plate at uniform surface temperature T_s. The free stream velocity is V_∞ and the Prandtl number is 0.1. Determine temperature gradient at the surface $dT(0)/dy$.

4.21 Fluid flows between two parallel plates. It enters with uniform velocity V_∞ and temperature T_∞. The plates are maintained at uniform surface temperature T_s. Assume laminar boundary layer flow at the entrance. Can Pohlhausen solution be applied to determine the heat transfer coefficient? Explain.

4.22 Two identical rectangles, A and B, of dimensions $L_1 \times L_2$ are drawn on the surface of a semi-infinite flat plate as shown. Rectangle A is oriented with side L_1 along the leading edge while rectangle B is oriented with side L_2 along the edge. The plate is maintained at uniform surface temperature.

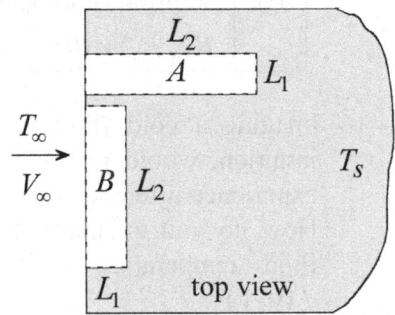

[a] If the flow over rectangle A is laminar, what is it for B ?

[b] If the heat transfer rate from plate A is 435 W, what is the rate from plate B ?

4.23 A semi-infinite plate is divided into four equal sections of one centimeter long each. Free stream temperature and velocity are uniform and the flow is laminar. The surface is maintained at uniform temperature. Determine the ratio of the heat transfer rate from the third section to that from the second section.

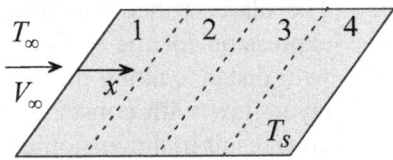

4.24 A fluid at a uniform velocity and temperature flows over a semi-infinite flat plate. The surface temperature is uniform. Assume laminar boundary layer flow.

[a] What will be the percent change in the local heat transfer coefficient if the free stream velocity is reduced by a factor of two?

[b] What will be the percent change in the local heat transfer coefficient if the distance from the leading edge is reduced by a

factor of two?

4.25 Use Pohlhausen's solution to derive an expression for the ratio of the thermal boundary layer thickness for two fluids. The Prandtl number of one fluid is 1.0 and its kinematic viscosity is $0.12 \times 10^{-6}\, m^2\!/s$. The Prandtl number of the second fluid is 100 and its kinematic viscosity is $6.8 \times 10^{-6}\, m^2\!/s$.

4.26 Water at 25°C flows over a flat plate with a uniform velocity of 2 m/s. The plate is maintained at 85°C. Determine the following:

[a] The thermal boundary layer thickness at a distance of 8 cm from the leading edge.

[b] The heat flux at this location.

[c] The total heat transfer from the first 8 cm of the plate.

[d] Whether Pohlhausen's solution can be used to find the heat flux at a distance of 80 cm from the leading edge.

4.27 The cap of an electronic package is cooled by forced convection. The free stream temperature is 25°C. The Reynolds number at the downstream end of the cap is 110,000. Surface temperature was found to be 145°C. However, reliability requires that surface temperature does not exceed 83°C. One possible solution to this design problem is to increase the free stream velocity by a factor of 3. You are asked to determine if surface the temperature under this plan will meet design specification.

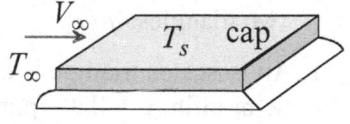

4.28 The back of the dinosaur Stegosaurus has two rows of fins. Each row is made up of several fins arranged in line and separated by a space. One theory suggests that providing a space between neighboring fins reduces the weight on the back of the dinosaur when compared with a single long fin along the back. On the other hand, having a space between neighboring fins reduces the total surface area. This may result in a reduction in the total heat loss.

Model the fins as rectangular plates positioned in line as shown. The length of each plate is L and its height is H. Consider two fins separated by a distance L. Compare the heat loss from the

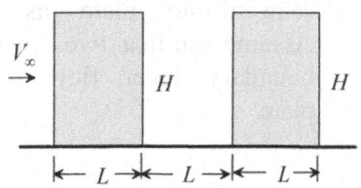

two fins with that of a single fin of length $3L$ and height H. Does your result support the argument that spaced fins result in a reduction in heat loss? To simplify the analysis assume laminar flow.

4.29   A fluid with Prandtl number 0.098 flows over a semi-infinite flat plate. The free stream temperature is T_∞ and the free stream velocity is V_∞. The surface of the plate is maintained at uniform temperature T_s. Assume laminar flow.

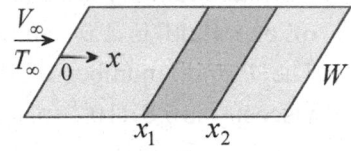

[a] Derive an equation for the local Nusselt number.

[b] Determine the heat transfer rate from a section of the plate between x_1 and x_2. The width of the plate is W.

[c] Derive an equation for the thermal boundary layer thickness $\delta_t(x)$.

4.30   Two identical triangles are drawn on the surface of a flat plate as shown. The plate, which is maintained at uniform surface temperature, is cooled by laminar forced convection. Determine the ratio of the heat transfer rate from the two triangles, q_1/q_2.

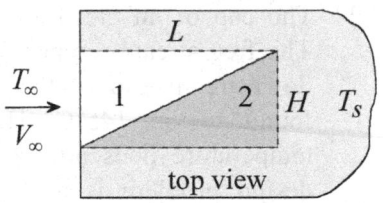

4.31   An isosceles triangle is drawn on a semi-infinite flat plate at a uniform surface temperature T_s. Consider laminar uniform flow of constant properties fluid over the plate. Determine the rate of heat transfer between the triangular area and the fluid

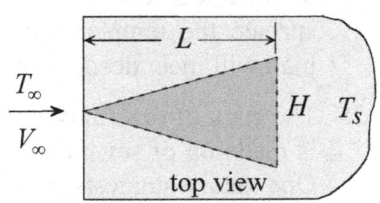

4.32   Determine the total heat transfer rate from a half circle drawn on a semi-infinite plate as shown. Assume laminar two-dimensional boundary layer flow over the plate.

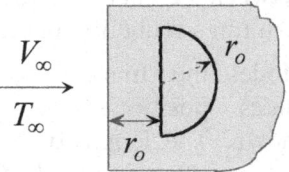

4.33 Consider steady, two-dimensional, laminar boundary layer flow over a semi-infinite plate. The surface is maintained at uniform temperature T_s. Determine the total heat transfer rate from the surface area described by $y(x) = H\sqrt{x/L}$ as shown.

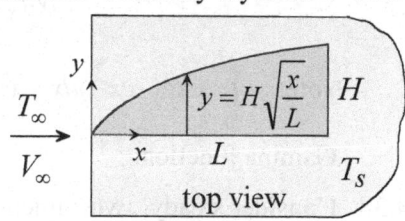

4.34 Fluid flows over a semi-infinite flat plate which is maintained at uniform surface temperature. It is desired to double the rate of heat transfer from a circular area of radius R_1 by increasing its radius to R_2. Determine the percent increase in radius needed to accomplish this change. In both cases the circle is tangent to the leading edge. Assume laminar boundary layer flow with constant properties.

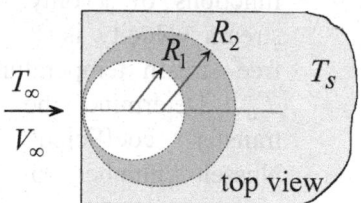

4.35 Liquid potassium ($Pr \ll 1$) flows over a semi-infinite plate. Assume laminar boundary layer flow. Suggest a simplified velocity profile for solving the energy equation.

4.36 For very low Prandtl numbers the thermal boundary layer is much thicker than the viscous boundary layer. Thus little error is introduced if the velocity everywhere in the thermal boundary layer is assumed to be the free stream velocity V_∞. Show that for laminar boundary layer flow over a flat plate at low Prandtl numbers, the local Nusselt number is given by

$$Nu_x = 0.564 Pr^{1/2} Re^{1/2}.$$

How does this result compare with scaling prediction?

4.37 Consider laminar boundary layer flow over a flat plate at a uniform temperature T_s. When the Prandtl number is very high the viscous boundary layer is much thicker than the thermal boundary layer. Assume that the thermal boundary layer is entirely within the part of the velocity boundary layer in which the velocity profile is approximately linear. Show that for such approximation the Nusselt number is given by

$$Nu_x = 0.339 Pr^{1/3} Re^{1/2}.$$

Note: $\displaystyle\int_0^\infty \exp(-cx^3)dx = (1/3)c^{-1/3}\Gamma(1/3)$, where Γ is the Gamma function.

4.38 Consider steady, two-dimensional, laminar boundary layer flow over a porous flat plate at uniform surface temperature. The plate is subject to a uniform suction $v(x,0) = -v_0$. Far away downstream both the axial velocity and the temperature may be assumed to be functions of y only. Free stream velocity is V_∞ and free stream temperature is T_∞. Determine the heat transfer coefficient and Nusselt number in this region.

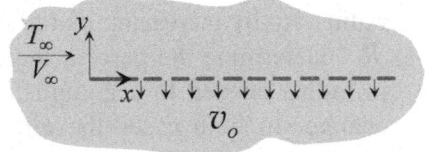

4.39 A semi infinite plate is heated with uniform flux q'' along its length. The free stream temperature is T_∞ and free stream velocity is V_∞. Since the heat transfer coefficient varies with distance along the plate, Newton's law of cooling requires that surface temperature must also vary to maintain uniform heat flux. Consider the case of laminar boundary layer flow over a plate whose surface temperature varies according to

$$T_s(x) - T_\infty = Cx^n.$$

Working with the solution to this case, show that $n = 1/2$ corresponds to a plate with uniform surface flux.

4.40 Water flows over a semi-infinite flat plate which is maintained at a variable surface temperature T_s given by

$$T_s(x) - T_\infty = Cx^{0.75},$$

where

$C = 54.27\,°C\,/\,m^{0.75}$
$T_\infty = $ free stream temperature $= 3°\,C.$
$x = $ distance from the leading edge, m

Determine the average heat transfer coefficient for a plate if length L = 0.3 m. The free stream velocity is 1.2 m/s.

4.41 Air flows over a plate which is heated non-uniformly such that its surface temperature increases linearly as the distance from the leading edge is increased according to

$$T_s(x) = T_\infty + Cx$$

where

$C = 24\ {}^\circ C\,/m$

T_∞ = free stream temperature = $20^\circ C$

x = distance from the leading edge, m

Determine the total heat transfer rate from a square plate 10 cm × 10 cm. The free stream velocity is 3.2 m/s.

4.42 The surface temperature of a plate varies with distance from the leading edge according to

$$T_s(x) = T_\infty + Cx^{0.8}$$

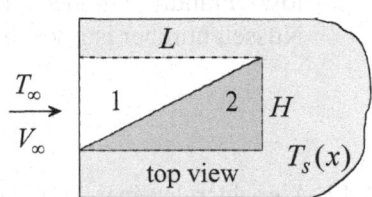

Two identical triangles are drawn on the surface as shown. Fluid at uniform upstream temperature T_∞ and uniform upstream velocity V_∞ flows over the plate. Assume laminar boundary layer flow. Determine the ratio of the heat transfer rate from the two triangles, q_1/q_2.

4.43 Construct a plot showing the variation of $Nu_x / \sqrt{Re_x}$ with wedge angle. Nu_x is the local Nusselt number and Re_x is the local Reynolds number. Assume laminar boundary layer flow of air.

4.44 Consider laminar boundary layer flow over a wedge. Show that the average Nusselt number $\overline{Nu}$ for a wedge of length L is given by

$$\overline{Nu} = \frac{2}{m+1}\frac{d\theta(0)}{d\eta}Re_L$$

where the Reynolds number is defined as $Re_L = \dfrac{LV_\infty(L)}{\nu}$.

4.45   Compare the total heat transfer rate from a $90°$ wedge, q_w, with that from a flat plate, q_p, of the same length. Construct a plot of q_w / q_p as a function of Prandtl number.

4.46   For very low Prandtl numbers the thermal boundary layer is much thicker than the viscous boundary layer. Thus little error is introduced if the velocity everywhere in the thermal boundary layer is assumed to be the free stream velocity V_∞. Show that for laminar boundary flow over a wedge at low Prandtl numbers the local Nusselt number is given by

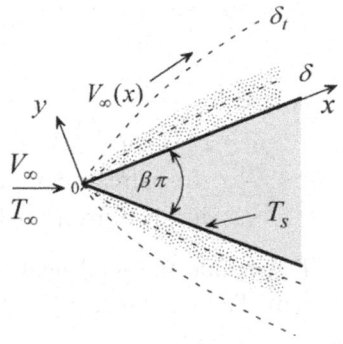

$$Nu_x = \sqrt{\frac{(m+1)Pr}{\pi} Re_x} \ .$$

4.47   Consider laminar boundary layer flow over a wedge at a uniform temperature T_s. When the Prandtl number is very high the viscous boundary layer is much thicker than the thermal boundary layer. Assume that the velocity profile within the thermal boundary layer is approximately linear. Show that for such approximation the local Nusselt number is given by

$$Nu_x = 0.489\big[(m+1)F''(0)Pr\big]^{1/3} Re^{1/2} \ .$$

Note: $\displaystyle\int_0^\infty \exp(-cx^3)dx = (1/3)c^{-1/3}\Gamma(1/3)$, where Γ is the Gamma function.

5

APPROXIMATE SOLUTIONS:
THE INTEGRAL METHOD

5.1 Introduction

There are various situations where it is desirable to obtain approximate analytic solutions. An obvious case is when an exact solution is not available or can not be easily obtained. Approximate solutions are also obtained when the form of the exact solution is not convenient to use. Examples include solutions that are too complex, implicit or require numerical integration. The *integral method* is used extensively in fluid flow, heat transfer and mass transfer. Because of the mathematical simplifications associated with this method, it can deal with such complicating factors as turbulent flow, temperature dependent properties and non-linearity.

5.2 Differential vs. Integral Formulation

To appreciate the basic approximation and simplification associated with the integral method, we consider the boundary layer flow shown in Fig. 5.1. In *differential formulation*, Fig. 5.1a, a differential element measuring $dx \times dy$ is selected. The three basic laws are formulated for this element.

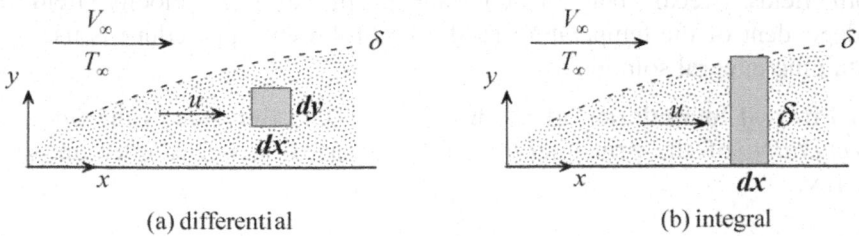

(a) differential   (b) integral

Fig. 5.1

The resulting equations thus apply to any point in the region and the solutions to these equations satisfy the basic laws everywhere exactly. Note that the same approach is used in three-dimensional transient problems. Here the basic laws are formulated for an element measuring $dx \times dy \times dz$ during an infinitesimal time dt. In *integral formulation*, Fig. 5.1b, a differential element measuring $dx \times \delta$ is selected. Note that this element is infinitesimal in x but finite in y. The three basic laws are formulated for this element. Here the resulting equations satisfy the basic laws for an entire cross section δ and not at every point. Thus solutions to this type of formulation are approximate in the sense that they do not satisfy the basic laws at every point.

5.3 Integral Method Approximation: Mathematical Simplification

Although integral solutions do not satisfy the basic laws at every point, they provide significant mathematical simplifications. A key simplification is a reduction in the number of independent variables. For example, for two-dimensional problems, instead of solving a partial differential equation in differential formulation, one solves an ordinary differential equation in integral formulation. In addition, an accompanying reduction of the order of the governing differential equation may result. Thus, major mathematical simplifications are associated with this approach. This explains why it is extensively used to solve a wide range of problems in fluid flow, heat transfer and mass transfer. In this chapter, the integral method is applied to boundary layer convection problems.

5.4 Procedure

Since convection heat transfer depends on fluid motion as well as temperature distribution, solutions require the determination of the velocity and temperature fields. The integral method is used in the determination of both fields. Recall that for constant properties the velocity field is independent of the temperature field. The following procedure is used in obtaining integral solutions:

(1) Integral formulation of the basic laws. The first step is the *integral formulation* of the principles of conservation of mass, momentum and energy.

(2) Assumed velocity and temperature profiles. Appropriate velocity and temperature profiles are assumed which satisfy known boundary conditions. An assumed profile can take on different forms. However, a polynomial is usually used in Cartesian coordinates. An assumed profile is expressed in terms of a single unknown parameter or variable which must be determined.

(3) Determination of the unknown parameter or variable. Substituting the assumed velocity profile into the integral form of conservation of momentum and solving the resulting equation gives the unknown parameter. Similarly, substituting the assumed velocity and temperature profiles into the integral form of conservation of energy yields an equation whose solution gives the unknown parameter in the temperature profile.

5.5 Accuracy of the Integral Method

Since basic laws are satisfied in an average sense, integral solutions are inherently approximate. The following observations are made regarding the accuracy of this method:

(1) Since an assumed profile is not unique (several forms are possible), the accuracy of integral solutions depends on the form of the assumed profile. In general, errors involved in this method are acceptable in typical engineering applications.

(2) The accuracy is not very sensitive to the form of an assumed profile.

(3) While there are general guidelines for improving the accuracy, no procedure is available for identifying assumed profiles that will result in the most accurate solutions.

(4) An assumed profile which satisfies conditions at a boundary yields more accurate information at that boundary than elsewhere.

5.6 Integral Formulation of the Basic Laws

5.6.1 Conservation of Mass

Consider boundary layer flow over a curved porous surface shown in Fig. 5.2. Fluid is injected into the boundary layer with velocity v_o through the porous surface. It is important to recognize that the edge

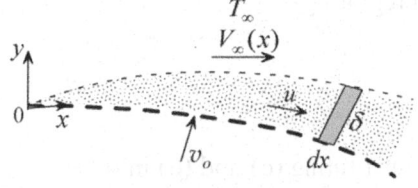

Fig. 5.2

of the viscous boundary layer does not coincide with a streamline. Thus, mass can enter the boundary layer from the external flow. Application of conservation of mass to the element $\delta \times dx$, shown in Fig. 5.2 and enlarged in Fig. 5.3, gives

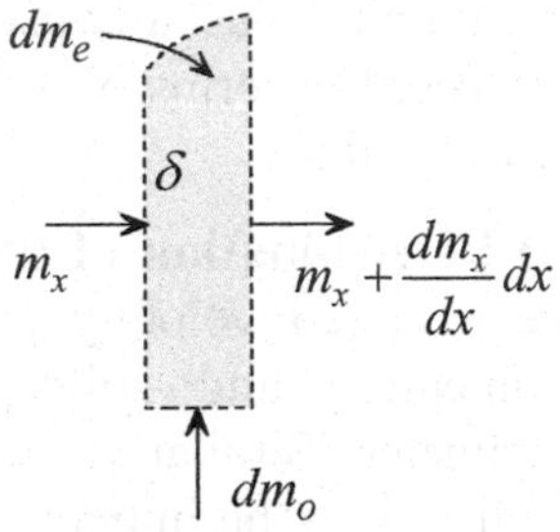

$$m_x + dm_o + dm_e = m_x + \frac{dm_x}{dx} dx,$$

or

$$dm_e = \frac{dm_x}{dx} dx - dm_o, \qquad \text{(a)}$$

where

Fig. 5.3

dm_e = mass flow rate supplied to element from the external flow
dm_o = mass flow rate supplied to element through porous wall
m_x = mass flow rate entering element at x

To formulate expressions for dm_o and m_x we apply the one-dimensional mass flow rate equation

$$m = \rho V A, \qquad \text{(b)}$$

where A is area, V is velocity normal to A, and ρ is density. Applying (b) to the porous side of the element and assuming that the injected fluid is identical to the external fluid, gives

$$dm_o = \rho v_o P dx, \qquad \text{(c)}$$

where P is wall porosity. To determine the rate of mass entering the element at section x, we note that the flow rate varies along y due to variations in velocity and density. Applying (b) to an infinitesimal distance dy gives

$$dm_x = \rho u dy.$$

Integrating

$$m_x = \int_0^{\delta(x)} \rho u dy. \qquad \text{(d)}$$

Substituting (c) and (d) into (a)

$$dm_e = \frac{d}{dx}\left[\int_0^{\delta(x)} \rho u \, dy\right] dx - \rho v_o P dx.$$ (5.1)

Equation (5.1) gives the mass supplied to the boundary layer from the external flow in terms of boundary layer variables and injected fluid. This result is needed in the integral formulation of the momentum and energy equations.

5.6.2 Conservation of Momentum

Application of the momentum theorem in the x-direction to the element $\delta \times dx$ shown in Fig. 5.2, gives

$$\sum F_x = M_x(\text{out}) - M_x(\text{in}),$$ (a)

where

$\sum F_x$ = sum of external forces acting on element in the x-direction

$M_x(\text{in})$ = x-momentum of the fluid entering element

$M_x(\text{out})$ = x-momentum of the fluid leaving element

Fig. 5.4a shows all external forces acting on the element in the x-direction. Fig. 5.4b shows the x-momentum of the fluid entering and leaving the element. Applying equation (a) and using the notations in Fig. 5.4, we obtain

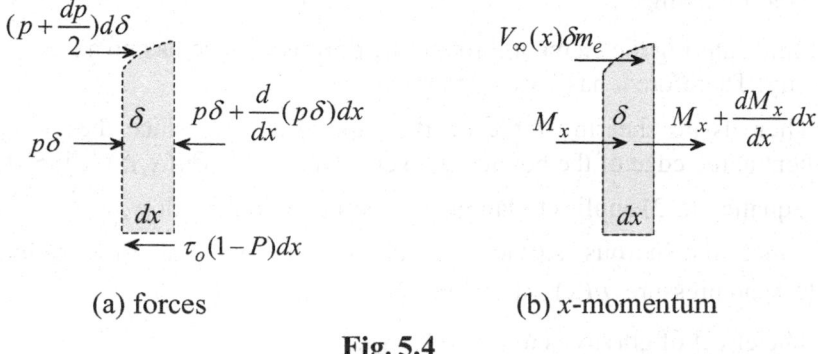

(a) forces   (b) x-momentum

Fig. 5.4

$$p\delta + \left(p + \frac{dp}{2}\right)d\delta - p\delta - \frac{d}{dx}(p\delta)dx - \tau_o(1-P)dx$$

$$= \left(M_x + \frac{dM_x}{dx}dx\right) - M_x - V_\infty(x)dm_e, \tag{b}$$

where

$M_x = x$-momentum
p = pressure
$V_\infty(x)$ = local fluid velocity at the edge of the boundary layer
τ_o = wall shearing stress

However

$$M_x = \int_0^{\delta(x)} \rho u^2 \, dy, \tag{c}$$

and

$$\tau_o = \mu \frac{\partial u(x,0)}{\partial y}. \tag{d}$$

Substituting (c) and (d) into (b) and neglecting higher order terms

$$-\delta \frac{dp}{dx} - \mu(1-P)\frac{\partial u(x,0)}{\partial y} = \frac{d}{dx}\int_0^{\delta(x)} \rho u^2 \, dy - V_\infty(x)\frac{d}{dx}\int_0^{\delta(x)} \rho u \, dy + V_\infty(x)\rho \, Pv_o$$

$$\tag{5.2}$$

Note the following:

(1) Fluid entering the element through the porous surface has no axial velocity. Therefore it has no x-momentum.

(2) There is no shearing force on the slanted surface since the velocity gradient at the edge of the boundary layer vanishes, i.e. $\partial u(x,\delta)/\partial y \approx 0$.

(3) Equation (5.2) applies to laminar as well as turbulent flow.

(4) Since the porous surface is curved, the external flow velocity $V_\infty(x)$ and pressure $p(x)$ vary along the surface.

(5) The effect of gravity is neglected in (5.2).

(6) Equation (5.2) represents the integral formulation of both conservation of momentum and mass.

(7) Although u is a function of x and y, once the integrals in (5.2) are evaluated one obtains a first order ordinary differential equation with x as the independent variable.

Special Cases:

Case 1: Incompressible fluid. Boundary layer approximation gives the axial pressure gradient as

$$\frac{dp}{dx} \approx \frac{dp_\infty}{dx}. \tag{4.12}$$

The x-momentum equation for boundary layer flow is

$$u\frac{\partial u}{\partial x} + v\frac{\partial u}{\partial y} = -\frac{1}{\rho}\frac{dp_\infty}{dx} + v\frac{\partial^2 u}{\partial y^2}. \tag{4.5}$$

Applying equation (4.5) at the edge of the boundary layer, $y = \delta$, where $u \approx V_\infty$ and $\partial u / \partial y \approx dV_\infty / dy \approx 0$, gives

$$\frac{dp}{dx} \approx \frac{dp_\infty}{dx} = -\rho V_\infty(x)\frac{dV_\infty}{dx}. \tag{5.3}$$

Substituting (5.3) into (5.2) and noting that ρ is constant

$$\delta V_\infty(x)\frac{dV_\infty}{dx} - v(1-P)\frac{\partial u(x,0)}{\partial y} = \frac{d}{dx}\int_0^{\delta(x)} u^2 dy - V_\infty(x)\frac{d}{dx}\int_0^{\delta(x)} u\,dy - V_\infty(x)Pv_0.$$

$$\tag{5.4}$$

Case 2: Incompressible fluid and impermeable flat plate. At the edge of boundary layer flow the fluid is assumed inviscid. Neglecting boundary layer thickness and viscous effects for the special case of a flat plate means that the external flow experiences no changes. It follows from (5.3) that

$$\frac{dV_\infty}{dx} = \frac{dp}{dx} \approx \frac{dp_\infty}{dx} = 0. \tag{e}$$

For an impermeable plate

$$v_o = 0, \quad P = 0, \tag{f}$$

Substituting (e) and (f) into (5.4)

$$\nu \frac{\partial u(x,0)}{\partial y} = V_\infty \frac{d}{dx} \int_0^{\delta(x)} u\,dy - \frac{d}{dx} \int_0^{\delta(x)} u^2\,dy, \qquad (5.5)$$

where ν is kinematic viscosity.

5.6.3 Conservation of Energy

Consider the flow of fluid at temperature T_∞ over a porous surface. The surface is maintained at a different temperature and thus heat exchange takes place. At high Reynolds and Peclet numbers temperature and velocity boundary layers form over the surface. Fluid at temperature T_o is injected into the boundary layer with velocity v_o. Conservation of energy is applied to the element $\delta_t \times dx$, shown in Fig. 5.5 and enlarged in Fig. 5.6. We neglect:

(1) Changes in kinetic and potential energy
(2) Axial conduction
(3) Dissipation

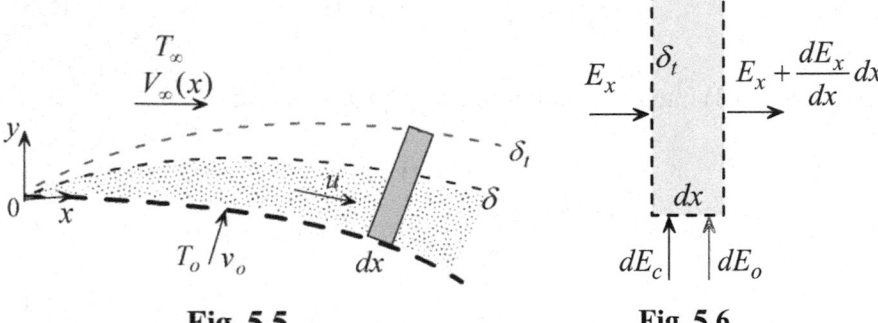

Fig. 5.5   **Fig. 5.6**

Based on these assumptions, conservation of energy for the element gives

$$E_x + dE_c + dE_o + dE_e = E_x + \frac{dE_x}{dx}\,dx.$$

Rearranging

$$dE_c = \frac{dE_x}{dx}\,dx - dE_e - dE_o, \qquad (a)$$

where

dE_c = energy added at surface by conduction

dE_e = energy added by the external mass

dE_o = energy added by the injected mass

E_x = energy convected with the boundary layer flow

Heat conduction at the porous surface is determined using Fourier's law

$$dE_c = -k(1-P)\frac{\partial T(x,0)}{\partial y}dx . \qquad (b)$$

Mass entering the element from the external flow, dm_e, is at the free stream temperature T_∞. Thus energy carried with this mass, dE_e, is

$$dE_e = h_\infty\, dm_e .$$

where h_∞ is the free stream enthalpy. Using (5.1) for dm_e

$$dE_e = h_\infty \frac{d}{dx}\left[\int_0^{\delta_t(x)} \rho u\, dy\right]dx - h_\infty \rho v_o P dx . \qquad (c)$$

Note that the upper limit of the integral in (c) is δ_t since the element extends to the edge of the thermal boundary layer. Neglecting conduction in the injected fluid, energy convected through the pores is

$$dE_o = \rho h_o v_o P dx . \qquad (d)$$

where h_o is the enthalpy of the injected fluid. Energy convected with fluid flow within the boundary layer, E_x, depends on the local axial velocity u and temperature T. Integration across the thermal boundary layer thickness gives the total convected energy

$$E_x = \int_0^{\delta_t(x)} \rho u h\, dy . \qquad (e)$$

where h is the local enthalpy. Substituting (b)-(e) into (a) and rearranging

$$-k(1-P)\frac{\partial T(x,0)}{\partial y} = \frac{d}{dx}\int_0^{\delta_t(x)} \rho u(h-h_\infty)dy - \rho v_o P(h_o - h_\infty). \qquad (5.6a)$$

Over small temperature differences the change in enthalpy can be expressed in term of temperature difference as $h - h_\infty \approx c_p (T - T_\infty)$ and $h_o - h_\infty \approx c_p (T_o - T_\infty)$. For this case (5.6a) becomes

$$-k(1-P)\frac{\partial T(x,0)}{\partial y} = \frac{d}{dx}\int_0^{\delta_t(x)} \rho u(T - T_\infty)dy - \rho v_o P(T_o - T_\infty). \quad (5.6b)$$

Note the following regarding this result:

(1) Equation (5.6) represents integral formulation of both conservation of mass and energy. (2) Although u and T are functions of x and y, once the integrals in (5.6) are evaluated one obtains a first order ordinary differential equation with x as the independent variable.

Special Case: Constant properties and impermeable flat plate

Setting $P = 0$ and assuming constant density and specific heat, equation (5.6b) simplifies to

$$-\alpha \frac{\partial T(x,0)}{\partial y} = \frac{d}{dx}\int_0^{\delta_t(x)} u(T - T_\infty)dy, \quad (5.7)$$

where α is thermal diffusivity.

5.7 Integral Solutions

To obtain solutions to the temperature distribution using the integral method, the velocity distribution u must be determined first. This is evident in equations (5.6) and (5.7) where the variable u appears in the integrands.

5.7.1 Flow Field Solution: Uniform Flow over a Semi-Infinite Plate

The integral method will be applied to obtain a solution to Blasius laminar flow problem, shown in Fig. 5.7. Equation (5.5) gives the integral formulation of momentum for this problem

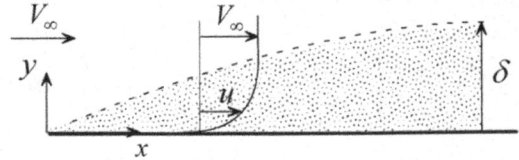

Fig. 5.7

$$v \frac{\partial u(x,0)}{\partial y} = V_\infty \frac{d}{dx} \int_0^{\delta(x)} u \, dy - \frac{d}{dx} \int_0^{\delta(x)} u^2 \, dy. \tag{5.5}$$

As pointed out in the procedure of Section 5.4, the next step is the introduction of an assumed velocity profile $u(x,y)$ to be used in equation (5.5). An assumed profile is usually based on some knowledge of the general flow characteristics.   For laminar flow over a flat plate, a polynomial is a reasonable representation of the velocity profile. Thus

$$u(x, y) = \sum_{n=0}^{N} a_n(x) y^n , \tag{5.8}$$

where N is the degree of the polynomial. As an example, we assume a third degree polynomial

$$u(x, y) = a_0(x) + a_1(x)y + a_2(x)y^2 + a_3(x)y^3 . \tag{a}$$

The coefficients $a_n(x)$ are determined using the following known exact and approximate boundary conditions on the velocity

(1) $u(x,0) = 0$,

(2) $u(x,\delta) \approx V_\infty$,

(3) $\dfrac{\partial u(x,\delta)}{\partial y} \approx 0$,

(4) $\dfrac{\partial^2 u(x,0)}{\partial y^2} = 0$.

Note the following regarding the above conditions:

(1) The second and third conditions are approximate since the edge of the boundary layer is not uniquely defined.

(2) Condition (4) is obtained by setting $y = 0$ in the x-component of the Navier-Stokes equations of motion (2.10x).

Equation (a) and the four boundary conditions give the coefficients $a_n(x)$

$$a_0 = a_2 = 0, \quad a_1 = \frac{3}{2}\frac{V_\infty}{\delta}, \quad a_3 = -\frac{1}{2}\frac{V_\infty}{\delta^3}.$$

Substituting the above into (a)

$$\frac{u}{V_\infty} = \frac{3}{2}\left(\frac{y}{\delta}\right) - \frac{1}{2}\left(\frac{y}{\delta}\right)^3. \tag{5.9}$$

Thus the assumed velocity is expressed in terms of the unknown variable $\delta(x)$. This variable is determined using the integral form of the momentum equation, (5.5). Substituting (5.9) into (5.5) and evaluating the integrals, gives

$$\frac{3}{2}vV_\infty\frac{1}{\delta} = \frac{39}{280}V_\infty^2\frac{d\delta}{dx}. \tag{b}$$

This is a first order ordinary differential equation in $\delta(x)$. Separating variables

$$\delta d\delta = \frac{140}{13}\frac{v}{V_\infty}dx.$$

Integrating and noting that $\delta(0) = 0$

$$\int_0^\delta \delta d\delta = \frac{140}{13}\frac{v}{V_\infty}\int_0^x dx.$$

Evaluating the integrals and rearranging

$$\frac{\delta}{x} = \frac{\sqrt{280/13}}{\sqrt{Re_x}} = \frac{4.64}{\sqrt{Re_x}}. \tag{5.10}$$

Substituting (5.10) into (5.9) gives the velocity u as a function of x and y. With the velocity distribution determined, friction coefficient C_f is obtained using (4.36) and (4.37a)

$$C_f = \frac{\tau_o}{\rho V_\infty^2 / 2} = \frac{\mu \dfrac{\partial u}{\partial y}(x,0)}{\rho V_\infty^2 / 2} = \frac{3v}{V_\infty \delta(x)}.$$

Using (5.10) to eliminate $\delta(x)$ in the above

$$C_f = \frac{0.646}{\sqrt{Re_x}}. \tag{5.11}$$

We are now in a position to examine the accuracy of the integral solution by comparing it with Blasius solution for $\delta(x)$ and C_f, equations (4.46) and (4.48):

$$\frac{\delta}{x} = \frac{5.2}{\sqrt{Re_x}}, \quad \text{Blasius solution}, \tag{4.46}$$

and

$$C_f = \frac{0.664}{\sqrt{Re_x}}, \quad \text{Blasius solution}. \tag{4.48}$$

The following observations are made:

(1) The integral and Blasius solutions for $\delta(x)$ and C_f have the same form.

(2) The constant 5.2 in Blasius solution for $\delta(x)$ differs by 10.8% from the corresponding integral solution of 4.64. However, it must be kept in mind that the constant in Blasius solution for $\delta(x)$ is not unique. It depends on how $\delta(x)$ is defined.

(3) The error in C_f is 2.7%.

(4) Predicting C_f accurately is more important than predicting $\delta(x)$.

5.7.2 Temperature Solution and Nusselt Number: Flow over a Semi-Infinite Plate

(i) Temperature Distribution

Consider uniform boundary layer flow over a semi-infinite plate shown in Fig. 5.8. A leading section of the plate of length x_o is insulat-ed and the remaining part is at uniform temperature T_s. Assume laminar, steady, two-

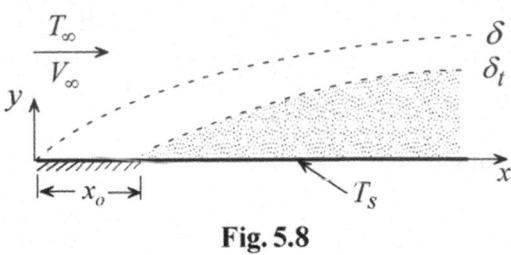

Fig. 5.8

dimensional, constant properties boundary layer flow and neglect axial conduction and dissipation. Of interest is the determination of the thermal boundary layer thickness, local heat transfer coefficient, and Nusselt number. This requires determining the temperature distribution. Since the velocity field is independent of temperature, the integral solution for the velocity $u(x, y)$ and boundary layer thickness $\delta(x)$ obtained in Section 5.7.1 is applicable to this case. Equation (5.7) gives the integral formulation of conservation of energy for this problem

$$-\alpha \frac{\partial T(x,0)}{\partial y} = \frac{d}{dx} \int_0^{\delta_t(x)} u(T - T_\infty)dy, \tag{5.7}$$

where $u(x, y)$ is given by equation (5.9). The next step is the introduction of an assumed temperature profile $T(x, y)$ to be used in equation (5.7). For laminar flow over a flat plate a polynomial is a reasonable representation for the temperature profile. Thus

$$T(x, y) = \sum_{n=0}^{N} b_n(x)y^n. \tag{5.12}$$

Following the procedure used in Section 5.7.1, we assume a third degree polynomial

$$T(x, y) = b_0(x) + b_1(x)y + b_2(x)y^2 + b_3(x)y^3. \tag{a}$$

The coefficients $b_n(x)$ are determined using the following known exact and approximate boundary conditions on the temperature

(1) $T(x,0) = T_s$,

(2) $T(x, \delta_t) \approx T_\infty$,

(3) $\dfrac{\partial T(x, \delta_t)}{\partial y} \approx 0$,

(4) $\dfrac{\partial^2 T(x, 0)}{\partial y^2} = 0$.

Note that the second and third conditions are approximate since the edge of the thermal boundary layer is not uniquely defined. The fourth condition is obtained by setting $y = 0$ in the energy equation (2.19). Equation (a) and the four boundary conditions give the coefficients $b_n(x)$

$$b_0 = T_s, \quad b_1 = \frac{3}{2}(T_\infty - T_s)\frac{1}{\delta_t}, \quad b_2 = 0, \quad b_3 = -\frac{1}{2}(T_\infty - T_s)\frac{1}{\delta_t^3}.$$

Substituting the above into (a)

$$T(x, y) = T_s + (T_\infty - T_s)\left[\frac{3}{2}\frac{y}{\delta_t} - \frac{1}{2}\frac{y^3}{\delta_t^3}\right]. \qquad (5.13)$$

Substituting (5.9) and (5.13) into (5.7) and evaluating the integral, gives

$$\frac{3}{2}\alpha\frac{T_\infty - T_s}{\delta_t} = \frac{d}{dx}\left\{(T_\infty - T_s)V_\infty\delta\left[\frac{3}{20}\left(\frac{\delta_t}{\delta}\right)^2 - \frac{3}{280}\left(\frac{\delta_t}{\delta}\right)^4\right]\right\}, \qquad (5.14)$$

where $\delta(x)$ is given in (5.10). Eliminating $\delta(x)$ in the above gives a first order ordinary differential equation for δ_t. However, equation (5.14) is simplified first. For Prandtl numbers greater than unity the thermal boundary layer is smaller than the viscous boundary layer. That is

$$\frac{\delta_t}{\delta} < 1, \quad \text{for } Pr > 1. \qquad (5.15)$$

Based on this restriction the last term in (5.14) can be neglected

$$\frac{3}{280}\left(\frac{\delta_t}{\delta}\right)^4 << \frac{3}{20}\left(\frac{\delta_t}{\delta}\right)^2 .$$

Equation (5.14) simplifies to

$$10\frac{\alpha}{\delta_t} = V_\infty \frac{d}{dx}\left[\delta\left(\frac{\delta_t}{\delta}\right)^2\right] . \tag{b}$$

To solve (b) for δ_t we use the integral solution to δ . Rewriting (5.10)

$$\delta = \sqrt{\frac{280}{13}}\sqrt{\frac{vx}{V_\infty}} . \tag{c}$$

Substitute (c) into (b) and rearrange

$$\left(\frac{\delta_t}{\delta}\right)^3 + 4x\left(\frac{\delta_t}{\delta}\right)^2 \frac{d}{dx}\left(\frac{\delta_t}{\delta}\right) = \frac{13}{14}\frac{1}{Pr} . \tag{d}$$

Equation (d) is solved for δ_t / δ by introducing the following definition:

$$r = \left(\frac{\delta_t}{\delta}\right)^3 . \tag{e}$$

Substitute (e) into (d)

$$r + \frac{4}{3}x\frac{dr}{dx} = \frac{13}{14}\frac{1}{Pr} . \tag{f}$$

This is a first order differential equation for r. Separating variables and integrating

$$r = \left(\frac{\delta_t}{\delta}\right)^3 = C(x)^{-3/4} + \frac{13}{14}\frac{1}{Pr} , \tag{g}$$

where C is constant of integration determined from the boundary condition on δ_t

$$\delta_t(x_o) = 0 . \tag{h}$$

Applying (h) to (g) gives the constant C

$$C = -\frac{13}{14}\frac{1}{Pr}x_o^{3/4}.$$ (i)

Substituting (i) into (g) and rearranging

$$\frac{\delta_t}{\delta} = \left\{\frac{13}{14}\frac{1}{Pr}\left[1 - \left(\frac{x_o}{x}\right)^{3/4}\right]\right\}^{1/3}.$$ (5.16)

Using (c) to eliminate δ in (5.16)

$$\delta_t = \left\{\frac{13}{14}\frac{1}{Pr}\left[1 - \left(\frac{x_o}{x}\right)^{3/4}\right]\right\}^{1/3}\sqrt{\frac{280}{13}}\sqrt{\frac{vx}{V_\infty}},$$ (5.17a)

or

$$\frac{\delta_t}{x} = \frac{4.528}{Pr^{1/3}Re_x^{1/2}}\left\{\left[1 - \left(\frac{x_o}{x}\right)^{3/4}\right]\right\}^{1/3},$$ (5.17b)

where Re_x is the local Reynolds number defined as

$$Re_x = \frac{V_\infty x}{v}.$$ (5.18)

(ii) Nusselt Number

The local Nusselt number is defined as

$$Nu_x = \frac{hx}{k},$$ (j)

where h is the local heat transfer coefficient given by

$$h = \frac{-k\dfrac{\partial T(x,0)}{\partial y}}{T_s - T_\infty}.$$ (k)

Using the temperature distribution (5.13) into (k)

$$h(x) = \frac{3}{2} \frac{k}{\delta_t} .$$ (5.19)

Eliminating δ_t by using (5.17b) gives the local heat transfer coefficient

$$h(x) = 0.331 \frac{k}{x} \left\{ 1 - \left(\frac{x_o}{x} \right)^{3/4} \right\}^{-1/3} Pr^{1/3} Re_x^{1/2} .$$ (5.20)

Substituting into (j)

$$Nu_x = 0.331 \left\{ 1 - \left(\frac{x_o}{x} \right)^{3/4} \right\}^{-1/3} Pr^{1/3} Re_x^{1/2} .$$ (5.21)

Special Case: Plate with no Insulated Section

Fig. 5.9 shows a flat plate which is maintained at uniform surface temperature. The plate has no insulated section. The solution to this case is obtained by setting $x_o = 0$ in the case of a plate with a leading insulated section presented above. The solution to the temperature distribution is given by equation (5.13).

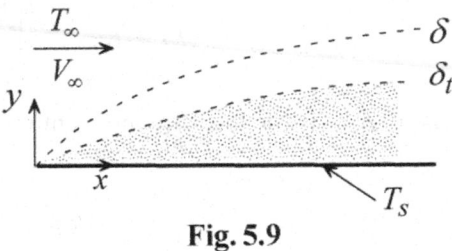

Fig. 5.9

Thermal boundary layer thickness, heat transfer coefficient, and Nusselt number are obtained by setting $x_o = 0$ in (5.16), (5.17), (5.20) and (5.21)

$$\frac{\delta_t}{\delta} = \left\{ \frac{13}{14} \frac{1}{Pr} \right\}^{1/3} = \frac{0.975}{Pr^{1/3}} ,$$ (5.22)

$$\frac{\delta_t}{x} = \frac{4.528}{Pr^{1/3} Re_x^{1/2}} ,$$ (5.23)

$$h(x) = 0.331 \frac{k}{x} Pr^{1/3} Re_x^{1/2}, \tag{5.24}$$

$$Nu_x = 0.331 Pr^{1/3} Re_x^{1/2}. \tag{5.25}$$

To examine the accuracy of the integral solution, comparison is made with Pohlhausen's results. For the limiting case of $Pr = 1$ the viscous and thermal boundary layers coincide, i.e. $\delta_t / \delta = 1$. Setting $Pr = 1$ in (5.22) gives

$$\frac{\delta_t}{\delta} = 0.975.$$

This has an error of 2.5%. We ext examine the accuracy of the local Nusselt number. For $Pr > 10$ equation (4.72c) gives Pohlhausen's solution

$$Nu_x = 0.339 Pr^{1/3} \sqrt{Re_x}, \quad \text{for } Pr > 10. \tag{4.72c}$$

Comparing this result with integral solution (5.25) gives an error of 2.4%.

Example 5.1: Laminar Boundary Layer Flow over a Flat Plate: Uniform Surface Temperature

Fluid flows with uniform velocity and temperature over a semi-infinite flat plate. The plate is maintained at uniform temperature T_s. A leading section of the plate of length x_o is insulated. Use the integral method to determine the local Nusselt number based on linear velocity and temperature profiles. Assume steady, two-dimensional, constant properties boundary layer flow and neglect dissipation.

(1) Observations. (i) The determination of the Nusselt number requires the determination of the velocity and temperature distribution. (ii) Results based on linear velocity and temperature profiles are less accurate than those using second or third degree polynomials. (iii) The velocity field is independent of temperature.

(2) Problem Definition. Determine the velocity and temperature distribution for boundary layer flow over a flat plate.

(3) Solution Plan. Start with equating Newton's law with Fourier's law to obtain an equation for the heat transfer coefficient h. Apply the integral form of the momentum equation using a linear velocity profile. Apply the integral form of the energy equation using a linear temperature profile

(4) Plan Execution.

(i) Assumptions. (1) Continuum, (2) Newtonian, (3) steady state, (4) constant properties, (5) two-dimensional, (6) laminar flow ($Re_x < 5 \times 10^5$), (7) viscous boundary layer flow ($Re_x > 100$), (8) thermal boundary layer ($Pe > 100$), (9) uniform upstream velocity and temperature, (10) flat plate, (11) uniform surface temperature, (12) negligible changes in kinetic and potential energy, (13) negligible axial conduction, (14) negligible dissipation and (15) no buoyancy ($\beta = 0$ or $g = 0$).

(ii) Analysis. The local Nusselt number is defined as

$$Nu_x = \frac{hx}{k}, \tag{a}$$

where the heat transfer coefficient h is given by equation (1.10)

$$h = \frac{-k\dfrac{\partial T(x,0)}{\partial y}}{T_s - T_\infty}. \tag{1.10}$$

Thus h depends on the temperature distribution $T(x, y)$. The integral form of the energy equation is used to determine the temperature distribution

$$-\alpha \frac{\partial T(x,0)}{\partial y} = \frac{d}{dx} \int_0^{\delta_t(x)} u(T - T_\infty)dy. \tag{5.7}$$

Before proceeding with the energy equation, axial velocity distribution $u(x,y)$ appearing in (5.7) must be determined. This is accomplished by applying the integral form of the momentum equation

$$v\frac{\partial u(x,0)}{\partial y} = V_\infty \frac{d}{dx} \int_0^{\delta(x)} udy - \frac{d}{dx} \int_0^{\delta(x)} u^2 dy. \tag{5.5}$$

Following the procedure outlined in Section 5.4, a velocity profile is assumed. As an example, assume a linear profile given by

$$u = a_0 + a_1 y. \tag{b}$$

Select the following two boundary conditions to determine the coefficients in (b)

(1) $u(x,0) = 0$,

(2) $u(x,\delta) \approx V_\infty$.

Applying these conditions to (b) gives

$$a_0 = 0, \quad a_1 = \frac{V_\infty}{\delta}.$$

Substituting into (b)

$$u = V_\infty \frac{y}{\delta}. \tag{c}$$

To determine $\delta(x)$ the assumed velocity (c) is substituted into (5.5)

$$v\frac{V_\infty}{\delta} = V_\infty \frac{d}{dx} \int_0^{\delta(x)} \frac{V_\infty}{\delta} y\, dy - \frac{d}{dx} \int_0^{\delta(x)} \frac{V_\infty^2}{\delta^2} y^2\, dy.$$

Evaluating the integrals

$$\frac{v}{\delta} = \frac{V_\infty}{2}\frac{d\delta}{dx} - \frac{V_\infty}{3}\frac{d\delta}{dx} = \frac{V_\infty}{6}\frac{d\delta}{dx}.$$

Separating variables

$$\delta\, d\delta = 6\frac{v}{V_\infty} dx.$$

Integrating and noting that $\delta(0) = 0$

$$\int_0^{\delta} \delta\, d\delta = 6\frac{v}{V_\infty} \int_0^{x} dx.$$

Evaluating the integrals and rearranging the result

$$\delta = \sqrt{\frac{12v}{V_\infty}} x, \tag{d}$$

or

$$\frac{\delta}{x} = \sqrt{\frac{12}{Re_x}} , \qquad (5.26)$$

where Re_x is the local Reynolds number defined as

$$Re_x = \frac{V_\infty x}{\nu}. \qquad (e)$$

Having determined the velocity $u(x, y)$ attention is focused on the determination of the temperature distribution. Assume, for example, a linear temperature profile

$$T = b_0 + b_1 y. \qquad (f)$$

Select the following two boundary conditions to determine the coefficients in (f)

(1) $T(x,0) = T_s$,

(2) $T(x, \delta_t) \approx T_\infty$.

Applying these conditions to (f) gives

$$b_0 = T_s, \quad b_1 = \frac{T_\infty - T_s}{\delta_t}.$$

Substituting into (f)

$$T = T_s + (T_\infty - T_s)\frac{y}{\delta_t}. \qquad (g)$$

Introducing (g) into (1.10)

$$h = \frac{k}{\delta_t}. \qquad (h)$$

Substituting (h) into (a)

$$Nu_x = \frac{x}{\delta_t}. \qquad (i)$$

Thus the key to the determination of the heat transfer coefficient and the Nusselt number is the solution to the thermal boundary layer thickness δ_t. The integral form of the energy equation (5.7) is used to determine δ_t. Substituting (c) and (g) into (5.7)

$$-\alpha \frac{T_\infty - T_s}{\delta_t} = V_\infty \frac{d}{dx} \int_0^{\delta_t(x)} \frac{V_\infty}{\delta} y \left[(T_s - T_\infty) + (T_\infty - T_s)(y/\delta_t) \right] dy \; .$$

Evaluating the integrals and rearranging the result

$$\frac{\alpha}{\delta_t} = \frac{V_\infty}{6} \frac{d}{dx} \left[\frac{\delta_t^2}{\delta} \right].$$

Rewriting the above

$$\frac{6\alpha}{V_\infty} \frac{1}{\delta} \left(\frac{\delta}{\delta_t} \right) = \frac{d}{dx} \left[\delta \left(\frac{\delta_t}{\delta} \right)^2 \right]. \tag{j}$$

To solve this equation for δ_t, let

$$r = \frac{\delta_t}{\delta}. \tag{k}$$

Substituting (k) into (j)

$$\frac{6\alpha}{V_\infty} \frac{1}{\delta} \frac{1}{r} = \frac{d}{dx} \left[\delta r^2 \right].$$

Using (d) to eliminate δ, the above becomes

$$\frac{6\alpha}{V_\infty} \sqrt{\frac{V_\infty}{12\nu}} \frac{1}{\sqrt{x}} \frac{1}{r} = \frac{d}{dx} \left[\sqrt{12\nu / V_\infty} \sqrt{x} \, r^2 \right].$$

Simplifying

$$\frac{1}{2} \frac{\alpha}{\nu} \frac{1}{\sqrt{x}} = r \frac{d}{dx} \left[\sqrt{x} \, r^2 \right],$$

Expanding and noting that $\dfrac{\nu}{\alpha} = Pr$, the above becomes

$$\frac{1}{2\,Pr}\frac{1}{\sqrt{x}} = r\left[2r\sqrt{x}\frac{dr}{dx} + \frac{r^2}{2\sqrt{x}}\right],$$

$$\frac{1}{Pr} = 4xr^2\frac{dr}{dx} + r^3.$$

Separating variables and integrating

$$\int_0^r \frac{4r^2\,dr}{(1/Pr) - r^3} = \int_{x_o}^x \frac{dx}{x}.$$
(1)

Note that the limits in (1) are based on the following boundary condition on δ_t

$$\delta_t = r = 0 \text{ at } x = x_o.$$
(m)

Evaluating the integrals in (1) and rearranging the results

$$r = \frac{\delta_t}{\delta} = \frac{1}{Pr^{1/3}}\left[1 - (x_o/x)^{3/4}\right]^{1/3}.$$
(n)

Using (d) to eliminate δ in the above

$$\delta_t = \frac{1}{Pr^{1/3}}\sqrt{\frac{12\nu}{V_\infty}}\sqrt{x}\left[1 - (x_o/x)^{3/4}\right]^{1/3}.$$
(o)

The local Nusselt number is determined by substituting (o) into (i) and using the definition of the local Reynolds number in (e) to obtain

$$Nu_x = 0.289\,Pr^{1/3}\sqrt{Re_x}\left[1 - (x_o/x)^{3/4}\right]^{-1/3}.$$
(5.27)

For the special case of a plate with no insulated section, setting $x_o = 0$ in (5.27) gives

$$Nu_x = 0.289\,Pr^{1/3}\sqrt{Re_x}.$$
(5.28)

(iii) **Checking.** *Dimensional check*: Solutions to δ/x and Nu_x are dimensionless. Units of δ_t in (o) are correct

Boundary conditions check: Assumed velocity and temperature profiles satisfy their respective boundary conditions.

Limiting check: For the special case of $Pr = 1$ an exact solution to the ratio $r = \delta_t/\delta$ should be unity for $x_o = 0$. Setting $Pr = 1$ in (n) gives the correct result.

(5) Comments. As might be expected, results based on assumed linear profiles for the velocity and temperature are less accurate than those based on third degree polynomials. Table 5.1 compares exact solutions for δ/x and $Nu_x / Pr^{1/3} Re_x^{1/2}$ with integral results for the case of a plate with no insulated section based on assumed linear and polynomial profiles. Equations (4.46) and (4.72c) give exact solutions, and equations (5.10), (5.25), (5.26) and (5.28) give integral results. Note that the integral method gives a more accurate prediction of Nusselt number than of the boundary layer thickness δ.

Table 5.1

Solution	$\dfrac{\delta}{x}\sqrt{Re_x}$	$\dfrac{Nu_x}{Pr^{1/3}Re^{1/2}}$
Exact (Blasius/ Pohlhausen)	5.2	0.332
3rd degree polynomial	4.64	0.331
Linear	3.46	0.289

5.7.3 Uniform Surface Flux

Figure 5.10 shows a flat plate with an insulated leading section of length x_o. The plate is heated with uniform flux q''_s along its surface $x \geq x_o$. We consider steady state, laminar, two-dimensional flow with constant properties. We wish to determine surface temperature distribution and the local Nusselt number. Application of Newton's law of cooling gives

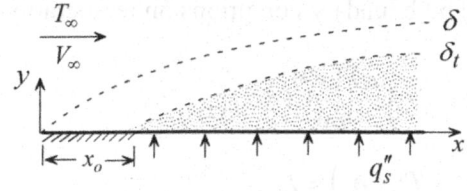

Fig. 5.10

$$q_s'' = h(x)\left[T_s(x) - T_\infty\right].$$ (a)

Solving (a) for $h(x)$

$$h(x) = \frac{q_s''}{T_s(x) - T_\infty}.$$

Introducing the definition of the Nusselt number, the above gives

$$Nu_x = \frac{q_s'' x}{k\left[T_s(x) - T_\infty\right]}.$$ (b)

Thus once surface temperature $T_s(x)$ is determined equation (b) gives the local Nusselt number. $T_s(x)$ is determined using the integral form of the energy equation

$$-\alpha \frac{\partial T(x,0)}{\partial y} = \frac{d}{dx} \int_0^{\delta_t(x)} u(T - T_\infty)dy.$$ (5.7)

For constant properties, the velocity distribution $u(x,y)$ in (5.7) is independent of temperature. Thus the integral solution to $u(x,y)$ for a third degree polynomial is given by (5.9)

$$\frac{u}{V_\infty} = \frac{3}{2}\left(\frac{y}{\delta}\right) - \frac{1}{2}\left(\frac{y}{\delta}\right)^3.$$ (5.9)

Assume a third degree polynomial for the temperature profile $T(x,y)$

$$T = b_0 + b_1 y + b_2 y^2 + b_3 y^3.$$ (c)

The boundary conditions on temperature are

(1) $-k\dfrac{\partial T(x,0)}{\partial y} = q_s''$,

(2) $T(x,\delta_t) \approx T_\infty$,

(3) $\dfrac{\partial T(x,\delta_t)}{\partial y} \approx 0$,

(4) $\dfrac{\partial^2 T(x,0)}{\partial y^2} = 0$.

Application of the boundary conditions gives the four coefficients. The temperature profile becomes

$$T(x,y) = T_\infty + \left[\frac{2}{3}\delta_t - y + \frac{1}{3}\frac{y^3}{\delta_t^2}\right]\frac{q_s''}{k}. \tag{5.29}$$

Surface temperature is obtained by setting $y = 0$ in the above

$$T_s(x) = T(x,0) = T_\infty + \frac{2}{3}\frac{q_s''}{k}\delta_t. \tag{5.30}$$

Substituting (5.30) into (b)

$$Nu_x = \frac{3}{2}\frac{x}{\delta_t(x)}. \tag{5.31}$$

Thus the problem reduces to determining δ_t. Substituting (5.9) and (5.29) into (5.7)

$$\alpha = V_\infty \frac{d}{dx}\left\{\int_0^{\delta_t}\left[\frac{3}{2}\frac{y}{\delta} - \frac{1}{2}\frac{y^3}{\delta^3}\right]\left[\frac{2}{3}\delta_t - y + \frac{1}{3}\frac{y^3}{\delta_t^2}\right]dy\right\}. \tag{d}$$

Evaluating the integrals

$$\frac{\alpha}{V_\infty} = \frac{d}{dx}\left\{\delta_t^2\left[\frac{1}{10}\frac{\delta_t}{\delta} - \frac{1}{140}\left(\frac{\delta_t}{\delta}\right)^3\right]\right\}. \tag{e}$$

For Prandtl numbers larger than unity, $\delta_t/\delta < 1$. Thus

$$\frac{1}{140}\left(\frac{\delta_t}{\delta}\right)^3 << \frac{1}{10}\frac{\delta_t}{\delta}.$$ (f)

Introducing (f) into (e), gives

$$10\frac{\alpha}{V_\infty} = \frac{d}{dx}\left[\frac{\delta_t^3}{\delta}\right].$$

Integrating the above

$$10\frac{\alpha}{V_\infty}x = \frac{\delta_t^3}{\delta} + C.$$ (g)

The boundary condition on δ_t is

$$\delta_t(x_o) = 0.$$ (h)

Applying (h) to (g)

$$C = 10\frac{\alpha}{V_\infty}x_o.$$ (i)

Equation (i) into (g)

$$\delta_t = \left[10\frac{\alpha}{V_\infty}(x - x_o)\delta\right]^{1/3}$$ (j)

Using (5.10) to eliminate δ in (j)

$$\delta_t = \left[10\frac{\alpha}{V_\infty}(x - x_o)\sqrt{\frac{280}{13}}\sqrt{\frac{1}{Re_x}}x\right]^{1/3}.$$

Introducing the definition of the Prandtl and Reynolds numbers and rearranging the above

$$\frac{\delta_t}{x} = \frac{3.594}{Pr^{1/3}Re_x^{1/2}}\left[1 - \frac{x_o}{x}\right]^{1/3}.$$ (5.32)

Surface temperature is obtained by substituting (5.32) into (5.30)

$$T_s(x) = T_\infty + 2.396\frac{q_s''}{k}\left[1 - \frac{x_o}{x}\right]^{1/3}\frac{x}{Pr^{1/3}Re_x^{1/2}}.$$ (5.33)

Substituting (5.32) into (5.31) gives the local Nusselt number

$$Nu_x = 0.417 \left[1 - \frac{x_o}{x} \right]^{-1/3} Pr^{1/3} Re_x^{1/2}. \tag{5.34}$$

For the special case of a plate with no insulated section, setting $x_o = 0$ in (5.33) and (5.34) gives

$$T_s(x) = T_\infty + 2.396 \frac{q_s''}{k} \frac{x}{Pr^{1/3} Re_x^{1/2}}, \tag{5.35}$$

$$Nu_x = 0.417 \, Pr^{1/3} Re_x^{1/2}. \tag{5.36}$$

This result is in good agreement with the more accurate differential formulation solution[1]

$$Nu_x = 0.453 \, Pr^{1/3} Re_x^{1/2}. \tag{5.37}$$

Examination of surface temperature (5.35) shows that it increases with distance along the plate according to $\sqrt{x}$.

Example 5.2: Laminar Boundary Layer Flow over a Flat Plate: Variable Surface Temperature

Consider uniform flow over a semi-infinite flat plate. The plate is maintained at a variable surface temperature given by

$$T_s(x) = T_\infty + C\sqrt{x}$$

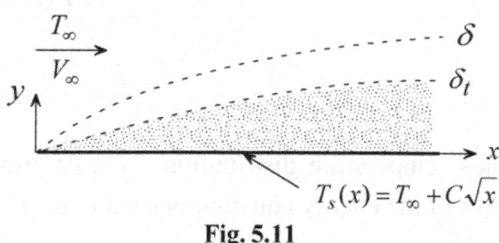

Fig. 5.11

where C is constant. Apply the integral method to determine the local Nusselt number using third degree polynomials for the velocity and temperature profiles. Assume steady, two-dimensional, constant properties boundary layer flow and neglect dissipation.

(1) Observations. (i) The determination of the Nusselt number requires the determination of the velocity and temperature distributions. (ii). Surface temperature is variable. (iii) For constant properties, velocity distribution is independent of temperature.

(2) Problem Definition. Determine the velocity and temperature distribution for laminar boundary layer flow over a flat plate.

(3) Solution Plan. Start with the definition of local Nusselt number and equation (1.10) for the heat transfer coefficient h. Apply the integral form of the energy equation to determine the temperature distribution

(4) Plan Execution.

   (i) Assumptions. (1) Continuum, (2) Newtonian, (3) steady state, (4) constant properties, (5) two-dimensional, (6) laminar flow ($Re_x < 5 \times 10^5$), (7) viscous boundary layer flow ($Re_x > 100$), (8) thermal boundary layer ($Pe > 100$), (9) uniform upstream velocity and temperature, (10) flat plate, (11) negligible changes in kinetic and potential energy, (12) negligible axial conduction, (13) negligible dissipation and (14) no buoyancy ($\beta = 0$ or $g = 0$).

   (ii) Analysis. The local Nusselt number is defined as

$$Nu_x = \frac{hx}{k}.$$
(a)

The heat transfer coefficient h is given by equation (1.10)

$$h = \frac{-k\dfrac{\partial T(x,0)}{\partial y}}{T_s(x) - T_\infty}.$$
(1.10)

Thus temperature distribution $T(x, y)$ must be determined. The integral form of the energy equation is used to determine temperature distribution

$$-\alpha \frac{\partial T(x,0)}{\partial y} = \frac{d}{dx} \int_0^{\delta_t(x)} u(T - T_\infty)dy.$$
(5.7)

The axial velocity distribution $u(x,y)$, based on an assumed third degree polynomial, was determined in Section 5.7.1 and is given by

$$\frac{u}{V_\infty} = \frac{3}{2}\left(\frac{y}{\delta}\right) - \frac{1}{2}\left(\frac{y}{\delta}\right)^3, \tag{5.9}$$

where

$$\delta = \frac{\sqrt{280/13}}{\sqrt{Re_x}} x = \sqrt{\frac{280}{13}\frac{xv}{V_\infty}}. \tag{5.10}$$

We assume a third degree temperature polynomial

$$T(x,y) = b_0(x) + b_1(x)y + b_2(x)y^2 + b_3(x)y^3. \tag{b}$$

The temperature boundary conditions are:

(1) $T(x,0) = T_s(x)$,

(2) $T(x,\delta_t) \approx T_\infty$,

(3) $\dfrac{\partial T(x,\delta_t)}{\partial y} \approx 0$,

(4) $\dfrac{\partial^2 T(x,0)}{\partial y^2} = 0$.

The four boundary conditions are used to determine the coefficients in (b). The assumed profile becomes

$$T(x,y) = T_s(x) + [T_\infty - T_s(x)]\left[\frac{3}{2}\frac{y}{\delta_t} - \frac{1}{2}\frac{y^3}{\delta_t^3}\right]. \tag{c}$$

Substituting (c) into (1.10)

$$h(x) = \frac{3}{2}\frac{k}{\delta_t}. \tag{d}$$

Introducing (d) into (a)

$$Nu_x = \frac{3}{2}\frac{x}{\delta_t}. \tag{e}$$

Thus the problem reduces to determining the thermal boundary layer thickness δ_t. This is accomplished by using the integral form of the energy equation (5.7). Substituting (5.9) and (c) into (5.7) and evaluating the integral, gives

$$\frac{3}{2}\alpha\frac{T_s(x)-T_\infty}{\delta_t}=\frac{d}{dx}\left\{[T_s(x)-T_\infty]V_\infty\delta\left[\frac{3}{20}\left(\frac{\delta_t}{\delta}\right)^2-\frac{3}{280}\left(\frac{\delta_t}{\delta}\right)^4\right]\right\}. \quad (f)$$

This equation is simplified for Prandtl numbers greater than unity. For this case

$$\frac{\delta_t}{\delta}<1,\quad \text{for}\ \ Pr>1, \quad (5.15)$$

Thus

$$\frac{3}{280}\left(\frac{\delta_t}{\delta}\right)^4<<\frac{3}{20}\left(\frac{\delta_t}{\delta}\right)^2.$$

Equation (f) simplifies to

$$10\frac{\alpha}{\delta_t}[T_s(x)-T_\infty]=V_\infty\frac{d}{dx}\left[[T_s(x)-T_\infty]\delta\left(\frac{\delta_t}{\delta}\right)^2\right]. \quad (g)$$

However

$$T_s(x)-T_\infty=C\sqrt{x}. \quad (h)$$

Substituting (5.10) and (h) into (g)

$$10\frac{\alpha}{\delta_t}[C\sqrt{x}]=V_\infty\frac{d}{dx}\left[C\sqrt{x}\sqrt{\frac{13}{280}\frac{V_\infty}{vx}}\delta_t^2\right].$$

Simplifying, rearranging and separating variables

$$5\sqrt{\frac{280}{13}\frac{\alpha}{v}}\ [v/V_\infty]^{3/2}\sqrt{x}\,dx=\delta_t^2\,d\delta_t. \quad (i)$$

The boundary condition on δ_t is

$$\delta_t(0)=0. \quad (j)$$

Integrating (i) using condition (j)

$$\delta_t = \left[10\sqrt{280/13} \, \right]^{1/3} (Pr)^{-1/3} (vx/V_\infty)^{1/2}. \tag{k}$$

Substituting into (e) and introducing the definition of the local Reynolds number

$$Nu_x = 0.417 \, Pr^{1/3} Re_x^{1/2}. \tag{5.38}$$

(5) Checking. *Dimensional check*: Equations (5.10), (c), (d) and (k) are dimensionally correct. Equations (d) and (5.38) are dimensionless.

Boundary conditions check: Assumed temperature profile satisfies the four boundary conditions.

(6) Comments. (i) The local Nusselt number given in (5.38) is identical to the result of Section 5.7.3 for the case of uniform surface flux shown in equation (5.36). This is not surprising since uniform surface flux results in a variable surface temperature given by

$$T_s(x) = T_\infty + 2.396 \frac{q_s''}{k} \frac{x}{Pr^{1/3} Re_x^{1/2}}. \tag{5.35}$$

Note that the above can be rewritten as

$$T_s(x) = T_\infty + C\sqrt{x}$$

This is identical to the surface temperature specified in this example.

(ii) The same procedure can be used to analyze plates with surface temperature distribution other than the above.

REFERENCES

[1] Kays, W.M. and M.E. Crawford, *Convection Heat Transfer, 3rd edition*, McGraw-Hill, 1993.

PROBLEMS

5.1   For fluids with $Pr \ll 1$ the thermal boundary layer thickness is much larger than the viscous boundary layer. That is $\delta_t / \delta \gg 1$. It is reasonable for such cases to assume that fluid velocity within the thermal layer is uniform equal to the free stream velocity. That is

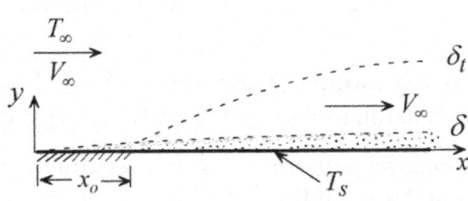

$$u \approx V_\infty.$$

Consider laminar boundary layer flow over a flat plate. The surface is maintained at uniform temperature T_s and has an insulated leading section of length x_0. Assume a third degree polynomial temperature profile. Show that the local Nusselt number is given by

$$Nu_x = 0.53 \left[1 - \frac{x_0}{x} \right]^{-1/2} Pr^{1/2} Re_x^{1/2},$$

where the local Reynolds number is $Re_x = V_\infty x / v$.

5.2   For fluids with $Pr \gg 1$ the thermal boundary layer thickness is much smaller than the viscous boundary layer. That is $\delta_t / \delta \ll 1$. It is reason-

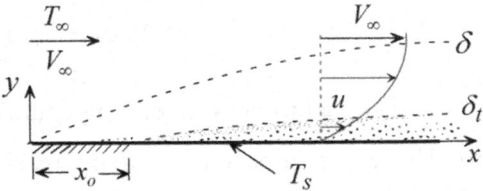

able for such cases to assume that fluid velocity within the thermal layer is linear given by

$$u = V_\infty \frac{y}{\delta}.$$

Consider uniform laminar boundary layer flow over a flat plate with an insulated leading section of length x_0. The plate is maintained at uniform surface temperature T_s. Assume a third degree polynomial temperature profile, show that the local Nusselt number is given by

$$Nu_x = 0.319 \left[1 - (x_0 / x)^{3/4} \right]^{-1/3} Pr^{1/3} Re_x^{1/2}.$$

5.3 A square array of chips is mounted flush on a flat plate. The array measures L cm $\times L$ cm. The forward edge of the array is at a distance x_o from the leading edge of the plate. The chips dissipate uniform surface flux q_s''. The plate is cooled by forced convection with uniform upstream velocity V_∞ and temperature T_∞. Assume laminar boundary layer flow. Assume further that the axial velocity within the thermal boundary layer is equal to the free stream velocity. Use a third degree polynomial temperature profile.

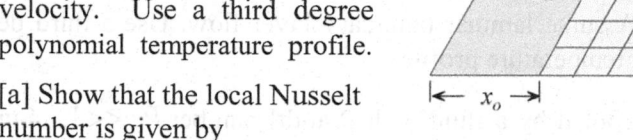

[a] Show that the local Nusselt number is given by

$$Nu_x = 0.75 \sqrt{\frac{Pr\, Re_x}{1-(x_o/x)}}.$$

[b] Determine the maximum surface temperature.

5.4 A liquid film of thickness H flows by gravity down an inclined surface. The axial velocity u is given by

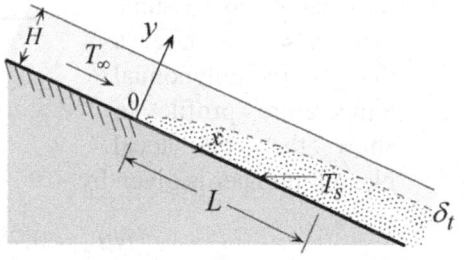

$$u = u_o \left[2\frac{y}{H} - \frac{y^2}{H^2} \right],$$

where u_o is the free surface velocity. At $x > 0$ the surface is maintained at uniform temperature T_s. Fluid temperature upstream of this section is T_∞. Assume laminar boundary layer flow and that $\delta_t / H < 1$. Determine the local Nusselt number and the total surface heat transfer from a section of width W and length L. Neglect heat loss from the free surface. Use a third degree polynomial temperature profile.

5.5 A thin liquid film flows under gravity down an inclined surface of width W. The film thickness is H and the angle of inclination is θ. The solution to the equations of motion gives the axial velocity u of the film as

$$u = \frac{gH^2 \sin\theta}{2\nu}\left[2\frac{y}{H} - \frac{y^2}{H^2}\right].$$

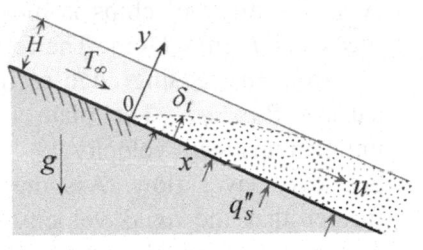

Heat is added to the film along the surface beginning at $x=0$ at uniform flux q''_s. Determine the total heat added from $x=0$ to the section where the thermal boundary layer penetrates half the film thickness. Assume laminar boundary layer flow. Use a third degree polynomial temperature profile.

5.6   A plate is cooled by a fluid with Prandtl number $Pr \ll 1$. Surface temperature varies with distance from the leading edge according to

$$T_s(x,0) = T_\infty + C\sqrt{x},$$

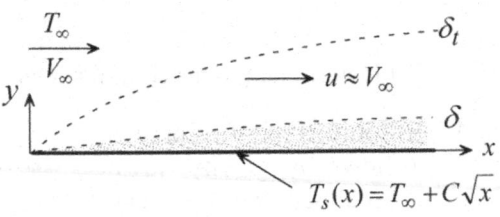

where C is constant. For such a fluid it is reasonable to assume that $u \approx V_\infty$. Use a third degree polynomial temperature profile to show that the local Nusselt number is given by

$$Nu_x = 0.75 \; Pr^{1/2} Re_x^{1/2},$$

and that surface heat flux is uniform. Assume laminar boundary layer flow.

5.7   A plate is cooled by a fluid with Prandtl number $Pr \gg 1$. Surface temperature varies with distance form the leading edge according to

$$T_s(x,0) = T_\infty + C\sqrt{x},$$

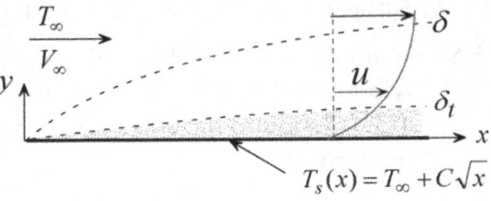

where C is constant. For $Pr \gg 1$ it is reasonable to assume that axial velocity within the thermal boundary layer is linear given by

$$u = V_\infty y/\delta.$$

Determine the local Nusselt number and show that surface heat flux is uniform. Use a third degree polynomial temperature profile and assume laminar boundary layer flow.

5.8 Surface temperature of a plate increases exponentially with distance from the leading edge according to

$$T_s(x,0) = T_\infty + C\exp(\beta x),$$

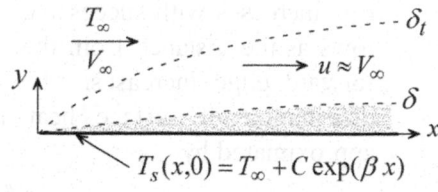

here C and β are constants. The plate is cooled with a low Prandtl number fluid ($Pr \ll 1$). Since for such fluids $\delta \ll \delta_t$, it is reasonable to assume uniform axial velocity within the thermal boundary layer. That is

$$u \approx V_\infty.$$

Use a third degree polynomial temperature profile and assume laminar boundary layer flow.

[a] Show that the local Nusselt number is given by

$$Nu_x = 0.75\sqrt{\frac{(\beta x)}{[1 - \exp(-2\beta x)]}PrRe_x}.$$

[b] Determine surface flux distribution.

5.9 A square array of chips of side L is mounted flush on a flat plate. The chips dissipate non-uniform surface flux according to

$$q''_x = \frac{C}{\sqrt{x}}.$$

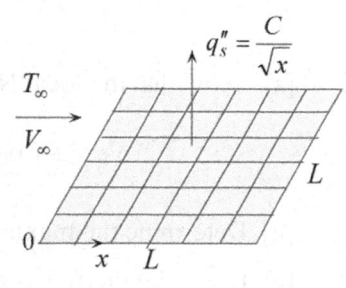

The plate is cooled by forced convection with uniform upstream velocity V_∞ and temperature T_∞. Assume laminar boundary layer flow with $\delta_t/\delta < 1$. Use third degree polynomials for the axial velocity and temperature.

[a] Show that the local Nusselt number is given by

$$Nu_x = 0.331 Pr^{1/3} Re^{1/2}.$$

[b] Show that surface temperature is uniform.

5.10 A square array of chips of side
L is mounted flush on a flat
plate. The forward edge of the
array is at a distance x_o from
the leading edge of the plate.
The heat dissipated in each
row increases with successive
rows as the distance from the
forward edge increases. The
distribution of surface heat flux for this arrangement may be
approximated by

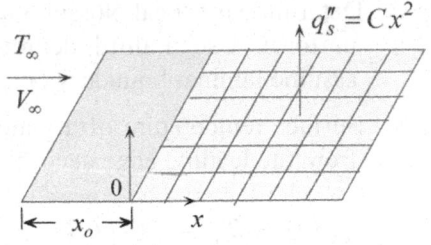

$$q''_s = Cx^2,$$

where C is constant. The plate is cooled by forced convection with
uniform upstream velocity V_∞ and temperature T_∞. Assume laminar
boundary layer flow. Assume further that the axial velocity within
the thermal boundary layer is equal to the free stream velocity,
$u \approx V_\infty$. Use a third degree polynomial temperature profile.

[a] Show that the local Nusselt number is given by

$$Nu_x = 1.3 \left[1 - (x_o / x)^3\right]^{-1/2} Pr^{1/2} Re^{1/2}.$$

[b] Determine the maximum surface temperature

[c] How should the rows be rearranged to reduce the maximum
surface temperature?

5.11 Repeat Problem 5.10 using a linear surface flux distribution

$$q''_s = Cx.$$

[a] Show that the local Nusselt number is given by

$$Nu_x = 1.06 \left[1 - (x_o / x)^2\right]^{-1/2} Pr^{1/2} Re^{1/2}.$$

[b] Determine the maximum surface temperature

[c] How should the rows be rearranged to reduce the maximum
surface temperature?

5.12 A fluid at temperature T_o and flow rate m_o is injected radially
between parallel plates. The spacing between the plates is H. The
upper plate is insulated and the lower plate is maintained at uniform

temperature T_s along $r \geq R_o$ and is insulated along $0 \leq r \leq R_o$. Consider laminar boundary layer flow and assume that the radial velocity u does not vary in the direction normal to the plates (slug flow).

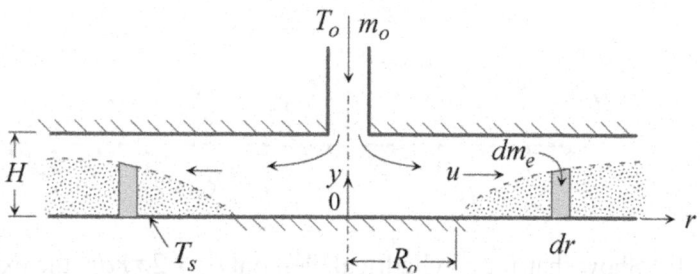

[a] Show that for a cylindrical element $\delta_t \times 2\pi r dr$ the external mass flow, dm_e, to the thermal boundary layer is

$$dm_e = 2\pi\rho \frac{d}{dr} \int_0^{\delta_t} u r dr = \frac{m_o}{H} d\delta_t.$$

[b] Show that the integral form of conservation of energy is

$$-kr\frac{\partial T(r,0)}{\partial r} = \frac{m_o c_p}{2\pi H}\frac{d}{dr}\int_0^{\delta_t}(T-T_o)dy.$$

[c] Assume a linear temperature profile, show that the local Nusselt number is

$$Nu_r = \frac{1}{\sqrt{2}}\left[1-(R_o/r)^2\right]^{-1/2}Pr^{1/2}Re_r^{1/2},$$

where

$$Re_r = \frac{\rho u r}{\mu} = \frac{m_o}{2\pi\mu H}.$$

5.13 The lower plate in Problem 5.12 is heated with uniform flux q_s'' along $r \geq R_o$ and insulated along $0 \leq r \leq R_o$.

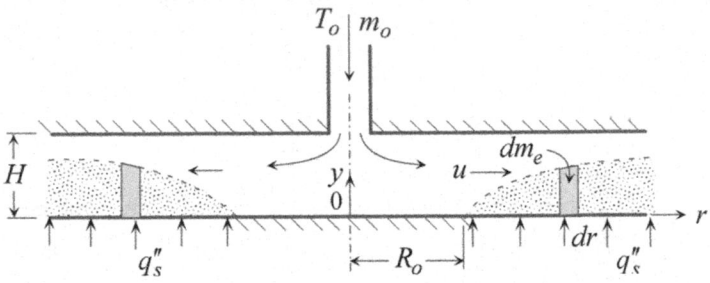

[a] Show that for a cylindrical element $\delta_t \times 2\pi r\, dr$ the external mass flow dm_e to the thermal boundary layer is

$$dm_e = 2\pi\rho \frac{d}{dr} \int_0^{\delta_t} ur\,dr = \frac{m_o}{H} d\delta_t .$$

[b] Show that the integral form of conservation of energy is

$$q_s'' = \frac{m_o c_p}{2\pi H} \frac{d}{dr} \int_0^{\delta_t} (T - T_o)\,dy .$$

[c] Assume a linear temperature profile, show that the local Nusselt number is

$$Nu_r = \left[1 - (R_o / r)^2\right]^{-1/2} Pr^{1/2} Re_r^{1/2} ,$$

where

$$Re_r = \frac{\rho u r}{\mu} = \frac{m_o}{2\pi \mu H} .$$

5.14 A porous plate with an impermeable and insulated leading section of length x_o is maintained at uniform temperature T_S along $x \geq x_o$. The plate is cooled by forced convection with a free stream velocity V_∞ and temperature T_∞. Fluid at temperature T_o is injected through the porous surface with uniform velocity v_o. The injected and free stream fluids are identical. Assume laminar boundary layer flow, introduce axial

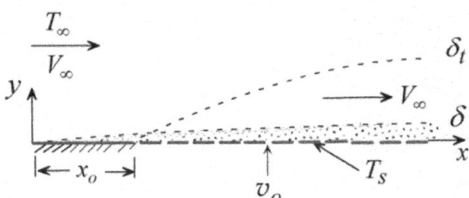

velocity simplification based on $Pr \ll 1$ and use a linear temperature pro-file to determine the local Nusselt number.

5.15 A porous plate with an impermeable and insulated leading section of length x_o is heated with uniform surface flux q_s'' along $x \geq x_o$. The plate is cooled by forced convection with a free stream velocity

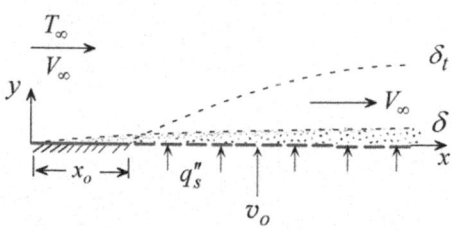

V_∞ and temperature T_∞. Fluid at temperature T_o is injected through the porous surface with uniform velocity v_o. The injected and free stream fluids are identical. Assume laminar boundary layer flow and introduce axial velocity simplification based on $Pr \ll 1$. Use a third degree polynomial temperature profile to determine the local Nusselt number.

5.16 Consider steady two-dimensional laminar flow in the inlet region of two parallel plates. The plates are separated by a distance H. The lower plate is maintained at uniform temperature T_o while heat is removed from the upper plate at uniform flux q_o''. The inlet temperature is T_i. Determine the distance from the inlet where the lower and upper thermal boundary layers meet. Use a linear temperature profile and assume that velocity is uniform equal to V_i. Express your result in terms of dimensionless quantities.

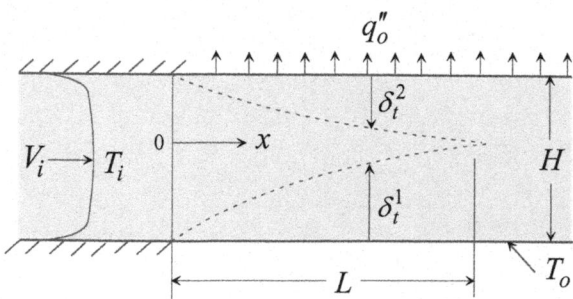

6

HEAT TRANSFER IN CHANNEL FLOW

6.1 Introduction

We consider internal flow through channels such as ducts, tubes and parallel plates. The following factors should be noted in analyzing heat transfer in internal flow.

(1) Laminar vs. turbulent flow. Transition from laminar to turbulent flow takes place when the Reynolds number reaches the transition value. For flow through tubes the experimentally determined *transition Reynolds number* Re_{D_t} is

$$Re_{D_t} = \frac{\overline{u}D}{v} \approx 2300,$$ (6.1)

where

D = tube diameter
$\overline{u}$ = mean velocity
v = kinematic viscosity

(2) Entrance vs. fully developed region. Based on velocity and temperature distribution, two regions are identified:

(i) Entrance region
(ii) Fully developed region

The length of the entrance region for velocity and temperature as well as the characteristics of these regions will be examined.

(3) Surface boundary conditions. Two common thermal boundary conditions will be considered:

(i) Uniform surface temperature

(ii) Uniform surface heat flux

(4) Objective. A common problem involves a fluid entering a channel with uniform velocity and temperature. The objective in analyzing internal flow heat transfer depends on the thermal boundary condition.

(i) Uniform surface temperature. In this class of problems we seek to determine axial variation of the following variables:

(1) Mean fluid temperature
(2) Heat transfer coefficient
(3) Surface heat flux

(ii) Uniform surface flux. For this class of problems the objective is to determine axial variation of the following variables:

(1) Mean fluid temperature
(2) Heat transfer coefficient
(3) Surface temperature

6.2 Hydrodynamic and Thermal Regions: General Features

We consider fluid entering a channel with uniform velocity V_i and temperature T_i. Velocity and temperature boundary layers form on the inside surface of the channel. The two boundary layers grow as the distance x from the entrance is increased. Two regions are identified for each of the velocity (hydrodynamic) and temperature (thermal) fields:

(1) Entrance region. This is also referred to as the developing region. It extends from the inlet to the section where the boundary layer thickness reaches the channel center.

(2) Fully developed region. This zone follows the entrance region.

Note that in general the lengths of the velocity and temperature entrance regions are not identical. The general features of velocity and temperature fields in the two regions will be examined in detail.

6.2.1 Flow Field

(1) Entrance Region (Developing Flow, $0 \le x \le L_h$ **).** Fig. 6.1 shows the developing velocity boundary layer in the entrance region of a tube. This region is called the *hydrodynamic entrance region*. Its length, L_h, is referred to as the *hydrodynamic entrance length*. This region is characterized by the following features:

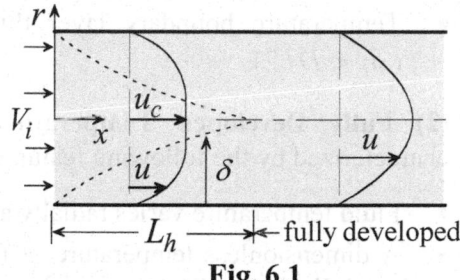

Fig. 6.1

- The radial velocity component does not vanish $(v_r \ne 0)$. Therefore, streamlines are not parallel.
- Core velocity u_c increases with axial distance x ($u_c \ne$ constant).
- Pressure decreases with axial distance ($dp/dx < 0$).
- Velocity boundary layer thickness is within tube radius ($\delta < D/2$).

(2) Fully Developed Flow Region. At $x \ge L_h$ the flow is described as *fully developed*. It is characterized by the following features:

- Streamlines are parallel $(v_r = 0)$.
- For two-dimensional incompressible flow the axial velocity u is invariant with axial distance x. That is $\partial u / \partial x = 0$.

6.2.2 Temperature Field

(1) Entrance Region (Developing Temperature, $0 \le x \le L_t$ **).** Fig. 6.2 shows fluid entering a tube with uniform velocity V_i and temperature T_i. The surface is at uniform temperature T_s. The region in which the temperature boundary layer forms and grows is referred to as the *thermal entrance region*. The length of this region, L_t, is called the *thermal entrance length*. At $x = L_t$ the thermal boundary layer thickness δ_t reaches the tube's center. This region is

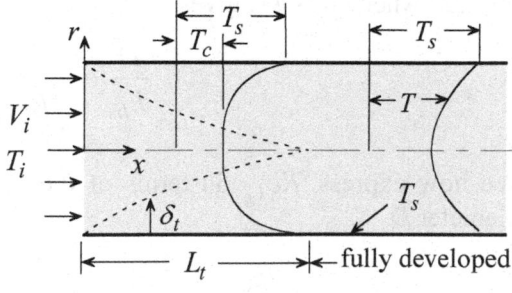

Fig. 6.2

characterized by the following features:

- Core temperature T_c is uniform equal to inlet temperature ($T_c = T_i$).
- Temperature boundary layer thickness is within the tube's radius ($\delta_t < D/2$).

(2) Fully Developed Temperature Region. The region $x \geq L_t$ is characterized by the following features:

- Fluid temperature varies radially and axially. Thus $\partial T / \partial x \neq 0$.
- A dimensionless temperature ϕ (to be defined later) is invariant with axial distance x. That is $\partial \phi / \partial x = 0$.

6.3 Hydrodynamic and Thermal Entrance Lengths

The determination of the two entrance lengths L_h and L_t is important in many applications. Scale analysis will be used to estimate the entrance lengths and results of analytic and numerical solutions will be summarized.

6.3.1 Scale Analysis

(1) Hydrodynamic Entrance Length L_h. Result of scale analysis for the velocity boundary layer thickness for external flow is given by (4.16)

$$\frac{\delta}{x} \sim \frac{1}{\sqrt{Re_x}}. \tag{4.16}$$

Applying (4.16) to the flow through a tube at the end of the entrance region $x = L_h$ where $\delta \sim D$, gives

$$\frac{D}{L_h} \sim \frac{1}{\sqrt{Re_{L_h}}}. \tag{a}$$

We now express Re_{L_h} in terms of the Reynolds number based on tube diameter D

$$Re_{L_h} = \frac{\overline{u}L_h}{\nu} = \frac{\overline{u}D}{\nu}\frac{L_h}{D} = Re_D \frac{L_h}{D}, \tag{b}$$

where $\overline{u}$ is mean or average velocity. Substituting (b) into (a) and rearranging

$$\left(\frac{L_h / D}{Re_D}\right)^{1/2} \sim 1. \tag{6.2}$$

(2) Thermal Entrance Length L_t. In internal flow both δ and δ_t increase with axial distance in the entrance region and eventually reach the centerline. Thus the scale for u for all Prandtl numbers is $u \sim \overline{u}$. This is unlike external flow where different scales are used depending on the Prandtl number. To scale δ_t we start with the external flow result, equation (4.24)

$$\delta_t \sim L Re_L^{-1/2} Pr^{-1/2}. \tag{4.24}$$

Applying (4.24) at $L = L_t$ where $\delta_t \sim D$

$$D \sim L_t Re_t^{-1/2} Pr^{-1/2}. \tag{a}$$

The Reynolds number for internal flow should more appropriately be based on the diameter D rather than length L_t. Thus

$$Re_{L_t} = \frac{\overline{u} L_t}{v} = \frac{\overline{u} D}{v} \frac{L_t}{D} = Re_D \frac{L_t}{D}. \tag{b}$$

Substituting (b) into (a) and rearranging

$$\left(\frac{L_t / D}{Re_D Pr}\right)^{1/2} \sim 1. \tag{6.3}$$

From (6.2) and (6.3) we obtain

$$\frac{L_t}{L_h} \sim Pr. \tag{6.4}$$

6.3.2 Analytic and Numerical Solutions: Laminar Flow

Solutions to the velocity and temperature distribution in the entrance region of various channel geometries have been obtained for laminar flow using analytic and numerical methods. Results provide information on L_h and

L_t. Since these lengths are not uniquely determined, their values depend on how they are defined.

(1) Hydrodynamic Entrance Length L_h. Results for L_h are expressed as

$$\frac{L_h}{D_e} = C_h Re_{D_e}, \tag{6.5}$$

where D_e is the equivalent diameter, defined as

$$D_e = \frac{4A_f}{P}.$$

Here A_f is channel flow area and P is channel perimeter. The coefficient C_h depends on channel geometry and is given in Table 6.1 [1]. Scaling prediction of C_h can now be evaluated using this table. Recall that scaling gives

$$\left(\frac{L_h / D}{Re_D}\right)^{1/2} \sim 1. \tag{6.2}$$

Table 6.1
Entrance length coefficients [1]

geometry	C_h	C_t	
		uniform surface flux	uniform surface temperature
○	0.056	0.043	0.033
b ▢ a $a = b$	0.09	0.066	0.041
b ▭ a $a = 2b$	0.085	0.057	0.049
$a = 4b$ b ▭ a $a/b = \infty$	0.075	0.042	0.054
▭	0.011	0.012	0.008

To compare this with the analytical results, equation (6.5) is rewritten as

$$\left(\frac{L_h / D_e}{Re_{D_e}}\right)^{1/2} = \left(C_h\right)^{1/2}. \tag{a}$$

As an example, for a rectangular channel of aspect ratio 2, Table 6.1 gives $C_h = 0.085$. Substituting this value into (a), gives

$$\left(\frac{L_h / D_e}{Re_{D_e}}\right)^{1/2} = \left(0.085\right)^{1/2} = 0.29. \tag{b}$$

Comparing (6.2) with (b) shows that scaling estimates the constant 0.29 to be unity.

(2) Thermal Entrance Length L_t. The length L_t depends on surface boundary conditions. Two cases are of special interest: uniform surface flux and uniform surface temperature. Solutions are expressed as

$$\frac{L_t}{D_e} = C_t PrRe_{D_e},$$ (6.6)

where C_t is a constant which depends on channel geometry as well as boundary conditions and is given in Table 6.1. To compare scaling prediction of L_t with the results of Table 6.1, equation (6.6) is rewritten as

$$\left(\frac{L_t / D_e}{PrRe_{D_e}}\right)^{1/2} = (C_t)^{1/2}.$$ (c)

Scaling gives

$$\left(\frac{L_t / D_e}{Re_D Pr}\right)^{1/2} \sim 1.$$ (6.3)

As an example, for a rectangular channel of aspect ratio 2 at uniform surface temperature, Table 6.1 gives $C_t = 0.049$. Substituting this value into (c), gives

$$\left(\frac{L_t / D_e}{PrRe_{D_e}}\right)^{1/2} = (0.049)^{1/2} = 0.22.$$ (d)

Comparing (6.3) with (d) shows that scaling estimates the constant 0.22 to be unity.

For turbulent flow, results for L_h and L_t are based on experimental data. In general, both lengths are much shorter than their corresponding laminar flow values. The following equation provides a guide for estimating the two lengths [2]

$$\frac{L}{D} \approx 10,$$ (6.7)

where $L = L_h = L_t$.

Example 6.1: Entrance Length for a Square Channel

In applications where entrance length is small compared to channel length, it is reasonable to neglect entrance length and assume fully developed conditions throughout. This approximation represents a significant simplification in obtaining analytic solutions. Consider the flow of water through a 0.75 cm × 0.75 cm square duct which is 2.5 m long. The duct is heated with uniform surface flux. The mean axial velocity is 0.12 cm/s. Is it justified to neglect entrance lengths? Evaluate water properties at $55^{\circ}C$.

(1) Observations. (i) This is an internal forced convection problem. (ii) The fluid is heated at uniform surface flux. (iii) The Reynolds number should be computed to establish if the flow is laminar or turbulent. (iv) If the flow is laminar, equations (6.5) and (6.6) can be used to determine entrance lengths L_h and L_t. (v) Velocity and temperature can be assumed fully developed if entrance lengths are small compared to channel length.

(2) Problem Definition. Determine entrance lengths L_h and L_t and compare them with total channel length.

(3) Solution Plan. Compute the Reynolds number to establish if the flow is laminar or turbulent. If laminar, apply (6.5) and (6.6) to determine entrance lengths L_h and L_t.

(4) Plan Execution.

(i)Assumptions. (1) Steady state, (2) continuum, (3) constant properties, (5) uniform surface flux, (6) negligible axial conduction, (7) negligible

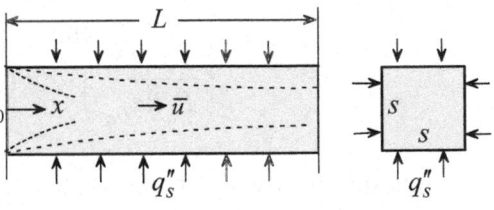

changes in kinetic and potential energy, and (8) negligible dissipation.

(ii) Analysis

The transition Reynolds number for flow through channels is

$$Re_{D_t} = \frac{\bar{u}D_e}{v} \approx 2300 . \qquad (a)$$

The Reynolds number for flow through a square channel is defined as

$$Re_{De} = \frac{\bar{u}D_e}{v} , \qquad (b)$$

where

D_e = equivalent diameter, m

Re_{D_e} = Reynolds number

$\bar{u}$ = mean velocity = 0.12 m/s

v = kinematic viscosity = 0.5116×10^{-6} m²/s

Water properties are evaluated at the mean temperature, $\overline{T}_m = 55^\circ$ C. The equivalent diameter for a square channel is defined as

$$D_e = 4\frac{A_f}{P} = 4\frac{S^2}{4S} = S, \tag{c}$$

where

A_f = channel flow area = S^2, m²

P = channel perimeter in contact with the fluid = $4S$, m

S = side dimension of the square channel = 0.0075 m

If the flow is laminar, L_h and L_t are determined using equations (6.5) and (6.6)

$$\frac{L_h}{D_e} = C_h Re_D, \tag{6.5}$$

and

$$\frac{L_t}{D_e} = C_t Pr Re_D, \tag{6.6}$$

where

C_h = hydrodynamic entrance length coefficient = 0.09, (Table 6.1)

C_t = thermal entrance length coefficient = 0.066, (Table 6.1)

L_h = hydrodynamic entrance length, m

L_t = thermal entrance length, m

Pr = Prandtl number = 3.27

Neglecting entrance lengths is justified if

$$\frac{L_h}{L} \ll 1 \text{ and } \frac{L_t}{L} \ll 1, \tag{d}$$

where

L = total channel length = 2.5 m

(iii) Computations. Substituting into (b)

$$Re_{De} = \frac{0.12(\text{m/s})0.0075(\text{m})}{0.5116 \times 10^{-6}(\text{m}^2/\text{s})} = 1759$$

Since the Reynolds number is smaller than 2300, the flow is laminar. Equations (6.5) and (6.6) give

$$L_h = 0.09(0.0075(\text{m})1759 = 1.19\,\text{m}$$

$$L_t = 0.066(0.0075(\text{m})(3.27)1759 = 2.85\,\text{m}$$

Thus

$$\frac{L_h}{L} = \frac{1.19(\text{m})}{2.5(\text{m})} = 0.48$$

$$\frac{L_t}{L} = \frac{2.85(\text{m})}{2.5(\text{m})} = 1.14$$

Comparison with (d) shows that the entrance lengths cannot be neglected.

(iv) Checking. *Dimensional check*: Computations showed that equations (b) and (d) are dimensionally consistent.

(5) Comments. (i) In general, the determination of the Reynolds number is an essential first step in analyzing internal flow. (ii) Fluids with Prandtl numbers greater than unity have longer thermal entrance lengths than hydrodynamic lengths. (iii) Entrance lengths can exceed channel length.

6.4 Channels with Uniform Surface Heat Flux q_s''

Consider a section of a channel shown in Fig. 6.3. Let the start of the section be at $x = 0$ and its end at $x = L$. The mean temperature at the inlet to this section is $T_{mi} = T_m(0)$. Heat is added at the surface at a uniform flux q_s''. We wish to determine the following:

(1) Total surface heat transfer rate q_s between $x = 0$ and location x along the channel.

(2) Mean temperature variation $T_m(x)$.

(3) Surface temperature variation $T_s(x)$.

Since the heat flux is uniform, the total heat transfer rate q_s is

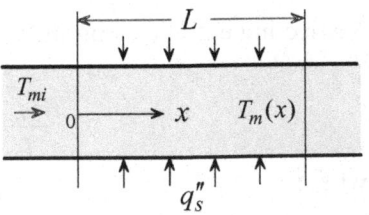

Fig. 6.3

$$q_s = q_s'' A_s = q_s'' P x, \qquad (6.8)$$

where A_s is channel surface area and P is perimeter.

The mean temperature $T_m(x)$ is obtained from energy conservation. Assume steady state, no energy generation, negligible changes in kinetic and potential energy and no axial conduction. Energy added at the surface must be equal to energy absorbed by the fluid. For constant properties, conservation of energy for a control volume between $x = 0$ and x gives

$$q_s = q_s'' P x = m c_p [T_m(x) - T_{mi}],$$

where m is mass flow rate and c_p is specific heat. Solving for $T_m(x)$, we obtain

$$T_m(x) = T_{mi} + \frac{q_s'' P}{m c_p} x. \qquad (6.9)$$

This equation gives the variation of the mean temperature along the channel. Note that no assumptions have been made regarding the region occupied by the section. That is, equations (6.8) and (6.9) are valid for the entrance region, fully developed region or a combination of the two. Furthermore, they apply to laminar as well as turbulent flow. The specific heat c_p should be evaluated at the average of the inlet and outlet mean temperatures.

Surface temperature $T_s(x)$ is determined using heat transfer analysis. Assume axisymmetric flow and neglect variations along the perimeter P, Newton's law of cooling gives

$$q_s'' = h(x)[T_s(x) - T_m(x)].$$

Solving for $T_s(x)$

$$T_s(x) = T_m(x) + \frac{q_s''}{h(x)}.$$

Using (6.9) to eliminate $T_m(x)$, we obtain

$$T_s(x) = T_{mi} + q_s'' \left[\frac{P x}{m c_p} + \frac{1}{h(x)} \right]. \qquad (6.10)$$

Thus, to determine surface temperature distribution the heat transfer coefficient $h(x)$ must be known. It is here that questions regarding the nature of the flow (laminar or turbulent) as well as the region of interest (entrance or fully developed) become crucial. Analytic and numerical solutions for h for laminar flow are available for both entrance and fully developed regions. For turbulent flow, empirical equations are used. The determination of h for laminar flow will be presented in Section 6.6.

Example 6.2: Maximum Surface Temperature

Water flows through a tube with a mean velocity of 0.2 m/s. The mean inlet and outlet temperatures are $20^\circ C$ and $80^\circ C$, respectively. The inside diameter of the tube is 0.5 cm. The surface is heated with uniform heat flux of 0.6 W/cm². If the flow is fully developed at the outlet the corresponding Nusselt number for laminar flow is given by

$$Nu_D = \frac{hD}{k} = 4.364 .$$

Determine the maximum surface temperature.

(1) Observations. (i) This is an internal forced convection problem in a tube. (ii) The surface is heated at uniform flux. (iii) Surface temperature changes along the tube. It is maximum at the outlet. (iv) The Reynolds number should be calculated to determine if the flow is laminar or turbulent. (v) If hydrodynamic and thermal entrance lengths are smaller than tube length, the flow can be assumed fully developed at the outlet. (vi) For fully developed flow, the heat transfer coefficient is uniform. (vii) Tube length is unknown.

(2) Problem Definition. (i) Find the required length to heat the water to a given temperature, and (ii) determine surface temperature at the outlet.

(3) Solution Plan. (i) Since surface flux, mean velocity, diameter, inlet and outlet temperatures are known, apply conservation of energy between the inlet and outlet to determine the required tube length. (ii) Compute the Reynolds number to determine if the flow is laminar or turbulent. (iii) Calculate the hydrodynamic and thermal entrance lengths and compare with tube length. (iv) Apply surface temperature solution for flow through a tube with constant surface flux.

(4) Plan Execution.

(i) Assumptions. (1) Continuum, (2) Newtonian, (3) steady state, (4) constant properties, (5) axisymmetric flow, (6) uniform surface heat flux, (7) negligible changes in kinetic and potential energy, (8) negligible axial conduction and (9) negligible dissipation.

(ii) Analysis. Application of conservation of energy between the inlet and outlet, gives

$$\pi DLq_s''' = mc_p(T_{mo} - T_{mi}),$$ (a)

where

c_p = specific heat, J/kg-°C
D = tube diameter = 0.5 cm = 0.005 m
L = tube length, m
m = mass flow rate, kg/s
T_{mi} = mean temperature at the inlet = 20°C
T_{mo} = mean temperature at the outlet = 80°C
q_s'' = surface heat flux = 0.6 W/cm² = 6000 W/m²

Solving (a) for L

$$L = \frac{mc_p(T_{mo} - T_{mi})}{\pi Dq_s''}.$$ (b)

The mass flow rate m is given by

$$m = (\pi/4)D^2\rho\bar{u},$$ (c)

where

$\bar{u}$ = mean flow velocity = 0.2 m/s
ρ = density, kg/m³

To determine surface temperature at the outlet, use the solution for surface temperature distribution for flow through a tube with uniform surface flux, given by equation (6.10)

$$T_s(x) = T_{mi} + q_s''\left[\frac{Px}{mc_p} + \frac{1}{h(x)}\right],$$ (6.10)

where

h = local heat transfer coefficient, W/m²-°C

P = tube perimeter, m
$T_s(x)$ = local surface temperature, °C
x = distance from inlet of heated section, m

The perimeter P is given by

$$P = \pi D. \tag{d}$$

Maximum surface temperature at the outlet, $T_s(L)$, is obtained by setting $x = L$ in (6.10)

$$T_s(L) = T_{mi} + q_s'' \left[\frac{PL}{mc_p} + \frac{1}{h(L)} \right]. \tag{e}$$

The determination of $h(L)$ requires establishing if the flow is laminar or turbulent and if it is fully developed at the outlet. Thus, the Reynolds number should be determined. It is defined as

$$Re_D = \frac{\bar{u}D}{\nu}, \tag{f}$$

where ν = kinematic viscosity, m²/s. Properties of water are determined at the mean temperature $\bar{T}_m$, defined as

$$\bar{T}_m = \frac{T_{mi} + T_{mo}}{2}. \tag{g}$$

Substituting into (g)

$$\bar{T} = \frac{(20 + 80)(°C)}{2} = 50°C$$

Properties of water at this temperature are

c_p = 4182 J/kg-°C
k = 0.6405 W/m-°C
Pr = 3.57
ν = 0.5537×10⁻⁶ m²/s
ρ = 988 kg/m³

Substituting into (f)

$$Re_D = \frac{0.2(\text{m/s})0.005(\text{m})}{0.5537 \times 10^{-6}(\text{m}^2/\text{s})} = 1806$$

Since the Reynolds number is less than 2300, the flow is laminar. The next step is calculating the hydrodynamic and thermal entrance lengths L_h and L_t to see if the flow is fully developed at the outlet. For laminar flow in a tube, the hydrodynamic and thermal lengths are given by (6.5) and (6.6)

$$\frac{L_h}{D_e} = C_h Re_D,$$ (6.5)

$$\frac{L_t}{D_e} = C_t Pr Re_D,$$ (6.6)

where

C_h = hydrodynamic entrance length coefficient = 0.056, (Table 6.1)
C_t = thermal entrance length coefficient = 0.043, (Table 6.1)
L_h = hydrodynamic entrance length, m
L_t = thermal entrance length, m

Substituting numerical values into (6.5) and (6.6)

L_h = 0.056 × 0.005 (m) × 1806 = 0.506 m

and

L_t = 0.043 × 0.005 (m) × 1806 × 3.57 = 1.386 m

If tube length L is larger than L_h and L_t, the flow is fully developed at the outlet. Thus, it is necessary to compute L using (b). The mass flow rate in (b) is given by (c)

m = 988(kg/m^3) 0.2(m/s)π (0.005)^2(m^2)/4 = 0.00388kg/s

Substituting into (b)

$$L = \frac{0.00388(\text{kg/s})4182(\text{J/kg}-^{\text{o}}\text{C})(80-20)(^{\text{o}}\text{C})}{\pi\, 0.005(\text{m})0.6(\text{W/cm}^2)10^4(\text{cm}^2/\text{m}^2)} = 10.33 \text{ m}$$

Since L is larger than both L_h and L_t, the flow is fully developed at the outlet. The heat transfer coefficient for fully developed laminar flow through a tube with uniform surface flux is given by

$$Nu_D = \frac{hD}{k} = 4.364.$$ (h)

(iii) Computations. The heat transfer coefficient at the outlet is computed using (h)

$$h(L) = 4.364 \frac{0.6405(\text{W/m}-^{\circ}\text{C})}{0.005(\text{m})} = 559 \ \text{W/m}^2\text{-}^{\circ}\text{C}$$

With L, m and $h(L)$ determined, equation (e) gives outlet surface temperature

$$T_s(L) = 20^{\circ}\text{C} + 6000(\text{W/m}^2) \left[\frac{\pi 0.005(\text{m})10.33(\text{m})}{0.00388(\text{kg/s})4182(\text{J/kg}-^{\circ}\text{C})} + \frac{1}{559(\text{W/m}^2 -^{\circ}\text{C})} \right]$$

$$= 90.7^{\circ}\text{C}$$

(iv) **Checking.** *Dimensional check*: Computations showed that equations (b), (c), (e), (f), (6.5), and (6.6) are dimensionally correct.

Quantitative checks: (1) Alternate approach to determining $T_s(L)$:
Application of Newton's law of cooling at the outlet gives

$$q_s'' = h(L)[T_s(L) - T_{mo}]. \tag{i}$$

Solving for $T_s(L)$

$$T_s(L) = T_{mo} + \frac{q_s''}{h} = 80 \ (^{\circ}\text{C}) + \frac{0.6(\text{W/cm}^2) \times 10^4 (\text{cm}^2/\text{m}^2)}{559(\text{W/m}^2 -^{\circ}\text{C})} = 90.7^{\circ}\text{C}$$

(2) The value of h is within the range reported in Table 1.1 for forced convection of liquids.

Limiting check: If $T_{mi} = T_{mo}$, the required length should vanish. Setting $T_{mi} = T_{mo}$ in (b) gives $L = 0$.

(5) **Comments.** (i) As long as the outlet is in the fully developed region, surface temperature at the outlet is determined entirely by the local heat transfer coefficient. Therefore, it is not necessary to justify neglecting entrance length to solve the problem.

(ii) In solving internal forced convection problems, it is important to establish if the flow is laminar or turbulent and if it is developing or fully developed.

6.5 Channels with Uniform Surface Temperature

Consider the same channel flow presented in the previous section with one important change. Instead of imposing uniform heat flux at the surface we specify uniform surface temperature as shown in Fig. 6.4.   We wish to determine the following:

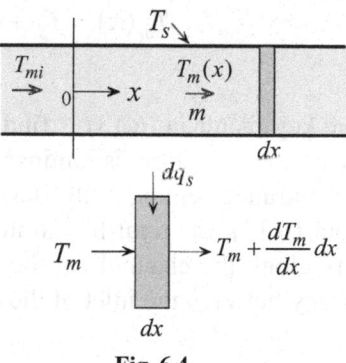

(1) Mean temperature variation $T_m(x)$.

(2) Total heat transfer rate q_s between $x = 0$ and location x along the channel.

(3) Surface heat flux variation $q''_s(x)$.

Fig. 6.4

Because surface flux is not uniform in this case, it is necessary to work with an infinitesimal element dx.   Applying conservation of energy to the element and invoking the assumptions used in the uniform surface flux case, we obtain

$$dq_s = m\, c_p\, dT_m.$$ (a)

Applying Newton's law of cooling to the element gives

$$dq_s = h(x)[T_s - T_m(x)]P dx.$$ (b)

Eliminating dq_s by combining the two equations and separating variables gives

$$\frac{dT_m}{T_s - T_m(x)} = \frac{P}{m\, c_p}h(x)dx.$$ (c)

Integrating (c) from $x = 0$, where $T_m = T_m(0) = T_{mi}$, to x where $T_m = T_m(x)$, gives

$$\ln\left[\frac{T_m(x) - T_s}{T_{mi} - T_s}\right] = -\frac{P}{m\, c_p}\int_0^x h(x)dx.$$ (6.11)

The integral in (6.11) cannot be evaluated unless $h(x)$ is known. We can rewrite this integral in terms of the average heat transfer coefficient $\bar{h}$ over the length x by applying the definition of $\bar{h}$

$$\bar{h} = \frac{1}{x}\int_0^x h(x)dx.$$ (6.12)

Introducing (6.12) into (6.11) and solving the resulting equation for $T_m(x)$

$$T_m(x) = T_s + (T_{mi} - T_s)\exp[-\frac{P\bar{h}}{mc_p}x]. \qquad (6.13)$$

The key factor in (6.13) is finding $h(x)$. The determination of $h(x)$ depends on whether the flow is laminar or turbulent and if the channel section is in the entrance region, fully developed region or both. With $T_m(x)$ known from (6.13), the total heat transfer rate and the variation of the local heat flux along the channel can be determined. Application of conservation of energy between the inlet of the channel and a section at location x gives

$$q_s = mc_p[T_m(x) - T_{mi}]. \qquad (6.14)$$

Application of Newton's law of cooling gives the heat flux $q_s''(x)$ at location x

$$q_s''(x) = h(x)[T_s - T_m(x)]. \qquad (6.15)$$

Properties such as kinematic viscosity, thermal conductivity, and specific heat should be evaluated at the average of the inlet and outlet mean temperatures.

Example 6.3: Required Tube Length

Air flows with a mean velocity of 2 m/s through a tube of diameter 1.0 cm. The mean temperature at a given section in the fully developed region is 35°C. The surface of the tube is maintained at a uniform temperature of 130°C. For fully developed laminar flow through tubes at uniform surface temperature, the Nusselt number is given by

$$Nu_D = \frac{hD}{k} = 3.657$$

Determine the length of the tube section needed to raise the mean temperature to 105°C.

(1) Observations. (i) This is an internal forced convection problem. (ii) The surface is maintained at uniform temperature. (iii) The Reynolds number should be checked to establish if the flow is laminar or turbulent. (iv) Since the Nusselt number for this flow is constant it follows that the heat transfer coefficient is uniform along the tube length.

(2) Problem Definition. Determine the tube length needed to raise the mean temperature to a specified level.

(3) **Solution Plan.** Use the analysis of flow in tubes at uniform surface temperature to determine the required tube length. Compute the Reynolds number to establish if the flow is laminar or turbulent.

(4) **Plan Execution.**

(i) **Assumptions.** (1) Continuum, (2) Newtonian, (3) steady state, (4) fully developed flow, (5) constant properties, (6) uniform surface temperature, (7) negligible changes in kinetic and potential energy, (8) negligible axial conduction, and (9) negligible dissipation.

(ii) **Analysis.** For flow through a tube at uniform surface temperature, conservation of energy and Newton's law of cooling lead to equation (6.13)

$$T_m(x) = T_s + (T_{mi} - T_s)\exp[-\frac{P\overline{h}}{m c_p}x],\qquad(6.13)$$

where

c_p = specific heat, J/kg–$^\circ$C
$\overline{h}$ = average heat transfer coefficient for a tube of length L, W/m^2–$^\circ$C
m = mass flow rate, kg/s
P = tube perimeter, m
$T_m(x)$ = mean temperature at x, $^\circ$C
T_{mi} = mean inlet temperature = $35\,^\circ$C
T_s = surface temperature = $130\,^\circ$C
x = distance from inlet of heated section, m

Applying (6.13) at the outlet of the heated section ($x = L$) and solving for L

$$L = \frac{m c_p}{P\overline{h}}\ln\frac{T_s - T_{mi}}{T_s - T_{mo}},\qquad(a)$$

where

T_{mo} = mean outlet temperature = $105\,^\circ$C

To compute L using (a), it is necessary to determine c_p, P, m, and $\overline{h}$. Air properties are determined at the mean temperature $\overline{T}_m$, defined as

$$\overline{T}_m = \frac{T_{mi} + T_{mo}}{2}.\qquad(b)$$

The perimeter P and flow rate m are given by

$$P = \pi D,\qquad(c)$$

and

$$m = \pi \frac{D^2}{4} \rho \bar{u} \,, \qquad\qquad (d)$$

where

   D – inside tube diameter = 1 cm = 0.01 m
   $\bar{u}$ = mean flow velocity = 2 m/s
   ρ = density, kg/m^3

The heat transfer coefficient for fully developed laminar flow is given by

$$Nu_D = \frac{hD}{k} = 3.657 \,, \qquad\qquad (e)$$

where

   h = heat transfer coefficient, W/m^2–$^\circ$C

   k = thermal conductivity of air, W/m–$^\circ$C

According to (e), h is uniform along the tube. Thus

$$h = \bar{h} = 3.657 \frac{k}{D} \,. \qquad\qquad (f)$$

To proceed, it is necessary to compute the Reynolds number to determine if the flow is laminar or turbulent. The Reynolds number for tube flow is defined as

$$Re_D = \frac{\bar{u} D}{v} \,, \qquad\qquad (g)$$

where

   Re_D = Reynolds number
   v = kinematic viscosity, m^2/s

The mean temperature is calculated using (b)

$$\bar{T}_m = \frac{(35 + 105)(^\circ C)}{2} = 70^\circ C$$

Properties of air at this temperature are

   c_p = 1008.7 J/kg–$^\circ$C
   k = 0.02922 W/m–$^\circ$C
   Pr = 0.707
   $v = 19.9 \times 10^{-6}$ m^2/s
   $\rho = 1.0287$ kg/m^3

Substituting into (g)

$$Re_D = \frac{2(\text{m/s})0.01(\text{m})}{19.9 \times 10^{-6}(\text{m}^2/\text{s})} = 1005$$

Since the Reynolds number is smaller than 2300, the flow is laminar. Thus (f) can be used to determine $\bar{h}$.

(iii) Computations. Substituting into (c), (d) and (f)

$$P = \pi 0.01(\text{m}) = 0.03142 \text{ m}$$

$$m = \pi \frac{(0.01)^2(\text{m}^2)}{4} 1.0287(\text{kg/m}^3)2(\text{m/s}) = 0.0001616\text{kg/s}$$

$$\bar{h} = 3.657 \frac{0.02922(\text{W/m}-^\circ\text{C})}{0.01(\text{m})} = 10.69 \text{ W/m}^2\text{-}^\circ\text{C}$$

Substituting into (a)

$$L = \frac{0.0001616(\text{kg/s})1008.7(\text{J/kg}-^\circ\text{C})}{0.03142(\text{m})10.69(\text{W/m}^2-^\circ\text{C})} \ln \frac{(130-35)(^\circ\text{C})}{(130-105)(^\circ\text{C})} = 0.65 \text{ m}$$

(iv) Checking. *Dimensional check:* Computations showed that equations (a)-(d), (f) and (g) are dimensionally consistent.

Limiting checks: (i) For the special case of $T_{mo} = T_{mi}$, the required length should vanish. Setting $T_{mo} = T_{mi}$ in (a) gives $L = 0$.

(ii) The required length for the outlet temperature to reach surface temperature is infinite. Setting $T_{mo} = T_s$ in (a) gives $L = \infty$.

Quantitative checks: (i) An approximate check can be made using conservation of energy and Newton's law of cooling. Conservation of energy is applied to the air between inlet and outlet

$$\text{Energy added at the surface} = \text{Energy gained by the air.} \quad\quad (h)$$

Assuming that air temperature in the tube is uniform equal to $\bar{T}_m$, Newton's law of cooling gives

$$\text{Energy added at surface} = \bar{h}\pi DL(T_s - \bar{T}_m). \quad\quad (i)$$

Neglecting axial conduction and changes in kinetic and potential energy, energy gained by the air is

$$\text{Energy gained by air} = mc_p(T_{mo} - T_{mi}). \quad\quad (j)$$

Substituting (j) and (i) into (h) and solving for the resulting equation for L

$$L = \frac{mc_p(T_{mo} - T_{mi})}{\bar{h}\,\pi D(T_s - \bar{T}_m)}. \tag{k}$$

Equation (k) gives

$$L = \frac{0.0001616(\text{kg/s})1008.7(\text{J/kg}-^\circ\text{C})(105-35)(^\circ\text{C})}{10.69(\text{W/m}^2-^\circ\text{C})\pi)\pi(0.01\text{m})(130-70)(^\circ\text{C})} = 0.57 \text{ m}$$

This is in reasonable agreement with the more exact answer obtained above.

(ii) The value of h appears to be low compared with typical values listed in Table 1.1 for forced convection of gases. However, it should be kept in mind that values of h in Table 1.1 are for typical applications. Exceptions should be expected.

(5) Comments. This problem is simplified by two conditions: fully developed and laminar flow.

6.6 Determination of Heat Transfer Coefficient $h(x)$ and Nusselt Number Nu_D

The heat transfer coefficient is critical in the analysis of channel flow heat transfer. Scale analysis will be presented first to obtain estimates of the heat transfer coefficient and Nusselt number. This will be followed by analytic solutions laminar flow in both entrance and fully developed regions.

6.6.1 Scale Analysis

Consider heat transfer in a tube of radius r_o shown in Fig. 6.5. Surface temperature is T_s and mean fluid temperature at a given section is T_m. Equating Fourier's law with Newton's law

$$h = \frac{-k\dfrac{\partial T(r_o, x)}{\partial r}}{T_m - T_s}. \tag{6.16}$$

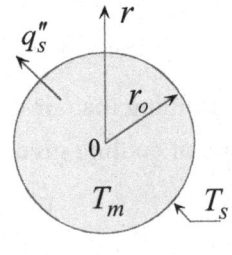

Fig. 6.5

A scale for r is

$$r \sim \delta_t.$$ (a)

A scale for the temperature gradient in (6.16) is

$$\frac{\partial T(r_o, x)}{\partial r} \sim \frac{T_m - T_s}{\delta_t}.$$ (b)

Substituting (a) and (b) into (6.16)

$$h \sim \frac{k \dfrac{(T_m - T_s)}{\delta_t}}{T_m - T_s},$$

or

$$h \sim \frac{k}{\delta_t}.$$ (6.17)

The Nusselt number is defined as

$$Nu_D = \frac{hD}{k}.$$

Introducing (6.17) into the above

$$Nu_D \sim \frac{D}{\delta_t}.$$ (6.18)

In the fully developed region where $\delta_t(x) \sim D$, equation (6.18) gives

$$Nu_D \sim 1 \quad \text{(fully developed)}.$$ (6.19)

This shows that in the fully developed region the Nusselt number is of order unity. However, in the entrance region where $\delta_t(x)$ grows from zero to r_o, the Nusselt number, according to (6.18), is greater than unity. To examine δ_t in the entrance region, we note that, unlike external flow, δ_t scales over the entire range of Prandtl numbers according to (4.24)

$$\delta_t \sim x \, Pr^{-1/2} Re_x^{-1/2}.$$ (4.24)

Substituting (4.24) into (6.18)

$$Nu_D \sim \frac{D}{x} Pr^{1/2} Re_x^{1/2}.$$ (c)

Expressing the Reynolds number Re_x in terms of diameter to form Re_D

$$Re_x = \frac{\overline{u}x}{v} = \frac{\overline{u}D}{v} \frac{x}{D} = Re_D \frac{x}{D}.$$ (d)

Substituting (d) into (c)

$$Nu_D \sim \left(\frac{D}{x}\right)^{1/2} Pr^{1/2} Re_D^{1/2}.$$ (6.20a)

The above is rewritten as

$$\frac{Nu_D}{\left(\dfrac{PrRe_D}{x/D}\right)^{1/2}} \sim 1.$$ (6.20b)

Scaling estimates (6.19) and (6.20) can be compared with the corresponding exact solutions presented in Section 6.6.2.

6.6.2 Basic Considerations for the Analytical Determination of Heat Flux, Heat Transfer Coefficient and Nusselt Number

The analytic determination of thermal characteristics such as heat transfer coefficient requires the determination of velocity and temperature distribution. An important simplification is the assumption of fully developed velocity. Neglecting axial conduction provides another major mathematical simplification. In this section we introduce basic definitions and present the governing equations for the analytic determination of surface heat flux, heat transfer coefficient and Nusselt number. In addition, the criterion for neglecting axial conduction will be identified.

(1) Fourier's law and Newton's law.   We return to Fig. 6.5 where heat flow is in the positive radial direction r. Fourier's law gives surface heat flux q_s''

$$q_s'' = -k \frac{\partial T(x,r_o)}{\partial r}.$$ (a)

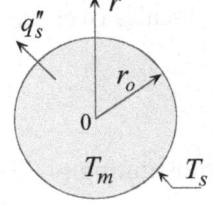

Fig. 6.5

Define dimensionless variables as

$$\theta = \frac{T - T_s}{T_i - T_s}, \ \xi = \frac{x/D}{Re_D Pr}, \ R = \frac{r}{r_o}, \ v_x^* = \frac{v_x}{\bar{u}}, \ v_r^* = \frac{v_r}{\bar{u}}, \ Re_D = \frac{\bar{u}D}{v},$$

$$(6.21)$$

where $\bar{u}$ is the mean axial velocity. Substituting into (a)

$$q_s''(\xi) = \frac{k}{r_o}(T_s - T_i)\frac{\partial\theta(\xi,1)}{\partial R}.$$

$$(6.22)$$

We define h using Newton's law of cooling

$$h(\xi) = \frac{q''_s}{T_m - T_s}.$$

$$(6.23)$$

Combining (6.22) and (6.23)

$$h(\xi) = \frac{k(T_s - T_i)}{r_o(T_m - T_s)}\frac{\partial\theta(\xi,1)}{\partial R} = -\frac{k}{r_o}\frac{1}{\theta_m(\xi)}\frac{\partial\theta(\xi,1)}{\partial R},$$

$$(6.24)$$

where the dimensionless mean temperature θ_m is defined as

$$\theta_m \equiv \frac{T_m - T_s}{T_i - T_s}.$$

$$(6.25)$$

The Nusselt number based on diameter is

$$Nu(\xi) = \frac{h(\xi)D}{k} = \frac{h(\xi)2r_o}{k}.$$

$$(6.26)$$

Substituting (6.24) into (6.26)

$$Nu(\xi) = \frac{-2}{\theta_m(\xi)}\frac{\partial\theta(\xi,1)}{\partial R}.$$

$$(6.27)$$

As can be seen from equations (6.22), 6.24) and (6.27), the key to determining $q_s''(\xi)$, $h(\xi)$ and $Nu(\xi)$ is the determination of the temperature distribution $\theta(\xi, R)$ which is obtained by solving the energy equation.

(2) The Energy Equation. Consider flow through a tube. Assume: (1) Continuum, (2) Newtonian, (3) steady state, (4) laminar flow, (5) axisymmetric, (6) negligible gravity, (7) negligible dissipation, (8) negligible changes in kinetic and potential energy and (9) constant properties. Based on these assumptions energy equation (2.24) gives

$$\rho c_p \left(v_r \frac{\partial T}{\partial r} + v_z \frac{\partial T}{\partial z} \right) = k \left[\frac{1}{r} \frac{\partial}{\partial r} \left(r \frac{\partial T}{\partial r} \right) + \frac{\partial^2 T}{\partial z^2} \right]. \tag{2.24}$$

Replacing the axial coordinate z by x, this equation is expressed in dimensionless form as

$$v_x^* \frac{\partial \theta}{\partial \xi} + 2 Re_D Pr\, v_r^* \frac{\partial \theta}{\partial R} = \frac{4}{R} \frac{\partial}{\partial R} \left(R \frac{\partial \theta}{\partial R} \right) + \frac{1}{(Re_D Pr)^2} \frac{\partial^2 \theta}{\partial \xi^2}. \tag{6.28}$$

The product of Reynolds and Prandtl numbers is called the *Peclet number*

$$Pe = Re_D Pr, \quad \text{Peclet number.} \tag{6.29}$$

Note that the first and second terms on the right-hand-side of (6.28) represent radial and axial conduction, respectively. Examination of the two terms suggests that for large values of the Peclet number, axial conduction may be neglected compared to radial conduction. Comparing solutions to (6.28) with and without the last term shows that axial conduction can be neglected for

$$Pe = Pr Re_D \geq 100. \tag{6.30}$$

Thus, under such conditions, (6.28) becomes

$$v_x^* \frac{\partial \theta}{\partial \xi} + 2 Re_D Pr\, v_r^* \frac{\partial \theta}{\partial R} = \frac{4}{R} \frac{\partial}{\partial R} \left(R \frac{\partial \theta}{\partial R} \right). \tag{6.31}$$

(3) Mean (Bulk) Temperature T_m. To determine h, a reference local temperature is needed. The mass average or mean temperature at a section of a channel is defined as

$$mc_p T_m = \int_0^{r_o} \rho c_p v_x T\, 2\pi r dr. \tag{a}$$

Each term in (a) represents the energy convected by the fluid. However, mass flow rate m is given by

$$m = \int_0^{r_o} \rho v_x 2\pi r dr .$$
(b)

Substituting (b) into (a) and assuming constant properties

$$T_m = \frac{\displaystyle\int_0^{r_o} v_x T r dr}{\displaystyle\int_0^{r_o} v_x r dr} .$$
(6.32a)

This result is expressed in dimensionless form as

$$\theta_m = \frac{T_m - T_s}{T_i - T_s} = \frac{\displaystyle\int_0^1 v_x^* \theta R dR}{\displaystyle\int_0^1 v_x^* R dR} .$$
(6.32b)

6.7 Heat Transfer Coefficient in the Fully Developed Temperature Region

As might be expected, analytical determination of the heat transfer coefficient in the fully developed region is simpler than that in the developing region. This section focuses on the fully developed region. Section 6.8 deals with the developing region.

6.7.1 Definition of Fully Developed Temperature Profile

Far away from the entrance of a channel ($x/d > 0.05 Re_D Pr$), temperature effect penetrates to the centerline and the temperature profile is said to be *fully developed*. This profile is not as easily visualized as a fully developed velocity profile. We introduce a dimensionless temperature ϕ defined as

$$\phi = \frac{T_s(x) - T(r,x)}{T_s(x) - T_m(x)} ,$$
(6.33)

where $T(r,x)$ is fluid temperature distribution. Note that this definition is applicable to a uniform as well as variable surface temperature. Since heat is added or removed from the fluid, it follows that its mean temperature varies with distance x along the channel. Fully developed temperature is defined as a profile in which ϕ is independent of x. That is

$$\phi = \phi(r). \qquad (6.34)$$

This definition means that a fully developed temperature profile has a single distribution in the radial direction at all locations x. It follows from (6.34) that

$$\frac{\partial \phi}{\partial x} = 0. \qquad (6.35)$$

Equations (6.33) and (6.35) give

$$\frac{\partial \phi}{\partial x} = \frac{\partial}{\partial x}\left[\frac{T_s(x) - T(r,x)}{T_s(x) - T_m(x)}\right] = 0. \qquad (6.36a)$$

Expanding and using the definition of ϕ in (6.33)

$$\frac{dT_s}{dx} - \frac{\partial T}{\partial x} - \phi(r)\left[\frac{dT_s}{dx} - \frac{dT_m}{dx}\right] = 0. \qquad (6.36b)$$

This result will be used in analyzing thermally developed flow in channels.

6.7.2 Heat Transfer Coefficient and Nusselt Number

We wish to examine the nature of h and Nu in the fully developed thermal region. Equating Fourier's with Newton's law, gives

$$h = \frac{-k\dfrac{\partial T(r_o,x)}{\partial r}}{T_m - T_s}. \qquad (6.16)$$

Using (6.33) to determine $\partial T(r_o,x)/\partial r$ and substituting into (6.16)

$$h = -k\frac{d\phi(r_o)}{dr} = \text{constant}. \qquad (6.37)$$

From this result we conclude that the heat transfer coefficient in the fully developed region is constant. This important conclusion is valid regardless of surface boundary conditions. Using (6.37) in the definition of the Nusselt number, gives

$$Nu_D = \frac{hD}{k} = -D\frac{d\phi(r_o)}{dr}.$$ (6.38)

In Section 6.6.1 scaling was used to estimate the Nusselt number in the entrance region. The result was used to examine the Nusselt number in the fully developed region as a limiting case of the entrance region. It was shown that

$$Nu_D \sim 1 \quad \text{(fully developed)}.$$ (6.19)

This result will now be arrived at using scale analysis of the fully developed temperature region where the thermal boundary layer fills the tube. A scale for the temperature gradient $\partial T(r_o, x)/\partial r$ is

$$\frac{\partial T(r_o, x)}{\partial r} \sim \frac{T_s - T_m}{D}.$$

Substituting into (6.16)

$$h \sim \frac{k}{D}.$$ (6.39)

Substituting (6.39) into the definition of the Nusselt number in (6.38)

$$Nu_D \sim 1 \quad \text{(fully developed)}.$$ (6.40)

6.7.3 Fully Developed Region for Tubes at Uniform Surface flux

Fig. 6.6 shows a tube section with uniform surface heat flux. We wish to determine the axial variation of surface temperature and heat transfer coefficient. We have shown in equation (6.37) that the heat

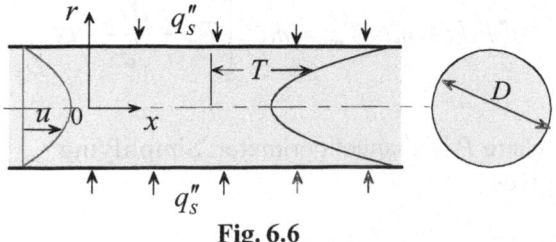

Fig. 6.6

transfer coefficient is uniform throughout the fully developed region, However, its value was not determined. Application of Newton's law of cooling gives

$$q_s'' = h[T_s(x) - T_m(x)].$$ (a)

Note that $T_s(x)$ and $T_m(x)$ are unknown. However, since q_s'' and h are constant it follows from (a) that

$$T_s(x) - T_m(x) = \text{constant}.$$ (b)

Differentiating (b)

$$\frac{dT_s}{dx} = \frac{dT_m}{dx}.$$ (c)

Substituting (c) into (6.36b)

$$\frac{\partial T}{\partial x} = \frac{dT_s}{dx}.$$ (d)

Combining (c) and (d)

$$\frac{\partial T}{\partial x} = \frac{dT_s}{dx} = \frac{dT_m}{dx} \quad \text{(for constant } q_s'' \text{).}$$ (6.41)

Note that $T(r,x)$, $T_m(x)$ and $T_s(x)$ are unknown. According to (6.16) and (6.38), these variables are needed to determine h and Nu_D. To determine the gradients in (6.41) an energy balance is made for an element dx of the tube shown in Fig. 6.7. Neglecting changes in kinetic and potential energy and assuming steady state and constant c_p, conservation of energy for the element gives

$$q_s'' P dx + m c_p T_m = m c_p \left[T_m + \frac{dT_m}{dx} dx \right],$$

where P is channel perimeter. Simplifying gives

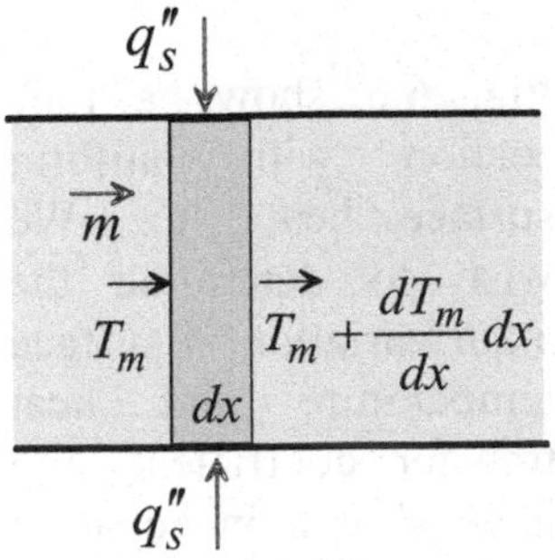

Fig. 6.7

$$\frac{dT_m}{dx} = \frac{q_s'' P}{mc_p} = \text{constant}. \tag{6.42}$$

This result shows that the axial gradient of the mean temperature is constant along the channel. Substituting (6.42) into (6.41)

$$\frac{\partial T}{\partial x} = \frac{dT_s}{dx} = \frac{dT_m}{dx} = \frac{q_s'' P}{mc_p} = \text{constant}. \tag{6.43}$$

Equation (6.42) shows that $T(x,r)$, $T_m(x)$ and $T_s(x)$ vary linearly with axial distance x. Integrating (6.43)

$$T_m(x) = \frac{q_s'' P}{mc_p} x + C_1, \tag{e}$$

where C_1 is constant of integration which is determined from inlet condition

$$T_m(0) = T_{mi}. \tag{f}$$

Application of (e) to (f) gives the $C_1 = T_{mi}$. Solution (e) becomes

$$T_m(x) = T_{mi} + \frac{q_s'' P}{mc_p} x. \tag{6.44}$$

Note that this result is identical to (6.9) which was obtained by applying conservation of energy to a finite tube section.

It remains to determine fluid temperature distribution $T(r,x)$ and surface temperature $T_s(x)$. This requires solving the differential form of the energy equation in the fully developed region. Neglecting axial conduction and dissipation, and noting that $v_r = 0$ for fully developed velocity, energy equation (2.24) simplifies to

$$\rho c_p v_x \frac{\partial T}{\partial x} = \frac{k}{r} \frac{\partial}{\partial r}\left(r \frac{\partial T}{\partial r}\right). \tag{6.45}$$

The axial velocity for fully developed flow is

$$v_x = 2\bar{u}\left[1 - \frac{r^2}{r_o^2}\right].$$

(6.46)

Substituting (6.43) and (6.46) into (6.45)

$$\rho c_p \, 2\bar{u}\left[1 - \frac{r^2}{r_o^2}\right]\frac{q_s'' P}{\dot{m} c_p} = \frac{k}{r}\frac{\partial}{\partial r}\left(r\frac{\partial T}{\partial r}\right).$$

(g)

However, $m = \pi r_o^2 \rho \bar{u}$ and $P = 2\pi r_o$, equation (g) becomes

$$\frac{4 q_s''}{r_o}\left[1 - \frac{r^2}{r_o^2}\right] = \frac{k}{r}\frac{\partial}{\partial r}\left(r\frac{\partial T}{\partial r}\right).$$

(6.47)

The boundary conditions are:

$$\frac{\partial T(0,x)}{\partial r} = 0,$$

(6.48a)

and

$$k\frac{\partial T(r_o,x)}{\partial r} = q_s''.$$

(6.48b)

Integrating (6.47) once with respect to r

$$\frac{4}{r_o}q_s''\left[\frac{r^2}{2} - \frac{r^4}{4r_o^2}\right] = kr\frac{\partial T}{\partial r} + f(x),$$

(h)

where $f(x)$ is "constant" of integration. Application of boundary condition (6.48a) gives $f(x) = 0$. Equation (h) becomes

$$\frac{\partial T}{\partial r} = \frac{4q_s''}{kr_o}\left[\frac{r}{2} - \frac{r^3}{4r_o^2}\right].$$

Integrating again

$$T(r,x) = \frac{4q_s''}{kr_o}\left[\frac{r^2}{4} - \frac{r^4}{16r_o^2}\right] + g(x).$$

(6.49)

The integration "constant" $g(x)$ represents the local centerline temperature $T(0, x)$. Boundary condition (6.48b) does not give $g(x)$ since (6.49) already satisfies this condition. The local mean temperature $T_m(x)$ is used to determine $g(x)$. Substituting (6.46) and (6.49) into (6.32a), gives

$$T_m(x) = \frac{7}{24} \frac{r_o q_s''}{k} + g(x).$$
(6.50)

Thus we have two equations for $T_m(x)$: (6.44) and (6.50). Equating the two gives $g(x)$

$$g(x) = T_{mi} - \frac{7}{24} \frac{r_o q_s''}{k} + \frac{P q_s''}{m c_p} x.$$
(6.51)

Substituting (6.51) into (6.49)

$$T(r, x) = T_{mi} + \frac{4 q_s''}{k r_o} \left[\frac{r^2}{4} - \frac{r^4}{16 r_o^2} \right] - \frac{7}{24} \frac{r_o q_s''}{k} + \frac{P q_s''}{m c_p} x.$$
(6.52)

Equation (6.52) satisfies the energy equation (6.45) and boundary conditions (6.48). Surface temperature $T_s(x)$ is obtained by setting $r = r_o$ in (6.52)

$$T_s(x) = T_{mi} + \frac{11}{24} \frac{r_o q_s''}{k} + \frac{P q_s''}{m c_p} x.$$
(6.53)

With $T(r, x)$, $T_m(x)$ and $T_s(x)$ determined, equation (6.33) gives $\phi(r)$ and (6.38) gives the Nusselt number. Substituting (6.44), (6.52) and (6.53) into (6.33) gives $\phi(r)$

$$\phi(r) = \frac{18}{11} - \frac{24}{11} \frac{1}{r_o^2} \left[r^2 - \frac{r^4}{4 r_o^2} \right].$$
(6.54)

Differentiating (6.54) and substituting into (6.38) gives the Nusselt number

$$Nu_D = \frac{48}{11} = 4.364.$$
(6.55)

The following observations are made regarding this result:

(1) Equation (6.55) applies to laminar fully developed velocity and temperature in tubes with uniform surface heat flux.

(2) Unlike other forced convection results, the Nusselt number for this case is independent of Reynolds and Prandtl numbers.

(4) Scaling prediction of the Nusselt number is given in equation (6.40) as

$$Nu_D \sim 1. \tag{6.40}$$

This compares favorably with (6.55).

6.7.4 Fully Developed Region for Tubes at Uniform Surface Temperature

We consider fully developed flow through a tube at uniform surface temperature T_s. Of interest is the determination of the Nusselt number. As shown in equation (6.38), the Nusselt number is constant throughout the fully developed region regardless of surface boundary condition. The determination of the Nusselt number requires solving the energy equation for the fully developed region. Neglecting axial conduction and dissipation and noting that $v_r = 0$ for fully developed velocity, energy equation (2.24) simplifies to

$$\rho c_p v_x \frac{\partial T}{\partial x} = \frac{k}{r} \frac{\partial}{\partial r} \left(r \frac{\partial T}{\partial r} \right). \tag{6.45}$$

The boundary conditions for this case are

$$\frac{\partial T(0,x)}{\partial r} = 0, \tag{6.56a}$$

and

$$T(r_o,x) = T_s. \tag{6.56b}$$

The axial velocity for fully developed flow is

$$v_x = 2\bar{u} \left[1 - \frac{r^2}{r_o^2} \right]. \tag{6.46}$$

The axial temperature gradient $\partial T / \partial x$ in equation (6.45) is eliminated using the definition of fully developed temperature profile, equation (6.36a)

$$\frac{\partial \phi}{\partial x} = \frac{\partial}{\partial x} \left[\frac{T_s(x) - T(r,x)}{T_s(x) - T_m(x)} \right] = 0. \tag{6.36a}$$

For uniform surface temperature, $T_s(x) = T_s$, (6.36a) give

$$\frac{\partial T}{\partial x} = \frac{T_s - T(r,x)}{T_s - T_m(x)} \frac{dT_m}{dx}. \tag{6.57}$$

Substituting (6.46) and (6.57) into (6.45)

$$2\rho c_p \overline{u} \left[1 - \frac{r^2}{r_o^2} \right] \frac{T_s - T(r,x)}{T_s - T_m(x)} \frac{dT_m}{dx} = \frac{k}{r} \frac{\partial}{\partial r} \left(r \frac{\partial T}{\partial r} \right). \tag{6.58}$$

Equation (6.58), subject to boundary conditions (6.56), was solved using an infinite power series [3]. The solution gives the Nusselt number as

$$Nu_D = 3.657. \tag{6.59}$$

6.7.5   Nusselt Number for Laminar Fully Developed Velocity and Temperature in Channels of Various Cross-Sections

Many internal flow applications involve channels with non-circular cross-sections. Analytical and numerical solutions for such cases have been obtained for various surface boundary conditions [3].   In all cases the Nusselt number in fully developed flow is uniform throughout. Table 6.2 [3] lists the Nusselt numbers for channels of various geometries for two surface conditions: (1) uniform heat flux, and (2) uniform temperature.   The Nusselt number for non-circular cross-sections is based on the equivalent diameter defined as

$$D_e = \frac{4A_f}{P} \tag{6.60}$$

where A_f the flow area and P is the perimeter. Note that the heat transfer coefficient for

Table 6.2

Nusselt number for laminar fully developed conditions in channels [3]

geometry	a/b	Nu_D	
		uniform surface flux	uniform surface temperature
		4.364	3.657
	1	3.408	2.976
	2	4.123	3.391
	4	5.331	4.439
	8	6.490	5.597
	∞	8.235	7.541
		3.102	2.460

non-circular channels varies along the periphery. However, Table 6.2 gives the average Nusselt number along the periphery. For uniform surface heat flux, surface temperature varies axially and along the periphery. The results shown in Table 6.2 are based on uniform periphery temperature. Note that in all cases the Nusselt number for uniform surface flux is greater than that for uniform surface temperature.

Scaling estimate of the Nusselt number in the fully developed region gives

$$Nu_D \sim 1 \quad \text{(fully developed).} \tag{6.40}$$

Examination of Table 6.2 shows that the Nusselt number ranges from 2.46 to 8.235. Thus scaling provides a reasonable estimate of the Nusselt number.

Example 6.4: Maximum Surface Temperature in an Air Duct

Air is heated in a 4 cm× 4 cm square duct from $40^{\circ}C$ to $120^{\circ}C$. A uniform heat flux of $590\ W/m^2$ is applied at the surface. The mean air velocity is 0.32 m/s. Neglecting entrance effects, determine the maximum surface temperature.

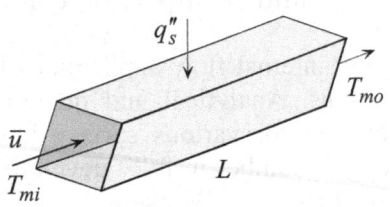

(1) Observations. (i) This is an internal forced convection problem in a square duct. (ii) The surface is heated at uniform flux. (iii) Surface temperature changes along the channel. It reaches a maximum value at the outlet. (iv) The Reynolds number should be checked to determine if the flow is laminar or turbulent. (v) Velocity and temperature profiles become fully developed far away from the inlet. (vi) The heat transfer coefficient is uniform for fully developed flow. (vii) Duct length is unknown. (viii) The fluid is air.

(2) Problem Definition. (i) Find the required length to heat the air to a given temperature and (ii) determine surface temperature at the outlet.

(3) Solution Plan. (i) Since surface flux, mean velocity, duct size, inlet and outlet temperatures are known, application of conservation of energy between the inlet and outlet gives the required duct length. (ii) Check the Reynolds number to determine if the flow is laminar or turbulent. (iii) Apply surface temperature solution for flow through a channel with constant surface flux. (iv) Use Table 6.2 to determine the heat transfer coefficient.

(4) Plan Execution.

(i) **Assumptions.** (1) Continuum, (2) steady state, (2) constant properties, (3) uniform surface heat flux, (5) negligible changes in kinetic and potential energy, (6) negligible axial conduction, and (7) negligible dissipation.

(ii) **Analysis.** Application of conservation of energy between the inlet and outlet gives the required channel length

$$PLq_s'' = mc_p(T_{mo} - T_{mi}),$$ (a)

where

c_p = specific heat, J/kg$-^\circ$C
L = channel length, m
m = mass flow rate, kg/s
P = perimeter, m
q_s'' = surface heat flux = 590 W/m^2
T_{mi} = 40°C
T_{mo} = 120°C

Solving (a) for L

$$L = \frac{mc_p(T_{mo} - T_{mi})}{Pq_s''}.$$ (b)

The mass flow rate and perimeter are given by

$$m = \rho S^2 \bar{u},$$ (c)

$$P = 4S,$$ (d)

where

S = duct side = 0.04 m
$\bar{u}$ = mean flow velocity = 0.32 m/s
ρ = density, kg/m^3

Substituting (c) and (d) into (b)

$$L = \frac{\rho S \bar{u} c_p(T_{mo} - T_{mi})}{4q_s''}.$$ (e)

To determine surface temperature at the outlet, use the solution for surface temperature distribution for channel flow with uniform surface flux, given by equation (6.10)

$$T_s(x) = T_{mi} + q_s'' \left(\frac{Px}{\dot{m} c_p} + \frac{1}{h(x)} \right),$$ (f)

where

$h(x)$ = local heat transfer coefficient, $W/m^2 - {}^\circ C$
$T_s(x)$ = local surface temperature, $^\circ C$
x = distance from inlet of heated section, m

Surface temperature at the outlet, $T_s(x)$, is obtained by setting $x = L$ in (f). Substituting (c) and (d) into (f)

$$T_s(L) = T_{mi} + q_s'' \left(\frac{4L}{\rho S \bar{u} c_p} + \frac{1}{h(L)} \right).$$ (g)

Finally, it remains to determine the heat transfer coefficient at the outlet, $h(L)$. This requires establishing whether the flow is laminar or turbulent. Thus, the Reynolds number should be determined. The Reynolds number for flow through a square channel is defined as

$$Re_{De} = \frac{\bar{u} D_e}{\nu},$$ (h)

where

D_e = equivalent diameter, m
ν = kinematic viscosity, m^2/s

The equivalent diameter for a square channel is defined as

$$D_e = 4\frac{A}{P} = 4\frac{S^2}{4S} = S.$$ (i)

Substituting (i) into (h)

$$Re_{De} = \frac{\bar{u} S}{\nu}.$$ (j)

Properties of air are determined at the mean temperature $\bar{T}_m$ defined as

$$\bar{T}_m = \frac{T_{mi} + T_{mo}}{2}.$$ (k)

Substituting into (k)

$$\overline{T}_m = \frac{(40 + 120)(^\circ C)}{2} = 80^\circ C$$

Properties of air at this temperature are:

$c_p = 1009.5$ J/kg$-^\circ$C
$k = 0.02991$ W/m$-^\circ$C
$Pr = 0.706$
$v = 20.92 \times 10^{-6}$ m^2/s
$\rho = 0.9996$ kg/m^3

Substituting into (j)

$$Re_{De} = \frac{0.32(\text{m/s})0.04(\text{m})}{20.92 \times 10^{-6}(\text{m}^2/\text{s})} = 611.9$$

Since the Reynolds number is smaller than 2300, the flow is laminar. The heat transfer coefficient for fully developed laminar flow through a square channel with uniform surface flux is constant. It is given by equation (6.55) and Table 6.2

$$\overline{Nu}_{D_e} = \frac{\overline{h}D_e}{k} = 3.608, \tag{l}$$

where $h = \overline{h}$. Solving (l) for $\overline{h}$

$$\overline{h} = 3.608\frac{k}{D_e}. \tag{m}$$

(iii) Computations. Substituting numerical values in (e) gives the channel length

$$L = \frac{0.9996(\text{kg/m}^3)0.04(\text{m})0.32(\text{m/s})1009.5(\text{J/kg-}^\circ C)(120 - 40)(^\circ C)}{(4)590(\text{W/m}^2)}$$

$$= 0.4378\,\text{m}$$

To determine surface temperature at the outlet, the heat transfer coefficient is computed using (m)

$$h(L) = \overline{h} = 3.608\frac{0.02991(\text{W/m-}^\circ C)}{0.04(\text{m})} = 2.7 \text{ W/m}^2 - ^\circ C$$

Equation (g) gives surface temperature at the outlet

$$T_s(L) = 40(^\circ C) + 590(W/m^2) \times$$

$$\left(\frac{4(0.4378)(m)}{0.9996(kg/m^3)\,0.04(m)0.32\,(m/s)\,1009.5(J/kg-^\circ C)} + \frac{1}{2.7(W/m^2-^\circ C)} \right)$$

$$T_s(L) = 338.5\,^\circ C$$

(iv) Checking. *Dimensional check*: Computations showed that equations (e), (g), (j), and (m) are dimensionally correct.

Quantitative checks: (i) Alternate approach to determining $T_s(x)$: Application of Newton's law of cooling at the outlet gives

$$q_s'' = h[T_s(L) - T_{mo}]. \tag{n}$$

Solving for $T_s(L)$

$$T_s(L) = T_{mo} + \frac{q_s''}{h} = 120(^\circ C) + \frac{590(W/m^2)}{2.7(W/m^2-^\circ C)} = 338.5\,^\circ C$$

(ii) The value of h is within the range reported in Table 1.1 for forced convection of gases.

Limiting check: If $T_{mo} = T_{mi}$, the required length should be zero. Setting $T_{mo} = T_{mi}$ into (e) gives $L = 0$.

(5) Comments. (i) As long as the outlet is in the fully developed region, surface temperature at the outlet is determined entirely by the local heat transfer coefficient.

(ii) In solving internal forced convection problems it is important to establish if the flow is laminar or turbulent and if it is developing or fully developed.

6.8 Thermal Entrance Region: Laminar Flow through Tubes

6.8.1 Uniform Surface Temperature: Graetz Solution

Consider laminar flow through a tube shown in Fig. 6.8. Fluid enters a heated or cooled section with a fully developed velocity. We neglect axial conduction ($Pe > 100$) and consider the case of uniform surface temperature T_s. It was shown in Chapter 3 that for two-dimensional fully developed flow the radial velocity vanishes

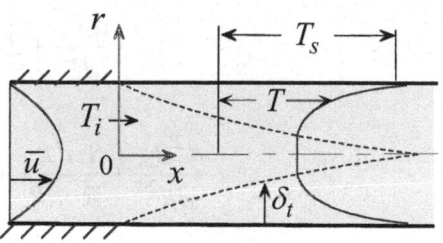

Fig. 6.8

$$v_r = 0. \qquad (3.1)$$

The axial velocity is

$$v_z = \frac{1}{4\mu}\frac{dp}{dz}(r^2 - r_o^2). \qquad (3.12)$$

Using the notation of this chapter, the axial velocity is expressed in dimensionless form as

$$v_x^* = \frac{v_x}{\bar{u}} = 2(1 - R^2). \qquad (6.61)$$

Substituting (3.1) and (6.61) into energy equation (6.31)

$$\frac{1}{2}\left(1 - R^2\right)\frac{\partial\theta}{\partial\xi} = \frac{1}{R}\frac{\partial}{\partial R}\left(R\frac{\partial\theta}{\partial R}\right). \qquad (6.62)$$

The boundary conditions for this case are

$$\frac{\partial\theta(\xi,0)}{\partial R} = 0, \qquad (6.63a)$$

$$\theta(\xi,1) = 0, \qquad (6.63b)$$

$$\theta(0, R) = 1. \qquad (6.63c)$$

Analytic and numerical solutions to this problem have been obtained [4, 5]. The following is a summary of the solution and results. Assume product solution of the form

$$\theta(\xi, R) = X(\xi)\mathcal{R}(R). \qquad (a)$$

Substituting (a) into (6.62), separating variables

$$\frac{dX_n}{d\xi} + 2\lambda_n^2 X_n = 0,$$ (b)

$$\frac{d^2 \mathcal{R}_n}{dR^2} + \frac{1}{R}\frac{d\mathcal{R}_n}{dR} + \lambda_n^2(1 - R^2)\mathcal{R}_n = 0.$$ (c)

where λ_n are the eigenvalues obtained from the boundary conditions. Solution $X_n(\xi)$ to (b) is exponential. However, solution $\mathcal{R}_n(R)$ to (c) is not available in terms of simple tabulated functions. Substituting the solutions to (b) and (c) into (a)

$$\theta(\xi, R) = \sum_{n=0}^{\infty} C_n \mathcal{R}_n(R)\exp(-2\lambda_n^2\xi),$$ (6.64)

where C_n is constant. With the temperature distribution given in (6.64), surface heat flux, mean temperature, local and average Nusselt numbers can be determined. Surface heat flux is given by

$$q_s''(\xi) = \frac{k}{r_o}\left(T_s - T_i\right)\frac{\partial\theta(\xi,1)}{\partial R}.$$ (6.22)

Surface temperature gradient $\partial\theta(\xi,1)/\partial R$ is obtained by differentiating (6.64)

$$\frac{\partial\theta(\xi,1)}{\partial R} = \sum_{n=0}^{\infty} C_n \frac{d\mathcal{R}_n(1)}{dR}\exp(-2\lambda_n^2\xi).$$ (d)

Defining the constant G_n as

$$G_n = -\frac{C_n}{2}\frac{d\mathcal{R}_n(1)}{dR}.$$ (e)

Substituting (d) and (e) into (6.22)

$$q_s''(\xi) = -\frac{2k}{r_o}\left(T_s - T_i\right)\sum_{n=0}^{\infty} G_n \exp\{-2\lambda_n^2\xi\}.$$ (6.65)

The local Nusselt number is given by

$$Nu(\xi) = \frac{-2}{\theta_m(\xi)} \frac{\partial \theta(\xi,1)}{\partial R}, \tag{6.27}$$

where the gradient $\partial \theta(\xi,1)/\partial R$ is given in (d). The local mean temperature $\theta_m(\xi)$ is obtained by substituting (6.61) and (6.64) into (6.32b), integrating by parts and using (e)

$$\theta_m(\xi) = 8 \sum_{n=0}^{\infty} \frac{G_n}{\lambda_n^2} \exp(-2\lambda_n^2 \xi). \tag{6.66}$$

Substituting (d), (e) and (6.66) into (6.27)

$$Nu(\xi) = \frac{\displaystyle\sum_{n=0}^{\infty} G_n \exp(-2\lambda_n^2 \xi)}{\displaystyle 2\sum_{n=0}^{\infty} \frac{G_n}{\lambda_n^2} \exp(-2\lambda_n^2 \xi)}. \tag{66.7}$$

The average Nusselt number for a tube of length ξ is defined as

$$\overline{Nu}(\xi) = \frac{\overline{h}(\xi)D}{k}. \tag{f}$$

where $\overline{h}(\xi)$ is determined by integrating the local heat transfer coefficient along a tube of length ξ. A simpler approach is to use equation (6.13) which contains the average heat transfer coefficient

$$T_m(x) = T_s + (T_{mi} - T_s)\exp[-\frac{P\overline{h}}{mc_p}x]. \tag{6.13}$$

Solving (6.13) for $\overline{h}$

$$\overline{h} = -\frac{mc_p}{Px}\ln\frac{T_m(x) - T_s}{T_{mi} - T_s}. \tag{g}$$

Substituting (g) into (f), noting that $m = \rho \bar{u} \pi D^2 / 4$, $P = \pi D$, and using the definitions of ξ, Re_D and θ_m in (6.21) and (6.25), gives

$$\overline{Nu}(\xi) = -\frac{1}{4\xi} \ln \theta_m(\xi) . \qquad (6.68)$$

The constants λ_n and G_n are needed to compute $q_s''(\xi)$, $\theta_m(\xi)$, $Nu(\xi)$ and $\overline{Nu}(\xi)$ in equations (6.65)-(6.68). Table 6.3 [4] lists values of λ_n and G_n for $0 \le n \le 10$. Equations (6.67) and (6.68) are used to plot and tabulate the local and average Nusselt numbers. Table 6.4 [5] gives $Nu(\xi)$ and $\overline{Nu}(\xi)$ at selected values of the axial distance ξ. Fig. 6.9 gives the variation of $Nu(\xi)$ and $\overline{Nu}(\xi)$ along a tube.

Table 6.3

Uniform surface temperature [4]

n	λ_n	G_n
0	2.70436	0.74877
1	6.67903	0.54383
2	10.67338	0.46286
3	14.67108	0.41542
4	18.66987	0.38292
5	22.66914	0.35869
6	26.66866	0.33962
7	30.66832	0.32406
8	34.66807	0.31101
9	38.66788	0.29984
10	42.66773	0.29012

Table 6.4

Local and average Nusselt number for tube at uniform surface temperature [5]

$\xi = \dfrac{x/D}{Re_D Pr}$	$Nu(\xi)$	$\overline{Nu}(\xi)$
0	∞	∞
0.0005	12.8	19.29
0.002	8.03	12.09
0.005	6.00	8.92
0.02	4.17	5.81
0.04	3.77	4.86
0.05	3.71	4.64
0.1	3.66	4.15
∞	3.66	3.66

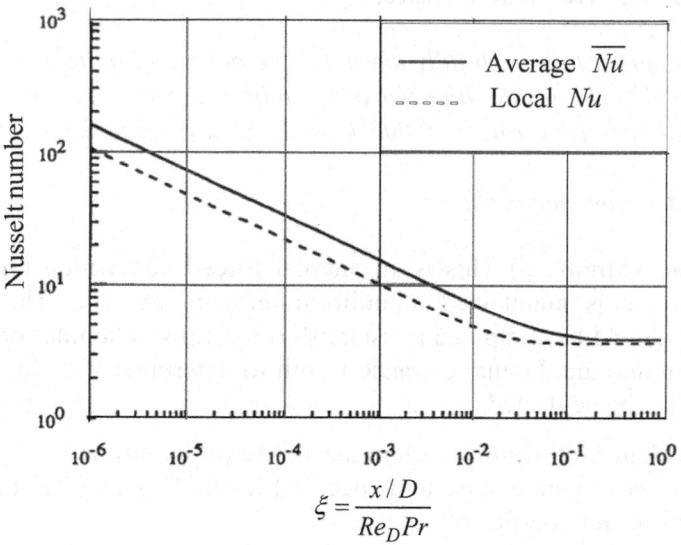

Fig. 6.9 Local and average Nusselt number for tube at uniform surface temepratu re [4]

The following observations are made:

(1) The local and average Nusselt numbers and heat transfer coefficient decrease as the distance ξ from the entrance is increased.

(2) At any given location ξ, the average Nusselt number is greater than the local Nusselt number.

(3) The Nusselt number reaches an asymptotic value of 3.657 at $\xi \approx 0.05$. As was shown in Section 6.7.4, this corresponds to the Nusselt number in the fully developed region. Thus

$$Nu(\infty) = 3.657.$$ (6.69)

(4) Surface heat flux and heat transfer coefficient depend on fluid properties such as thermal conductivity k, kinematic viscosity v and Prandtl number Pr. Properties in channel flow are evaluated at the average of inlet and outlet mean temperatures $\overline{T}_m$, defined as

$$\overline{T}_m = (T_{mi} + T_{mo})/2.$$ (6.70)

Note that in problems where the outlet temperature is not known a priori and must be determined as part of the solution, a trial and error procedure is used.

Example 6.5: Hot Water Heater

Water enters a tube with fully developed velocity and uniform temperature $T_i = 25°\,C$. The inside diameter of the tube is 1.5 cm and its length is 80 cm. The mass flow rate is 0.002 kg/s. It is desired to heat the water to $75°\,C$ by maintaining the surface at uniform temperature. Determine the required surface temperature.

(1) Observations. (i) This is an internal forced convection problem. (ii) The surface is maintained at uniform temperature. (iii) The Reynolds number should be computed to establish if the flow is laminar or turbulent. (iv) Compute the thermal entrance length to determine if thermal entrance effects can be neglected.

(2) Problem Definition. Determine the required surface temperature to raise the mean temperature to a specified level. This requires determining the heat transfer coefficient.

(3) Solution Plan. Use the analysis of flow in tubes at uniform surface temperature. Compute the Reynolds number to establish if the flow is laminar or turbulent. Compute the thermal entrance length to determine if entrance or fully developed analysis is required.

(4) Plan Execution.

   (i) Assumptions. (1) Continuum, (2) steady state, (3) constant properties, (4) uniform surface temperature, (5) negligible changes in kinetic and potential energy, (6) negligible axial conduction, (7) fully developed velocity, and (8) negligible dissipation.

   (ii) Analysis. For flow through a tube at uniform surface temperature, conservation of energy and Newton's law of cooling lead to equation (6.13)

$$T_m(x) = T_s + (T_{mi} - T_s)\exp[-\frac{P\overline{h}}{m c_p}x], \qquad (6.13)$$

where

   $\overline{h}$ = average heat transfer coefficient for a tube of length x, $\text{W/m}^2 - °\text{C}$
   m = mass flow rate = $0.002\,\text{kg/s}$
   T_{mi} = mean inlet temperature = $25°\,\text{C}$
   T_{mo} = mean outlet temperature = $75°\,\text{C}$

Applying (6.13) at the outlet, $x = L$, and solving for T_s

$$T_s = \frac{\left[T_{mi} - T_m(L)\exp(\bar{P}\bar{h}L/mc_p)\right]}{1-\exp(\bar{P}\bar{h}L/mc_p)}. \tag{a}$$

To compute T_s using (a), it is necessary to determine c_p, P, and $\bar{h}$. Water properties are determined at the mean temperature $\bar{T}_m$, defined as

$$\bar{T}_m = \frac{T_{mi}+T_{mo}}{2}. \tag{b}$$

The perimeter P is

$$P = \pi D, \tag{c}$$

where

D = inside tube diameter = 1.5 cm = 0.015 m

The determination of $\bar{h}$ requires computing the Reynolds number to establish if the flow is laminar or turbulent, and computing the thermal entrance length to determine if it is important. The Reynolds number is

$$Re_D = \frac{\bar{u}D}{\nu}, \tag{d}$$

where

ν = kinematic viscosity, m²/s
$\bar{u}$ = mean velocity, m/s

The flow rate gives the mean velocity

$$\bar{u} = \frac{4m}{\rho\pi D^2}, \tag{e}$$

where ρ is density. To determine water properties, (b) is used to compute $\bar{T}_m$

$$\bar{T}_m = \frac{(20+80)(^\circ C)}{2} = 50^\circ C$$

Properties of water at this temperature are

c_p = 4182 J/kg-°C
k = 0.6405 W/m-°C
Pr = 3.57
ν = 0.5537×10⁻⁶ m²/s
ρ = 988 kg/m³

Substituting into (e)

$$\bar{u} = \frac{4(0.002)(\text{kg/s})}{988(\text{kg/m}^3)\pi(0.015)^2(\text{m}^2)} = 0.01146 \text{ m/s}$$

Equation (d) gives

$$Re_D = \frac{0.01146(\text{m/s})0.015(\text{m})}{0.5537 \times 10^{-6}(\text{m}^2/\text{s})} = 310.5$$

Since the Reynolds number is less than 2300, the flow is laminar. The next step is to compute the thermal entrance length L_t. For laminar flow through channels, the thermal entrance length is given by (6.6)

$$\frac{L_t}{D} = C_t Pr Re_D, \qquad (6.6)$$

where

C_t = thermal entrance length coefficient = 0.033, (Table 6.1)
L_t = thermal entrance length, m

Substituting numerical values into (6.6)

$$L_t = 0.033 \times 0.015 \text{ (m)} \times 310.5 \times 3.57 = 0.55 \text{ m}$$

Since L_t is not negligible compared to tube length L, it follows that temperature entrance effects must be taken into consideration in determining $\bar{h}$. For laminar flow in the entrance region of a tube at fully developed velocity profile and uniform surface temperature, Graetz solution gives $\bar{h}$. Fig. 6.9 and Table 6.4 give the average Nusselt number $\overline{Nu}$ as a function of the dimensionless axial distance ξ, defined as

$$\xi = \frac{x/D}{Re_D Pr}. \qquad (f)$$

The average heat transfer coefficient $\bar{h}$ is given by

$$\bar{h} = \frac{k}{D}\overline{Nu}. \qquad (g)$$

(iii) Computation. Evaluating ξ at $x = L$

$$\xi = \frac{0.8(m)/0.015(m)}{310.5 \times 3.57} = 0.0481$$

At $\xi = 0.481$ Fig. 6.9 gives

$$\overline{Nu} \approx 4.6$$

Substituting into (g)

$$\overline{h} = \frac{0.6405(W/m-^{\circ}C)}{0.015(m)} 4.6 = 196.4 W/m^{2} -^{\circ}C$$

Equation (a) gives the required surface temperature. First, the exponent of the exponential is calculated

$$\frac{P\overline{h}L}{mc_p} = \frac{\pi(0.015)(m)(196.4(W/m^2-^{\circ}C)0.8(m)}{0.002(kg/s)4182(J/kg-^{\circ}C)} = 0.88524$$

Substituting into (a)

$$T_s = \frac{1}{1-\exp(0.88524)}\left[25(^{\circ}C)-75(^{\circ}C)\exp(0.88524)\right] = 110.1^{\circ}C$$

(iv) Checking. *Dimensional check*: (i) Computations showed that equations (a), (e), (g) and (6.6) are dimensionally consistent. (ii) The Reynolds number and the exponent of the exponential are dimensionless.

Limiting checks: (i) For the special case of $T_{mi} = T(L)$, the required surface temperature should be the same as inlet temperature. Setting $T_{mi} = T(L)$ in (a) gives $T_s = T_{mi}$.

(ii) The required surface temperature should be infinite if the length is zero. Setting $L = 0$ in (a) gives $T_s = \infty$.

Quantitative checks: (i) An approximate check can be made using conservation of energy and Newton's law of cooling. Conservation of energy is applied to the water between inlet and outlet

$$\textit{Energy added at the surface} = \textit{Energy gained by water.} \qquad (h)$$

Assuming that water temperature in the tube is uniform equal to $\overline{T}_m$, Newton's law of cooling gives

$$\text{Energy added at surface} = \bar{h}\pi DL(T_s - \bar{T}_m). \tag{i}$$

Neglecting axial conduction and changes in kinetic and potential energy, energy gained by water is

$$\text{Energy gained by water} = mc_p(T_{mo} - T_{mi}). \tag{j}$$

Substituting (i) and (j) into (h) and solving the resulting equation for T_s

$$T_s = \bar{T}_m + \frac{mc_p(T_{mo} - T_{mi})}{\bar{h}\pi DL}. \tag{k}$$

Equation (k) gives

$$T_s = 50(^\circ C) + \frac{0.002(\text{kg/s})4182(\text{J/kg}-^\circ C)(75-25)(^\circ C)}{196.4(\text{W/m}^2-^\circ C)\pi(0.015)(\text{m})(0.8)(\text{m})} = 106.5^\circ C$$

This is in reasonable agreement with the more exact answer obtained above.

(ii) The value of $\bar{h}$ is within the range listed in Table 1.1 for forced convection of liquids.

(5) Comments. (i) Using Fig. 6.9 to determine $\bar{h}$ introduces a small error.

(ii) If entrance effects are neglected and the flow is assumed fully developed throughout, the corresponding Nusselt number is 3.657. Using this value gives $\bar{h} = 156.3\text{W/m}^2-^\circ C$ and $T_s = 121\,^\circ C$.

6.8.2 Uniform Surface Heat Flux

We repeat Graetz entrance problem replacing the uniform surface temperature with uniform surface heat flux, as shown in Fig. 6.10. The fluid enters the heated or cooled section with fully developed velocity. The governing

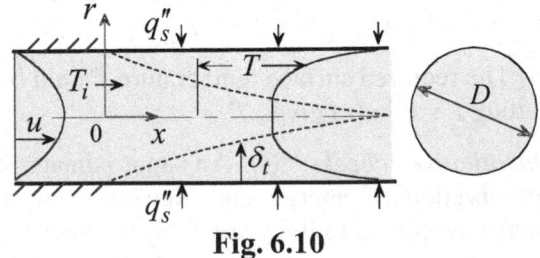

Fig. 6.10

energy equation is the same as that for the Graetz problem, given by (6.62)

$$\frac{1}{2}\left(1 - R^2\right)\frac{\partial \theta}{\partial \xi} = \frac{1}{R}\frac{\partial}{\partial R}\left(R\frac{\partial \theta}{\partial R}\right). \tag{6.62}$$

The boundary conditions, expressed in dimensionless form are:

$$\frac{\partial \theta(\xi,0)}{\partial R} = 0, \tag{6.71a}$$

$$\frac{\partial \theta(\xi,1)}{\partial R} = \frac{q_s'' r_o}{k(T_i - T_s)}, \tag{6.71b}$$

$$\theta(0,R) = 1. \tag{6.71c}$$

Analytic solution based on separation of variables as well as numerical solution to this problem is available [4]. The solution for the local Nusselt number is

	Table 6.5	
	Uniform surface flux [4]	
n	β_n^2	A_n
1	25.6796	0.198722
2	83.8618	0.069257
3	174.1667	0.036521
4	296.5363	0.023014
5	450.9472	0.016030
6	637.3874	0.011906
7	855.8495	0.009249
8	1106.3290	0.007427
9	1388.8226	0.006117
10	1703.3279	0.005141

$$Nu(\xi) = \frac{hx}{k} = \left[\frac{11}{48} + \frac{1}{2}\sum_{n=1}^{\infty} A_n \exp(-2\beta_n^2 \xi)\right]^{-1}. \tag{6.72}$$

The eigenvalues β_n^2 and the constant A_n are listed in Table 6.5 [4]. The average Nusselt number is given by

$$\overline{Nu}(\xi) = \frac{hx}{k} = \left[\frac{11}{48} + \frac{1}{2}\sum_{n=1}^{\infty} A_n \frac{1 - \exp(-2\beta_n^2 \xi)}{2\beta_n^2 \xi}\right]^{-1}. \tag{6.73}$$

The limiting case corresponding to $\xi = \infty$ gives the Nusselt number in the fully developed region. Setting $\xi = \infty$ in (6.72) or (6.73) gives

$$Nu(\infty) = \frac{48}{11} = 4.364. \tag{6.74}$$

This agrees with the solution of the fully developed region given in equation (6.55). The solutions to the local and average Nusselt numbers are presented graphically in Fig. 6.11 [4].

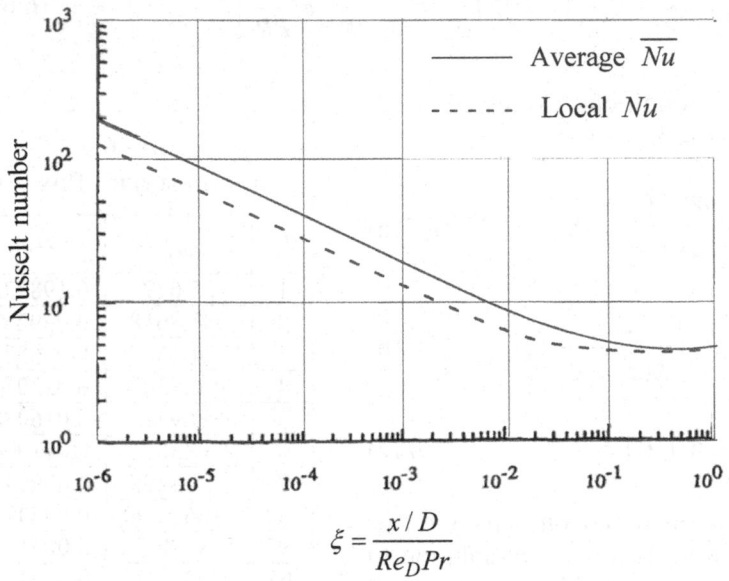

$$\xi = \frac{x / D}{Re_D Pr}$$

Fig. 6.11 Local and average Nusselt number for
tube at uniform surface heat flux [4]

REFERENCES

[1] Ozisik, M. N., *Heat Transfer*, McGraw-Hill, New York 1985.

[2] Bejan, A. Force Convection: Internal Flows, in *Heat Transfer Handbook*, A. Bejan and A. D. Kraus, eds., Wiley, New York, 2003.

[3] Kays, W.M. and M.E. Crawford, *Convection Heat Transfer, 3rd edition*, McGraw-Hill, 1993.

[4] Ebadian, M.A. and Z.F. Dong, Forced Convection, Internal Flow in Ducts, in *Handbook of Heat Transfer*, Third Edition, W.M. Rohsenow, J.P. Hartnett and Y.I. Cho, eds., McGraw-Hill, 1998.

[5] Shah, R.K. and M.S. Bhatti, "Laminar Convection Heat Transfer in Ducts," in, *Handbook of Single-Phase Convective Heat Transfer*, S. Kakac, R.K. Shah and W. Aung, eds., Wiley, 1987.

PROBLEMS

6.1   Use scaling to determine the ratio L_t / L_h. Compare scaling estimates with exact solutions.

6.2   Use scaling to estimate the hydrodynamic and thermal entrance lengths for the flow of air in a $3\,\text{cm} \times 3\,\text{cm}$ square duct. The mean velocity is 0.8 m/s. Compare scaling estimates with exact solutions. Evaluate properties at $50^\circ\,\text{C}$.

6.3   Far away from the entrance of a channel the velocity and temperature become fully developed. It can be shown that under such conditions the Nusselt number becomes constant. Consider air flowing with a mean velocity of 2 m/s through a long square duct of a 2.0 cm. The mean temperature at a section in the fully developed region is 40°C. The surface of the tube is maintained at a uniform temperature of 140°C. What is the length of the tube section needed for the mean temperature to reach 110°C? The Nusselt number for this case is given by

$$Nu_{D_e} = 2.976.$$

The equivalent diameter D_e is defined as $D_e = 4A / P$, where A is the flow area and P is the duct perimeter.

6.4   A fluid is heated in a long tube with uniform surface flux. The resulting surface temperature distribution is found to be higher than design specification. Two suggestions are made for lowering surface temperature without changing surface flux or flow rate: (1) increasing the diameter, (2) decreasing the diameter. You are asked to determine which suggestion to follow. The flow is laminar and fully developed. Under such conditions the Nusselt number is given by

$$Nu_D = 4.364.$$

6.5   Two identical tubes are heated with the same uniform flux at their surfaces. Air flows through one tube while water flows at the same rate through the other. The mean inlet temperature for both tubes is the same. Which tube will have a higher surface temperature distribution? Assume laminar flow and neglect entrance effects. For this case the Nusselt number is given by

$$Nu_D = 4.364.$$

6.6   Water flows through a tube with a mean velocity of 0.2 m/s. The mean inlet temperature is 20°C and the inside diameter of the tube is 0.5 cm. The water is heated to 80°C with uniform surface heat flux of 0.6 W/cm². Determine surface temperature at the outlet. If entrance effects can be neglected the Nusselt number for fully developed flow is constant given by

$$Nu_D = 4.364.$$

Is it justifiable to neglect entrance effects?

6.7   Fluid flows with a mean axial velocity $\overline{u}$ in a tube of diameter D. The mean inlet temperature is T_{mi}. The surface is maintained at uniform temperature T_s. Show that the average Nusselt number for a tube of length L is given by

$$\overline{Nu}_L = \frac{Re_D Pr}{4} \ln \frac{T_{mi} - T_s}{T_m(L) - T_s},$$

where $\overline{Nu}_L = \dfrac{\overline{h}_L L}{k}$, $Re_D = \dfrac{\overline{u}D}{\nu}$ and $\overline{h}_L$ is the average heat transfer coefficient over the length L.

6.8   Water flows through a $0.75\,\text{cm} \times 0.75\,\text{cm}$ square duct with a mean velocity of 0.12 m/s. The duct is heated with a uniform surface flux of 0.25 W/cm². The mean inlet temperature is 25°C. The maximum allowable surface temperature is 95°C. Justify neglecting entrance effects and determine the maximum outlet mean temperature.

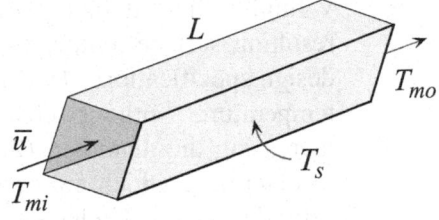

6.9   Two experiments were conducted on fully developed laminar flow through a tube. In both experiments surface temperature is $180°C$ and the mean inlet temperature is $20°C$. The mean outlet temperature for the first experiment is found to be $120°C$. In the second experiment the flow rate is reduced by a factor of 2. All other conditions remained the same. Determine:

[a] The outlet temperature of the second experiment.

[b] The ratio of heat transfer rate for the two experiments.

6.10 A long rectangular duct with a 4cm × 8cm cross section is used to heat air from −19.6°C to 339.6°C. The mean velocity in the duct is 0.2 m/s and surface temperature is 340 °C. Determine the required duct length. Is neglecting entrance effects justified?

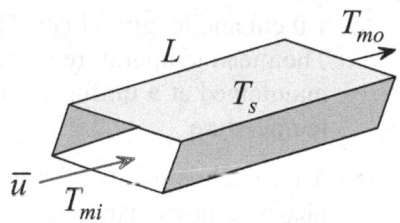

6.11 A rectangular duct with inside dimensions of 2 cm × 4 cm is used to heat water from $25\,^{\circ}C$ to $115\,^{\circ}C$. The mean water velocity is 0.018 m/s. The surface of the duct is maintained at $145\,^{\circ}C$. Determine the required duct length. Assume fully developed flow conditions throughout.

6.12 Air is heated in a 4cm × 4cm square duct at uniform surface flux of 590 W/m^2. The mean air velocity is 0.32 m/s. At a section far away from the inlet the mean temperature is $40\,^{\circ}C$. The mean outlet temperature is $120\,^{\circ}C$. Determine the required length and maximum surface temperature.

6.13 Consider fully developed laminar flow in two tubes having the same length. The flow rate, fluid, inlet temperature and surface temperature are the same for both tubes. However, the diameter of one tube is double that of the other. Determine the ratio of the heat transfer rate from the two tubes.

6.14 To evaluate the accuracy of scaling prediction of the thermal entrance length and Nusselt number, compare scaling estimates with the exact results of Graetz solution for flow through tubes.

6.15 Use scaling to estimate the heat transfer coefficient for plasma at a distance of 9 cm from the entrance of a vessel. The mean plasma velocity is 0.042 m/s and the vessel diameter is 2.2 mm. Properties of plasma are:

$c_p = 3900\ \text{J/kg-}^{\circ}C$, $k = 0.5\ \text{W/m-}^{\circ}C$, $v = 0.94 \times 10^{-6}\ \text{m}^2/\text{s}$, $\rho = 1040\ \text{kg/m}^3$

6.16 Air flows with a fully developed velocity through a tube of inside diameter 2.0 cm. The mean velocity of 1.2 m/s. The surface is maintained at a uniform temperature of 90°C. Inlet temperature is uniform equal 30 °C. Determine the length of tube needed to increase the mean temperature to 70°C.

6.17 Air flows with a mean velocity of 2 m/s through a tube of diameter 1.0 cm and length 14 cm. The velocity is fully developed throughout. The mean temperature at the inlet is 35°C. The surface of the tube is maintained at a uniform temperature of 130°C. Determine the outlet temperature.

6.18 A research apparatus for a pharmaceutical laboratory requires heating plasma in a tube 0.5 cm in diameter. The tube is heated by uniformly wrapping an electric element over its

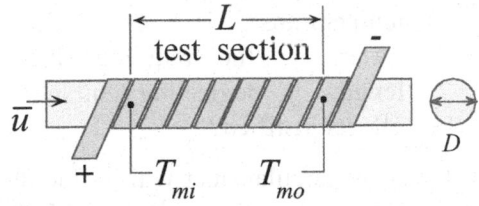

surface. This arrangement provides uniform surface heat flux. The plasma is monitored in a 15 cm long section. The mean inlet temperature to this section is $18°C$ and the mean velocity is 0.025 m/s. The maximum allowable temperature is $42°C$. You are asked to provide the designer of the apparatus with the outlet temperature and required power corresponding to the maximum temperature. Assume fully developed inlet velocity. Properties of plasma are:

$$c_p = 3900 \text{ J/kg}-°C, \ \nu = 0.94 \times 10^{-6} \text{ m}^2/\text{s}, \ \rho = 1040 \text{ kg/m}^3.$$

6.19 An experiment is designed to investigate heat transfer in rectangular ducts at uniform surface temperature. One method for providing heating at uniform surface temperature is based on wrapping a set of electric elements around the surface. Power supply to each element is individually adjusted to provide uniform surface temperature. This experiment uses air flowing in a $4\text{cm} \times 8\text{cm}$ rectangular duct 32 cm long. The air is to be heated from $22°C$ to $98°C$. The velocity is fully developed with a mean value of 0.15 m/s. Your task is to provide the designer of the experiment with the heat flux distribution along the surface. This data is needed to determine the power supplied to the individual elements.

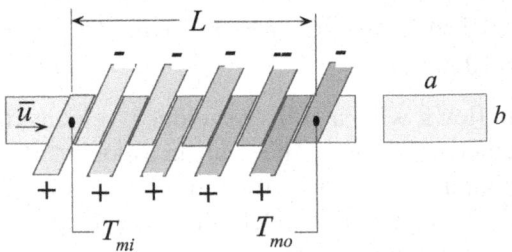

7

Free Convection

7.1 Introduction

Free convection is encountered in many situations. In fact, it is always present as long as fluid temperature is not uniform and there is an acceleration field such as gravity. In some applications, free convection heat transfer is small compared to other modes and therefore may be neglected. In others it is the dominant mechanism for heat transfer. There are situations where it is desirable to suppress free convection, such as in heat loss from steam pipes, windows, and solar collectors. On the other hand, one seeks to enhance the transfer of heat by free convection in cooling microelectronic components and packages.

7.2 Features and Parameters of Free Convection

(1) Driving Force. Fluid motion in free convection is driven by natural forces. In general, two conditions are required for fluids to be set in motion in free convection: (1) the presence of an *acceleration field*, and (2) a *density gradient* in the fluid. The most common acceleration field is *gravity*. Since all fluids undergo changes in density due to temperature changes, it follows that a *temperature gradient* will set up a *density gradient*. However, there are cases in which the presence of a density gradient in an acceleration field does not result in fluid motion. An example is a fluid which is contained between two horizontal plates with the upper plate at a higher temperature than the lower plate.

(2) Governing Parameters. Two parameters play a key role in the determination of the Nusselt number in free convection: (1) the *Grashof number*, and (2) the *Prandtl number*. The Grashof number is defined as

$$\text{Grashof number} = Gr_L = \frac{\beta g (T_s - T_\infty) L^3}{\nu^2}, \qquad (7.1)$$

where L is a characteristic dimension of the body. For a horizontal cylinder it is the diameter and for a vertical plate it is a dimension in the vertical direction. β is a fluid property defined in equation (2.16). It is called the *coefficient of thermal expansion,* also known as the *compressibility factor.* For ideal gases it is given by

$$\beta = \frac{1}{T}, \quad \text{for ideal gas}, \qquad (2.21)$$

where T is in absolute degrees. In some solutions, the Grashof number appears multiplied by the Prandtl number. This dimensionless product is called the *Rayleigh number*, defined as

$$Ra_L = Gr_L Pr = \frac{\beta g (T_s - T_\infty) L^3}{\nu^2} Pr = \frac{\beta g (T_s - T_\infty) L^3}{\nu \alpha}, \qquad (7.2)$$

where α is *thermal diffusivity*.

(3) Boundary Layer. As with forced convection, viscous and thermal boundary layers do exist in free convection. Furthermore the flow can be laminar, turbulent, or mixed. Boundary layer approximations for free convection are valid for $Ra_x > 10^4$.

(4) Transition from Laminar to Turbulent Flow. The criterion for transition from laminar to turbulent flow is expressed in terms of the Grashof or Rayleigh number. For vertical plates the *transition Rayleigh number*, Ra_{x_t}, is given by

$$Ra_{x_t} \approx 10^9. \qquad (7.3)$$

(5) External vs. Enclosure Free Convection. It is convenient to classify free convection as (i) external free convection, and (ii) enclosure free convection. In external free convection a surface is immersed in a fluid of infinite extent. Examples include free convection over vertical plates, horizontal cylinders, and spheres. Enclosure free convection takes place inside closed volumetric regions such as rectangular confines, concentric cylinders, and concentric spheres.

(6) Analytic Solutions. Since the velocity and temperature fields are coupled in free convection, analytic solutions require the simultaneous integration of the continuity, momentum, and energy equations. Even for the simplest case of laminar free convection over an isothermal vertical plate, the mathematical analysis is not elementary and results are obtained through numerical integration.

7.3 Governing Equations

The following approximations are used in the analysis of free convection:

(1) Density is assumed constant except in evaluating gravity forces.

(2) The Boussinesq approximation which relates density change to temperature change is used in formulating buoyancy force in the momentum equation.

(3) Dissipation effect is neglected in the energy equation.

Considering steady state, laminar, two-dimensional flow with constant properties, the continuity, momentum, and energy equations are obtained from equations (2.2), (2.29) and (2.19), respectively

$$\frac{\partial u}{\partial x} + \frac{\partial v}{\partial y} = 0 ,$$

$$(7.4)$$

$$u\frac{\partial u}{\partial x} + v\frac{\partial u}{\partial y} = \beta \, g(T - T_\infty) - \frac{1}{\rho_\infty}\frac{\partial}{\partial x}(p - p_\infty) + v\left(\frac{\partial^2 u}{\partial x^2} + \frac{\partial^2 u}{\partial y^2}\right), \quad (7.5)$$

$$u\frac{\partial v}{\partial x} + v\frac{\partial v}{\partial y} = -\frac{1}{\rho_\infty}\frac{\partial}{\partial y}(p - p_\infty) + v\left(\frac{\partial^2 v}{\partial x^2} + \frac{\partial^2 v}{\partial y^2}\right), \quad (7.6)$$

$$u\frac{\partial T}{\partial x} + v\frac{\partial T}{\partial y} = \alpha\left(\frac{\partial^2 T}{\partial x^2} + \frac{\partial^2 T}{\partial y^2}\right). \quad (7.7)$$

In equation (7.5) gravity points in the negative x-direction, as shown in Fig. 7.1. Additional simplifications will be introduced in the above equations based on boundary layer approximation.

7.3.1 Boundary Layer Equations

Boundary layer approximations used to simplify the governing equations in forced convection are applied to free convection. Fig. 7.1 shows the viscous and thermal boundary layers over a vertical plate. In boundary layer flow, the y-component of the Navier-Stokes equations, (7.6), reduces to

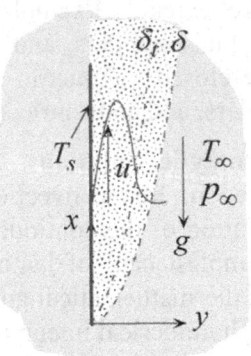

Fig. 7.1

$$\frac{\partial}{\partial y}(p - p_\infty) = 0. \tag{a}$$

Neglecting ambient pressure variation with distance x, it follows that

$$\frac{\partial p}{\partial x} = 0. \tag{b}$$

Furthermore, for boundary layer flow

$$\frac{\partial^2 u}{\partial x^2} << \frac{\partial^2 u}{\partial y^2}. \tag{c}$$

Thus the x-component of the Navier-Stokes equations simplifies to

$$u\frac{\partial u}{\partial x} + v\frac{\partial u}{\partial y} = \beta g(T - T_\infty) + v\frac{\partial^2 u}{\partial y^2}. \tag{7.8}$$

Similarly axial conduction is neglected compared to normal conduction

$$\frac{\partial^2 T}{\partial x^2} << \frac{\partial^2 T}{\partial y^2}. \tag{d}$$

Substituting (d) into energy equation (7.7)

$$u\frac{\partial T}{\partial x} + v\frac{\partial T}{\partial y} = \alpha\frac{\partial^2 T}{\partial y^2}. \tag{7.9}$$

Thus the governing equations for laminar boundary layer free convection are: continuity equation (7.4), x-momentum (7.8), and energy equation (7.9). These equations contain three unknowns: u, v, and T. However, it should be noted that the momentum and energy are coupled since both contain the variables u, v, and T.

7.4 Laminar Free Convection over a Vertical Plate: Uniform Surface Temperature

Consider the vertical plate at uniform temperature T_s shown in Fig. 7.1. The plate is submerged in an infinite fluid at temperature T_∞. Of primary interest is the velocity and temperature distribution in the fluid adjacent to the plate.

7.4.1 Assumptions. (1) Continuum, (2) Newtonian, (3) steady state, (4) laminar flow, (5) two-dimensional, (6) constant properties, (7) Boussinesq approximation, (8) uniform surface temperature, (9) uniform ambient temperature, (10) vertical plate, and (11) negligible dissipation.

7.4.2 Governing Equations. Based on the above assumptions the governing equations are: continuity (7.4), momentum (7.8), and energy (7.9)

$$\frac{\partial u}{\partial x} + \frac{\partial v}{\partial y} = 0, \tag{7.4}$$

$$u\frac{\partial u}{\partial x} + v\frac{\partial u}{\partial y} = \beta g(T - T_\infty) + v\frac{\partial^2 u}{\partial y^2}, \tag{7.8}$$

$$u\frac{\partial \theta}{\partial x} + v\frac{\partial \theta}{\partial y} = \alpha\frac{\partial^2 \theta}{\partial y^2}, \tag{7.10}$$

where θ is a dimensionless temperature defined as

$$\theta = \frac{T - T_\infty}{T_s - T_\infty}. \tag{7.11}$$

7.4.3 Boundary Conditions. The boundary conditions on velocity and temperature are:

Velocity:

(1) $u(x,0) = 0$,

(2) $v(x,0) = 0$,

(3) $u(x,\infty) = 0$,

(4) $u(0,y) = 0$.

Temperature:

(5) $\theta(x,0) = 1$,

(6) $\theta(x,\infty) = 0$,

(7) $\theta(0,y) = 0$.

7.4.4 Similarity Transformation [1]. Equations (7.4), (7.8), and (7.10) are solved simultaneously using the similarity method to transform the three partial differential equations to two ordinary differential equations. The resulting ordinary differential equations are solved numerically. The appropriate similarity variable $\eta(x,y)$ for this case takes the form

$$\eta(x,y) = C\frac{y}{x^{1/4}}, \tag{7.12}$$

where

$$C = \left[\frac{\beta g(T_s - T_\infty)}{4\nu^2}\right]^{\frac{1}{4}}. \tag{7.13}$$

Substituting (7.13) into (7.12)

$$\eta = \left(\frac{Gr_x}{4}\right)^{1/4}\frac{y}{x}, \tag{7.14}$$

where the local Grashof number based on (7.1) is defined as

$$Gr_x = \frac{\beta g(T_s - T_\infty)x^3}{\nu^2}. \tag{7.15}$$

We postulate that the dimensionless temperature θ depends on η. That is

$$\theta(x,y) = \theta(\eta). \tag{7.16}$$

Continuity equation (7.4) is satisfied by introducing a stream function ψ which gives the velocity components u and v as

$$u = \frac{\partial \psi}{\partial y}, \tag{7.17}$$

and

$$v = -\frac{\partial \psi}{\partial x}. \tag{7.18}$$

Using the stream function of Blasius solution for forced convection over a flat plate as a guide, the stream function for this problem is given by

$$\psi = 4v \left(\frac{Gr_x}{4} \right)^{\frac{1}{4}} \xi(\eta), \tag{7.19}$$

where $\xi(\eta)$ is an unknown function to be determined. Introducing (7.19) into (7.17) and (7.18) gives

$$u = 2v \frac{\sqrt{Gr_x}}{x} \frac{d\xi}{d\eta}, \tag{7.20}$$

and

$$v = \frac{v}{(4)^{1/4}} \frac{(Gr_x)^{1/4}}{x} \left[\eta \frac{d\xi}{d\eta} - 3\xi \right]. \tag{7.21}$$

Substituting (7.20) and (7.21) into (7.8) and (7.10) and using (7.11) and (7.16), gives

$$\frac{d^3\xi}{d\eta^3} + 3\xi \frac{d^2\xi}{d\eta^2} - 2 \left(\frac{d\xi}{d\eta} \right)^2 + \theta = 0, \tag{7.22}$$

$$\frac{d^2\theta}{d\eta^2} + 3Pr\xi \frac{d\theta}{d\eta} = 0. \tag{7.23}$$

Note that the original variables x and y do not appear in the transformed equations (7.22) and (7.23). They are replaced by the single independent variable η.

Using (7.14), (7.16), (7.20), and (7.21), the four boundary conditions on velocity and three conditions on temperature transform to:

Velocity:

(1) $\dfrac{d\xi(0)}{d\eta} = 0$,

(2) $\xi(0) = 0$,

(3) $\dfrac{d\xi(\infty)}{d\eta} = 0$,

(4) $\dfrac{d\xi(\infty)}{d\eta} = 0$.

Temperature:

(1) $\theta(0) = 1$,

(2) $\theta(\infty) = 0$,

(3) $\theta(\infty) = 0$.

The following observations are made regarding the above transformation:

(1) The three original partial differential equations, (7.4), (7.8), and (7.10) are transformed into two ordinary differential equations.

(2) Equation (7.22) is a third order non-linear ordinary differential equation requiring three boundary conditions.

(3) Equation (7.23) is a second order ordinary differential equation requiring two boundary conditions.

(4) The boundary conditions are transformed in terms of the similarity variable η.

(5) The original seven boundary conditions on u, v, and T are transformed into five conditions on ξ and θ.

(6) The problem is characterized by a single parameter which is the Prandtl number.

7.4.5 Solution.

Equations (7.22) and (7.23) and their five boundary conditions are solved numerically [1]. The solution is presented graphically in Figs. 7.2 and 7.3. Fig. 7.2 gives the axial velocity $u(x,y)$ and Fig. 7.3 gives the temperature distribution $T(x,y)$ for various Prandtl numbers.

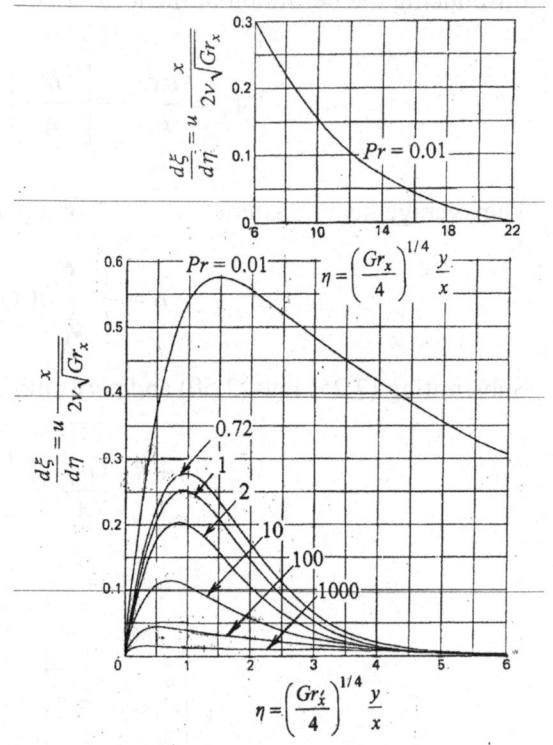

Fig. 7.2 Axial velocity distribution [1]

7.4.6 Heat Transfer Coefficient and Nusselt Number. The heat transfer coefficient h is based on Fourier's law and Newton's law. It is given in Sections 1.6 and 2.10.6 as

$$h = -k\frac{\partial T(x,0)}{\partial y}/(T_s - T_\infty). \tag{7.24}$$

Expressing the above in terms of the variables θ and η

$$h = \frac{-k}{T_s - T_\infty}\frac{dT}{d\theta}\frac{d\theta(0)}{d\eta}\frac{\partial \eta}{\partial y}.$$

Using (7.11) and (7.14), the above gives

$$h = \frac{-k}{x}\left[\frac{Gr_x}{4}\right]^{1/4}\frac{d\theta(0)}{d\eta}. \tag{7.25}$$

Introducing the definition of the local Nusselt number, the above becomes

$$Nu_x = \frac{hx}{k} = -\left[\frac{Gr_x}{4}\right]^{1/4} \frac{d\theta(0)}{d\eta}. \tag{7.26}$$

The average heat transfer coefficient for a plate of length L is defined as

$$\bar{h} = \frac{1}{L} \int_0^L h(x)dx. \tag{2.50}$$

Substituting (7.25) into (2.50) and performing the integration

$$\bar{h} = -\frac{4}{3}\frac{k}{L}\left(\frac{Gr_L}{4}\right)^{1/4} \frac{d\theta(0)}{d\eta}. \tag{7.27}$$

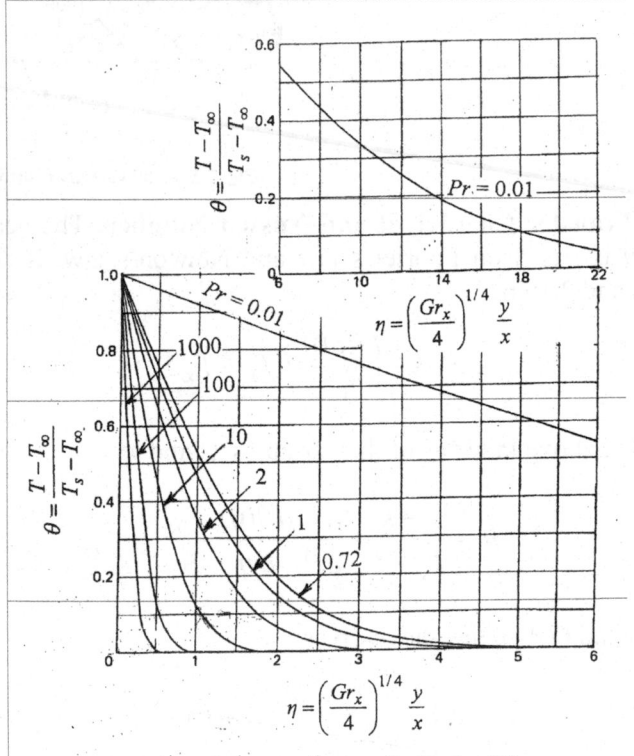

Fig. 7.3 Temperature distribution [1]

The average Nusselt number is given by

$$\overline{Nu}_L = \frac{\overline{h}L}{k} = -\frac{4}{3}\left(\frac{Gr_L}{4}\right)^{1/4} \frac{d\theta(0)}{d\eta}. \quad (7.28)$$

Surface temperature gradient, $d\theta(0)/d\eta$, which appears in the above equations is obtained from the numerical solution of equations (7.22) and (7.23). This important factor depends on the Prandtl number only and is listed in Table 7.1 for selected values of the Prandtl number. Also listed in this table is $d^2\xi(0)/d\eta^2$. This constant is needed to determine surface velocity gradient and shearing stress.

Table 7.1 [1,2]

Pr	$-\dfrac{d\theta(0)}{d\eta}$	$\dfrac{d^2\xi(0)}{d\eta^2}$
0.01	0.0806	0.9862
0.03	0.136	
0.09	0.219	
0.5	0.442	
0.72	0.5045	0.676
0.733	0.508	0.6741
1.0	0.5671	0.6421
1.5	0.6515	
2.0	0.7165	0.5713
3.5	0.8558	
5.0	0.954	
7.0	1.0542	
10	1.1649	0.4192
100	2.191	0.2517
1000	3.9660	0.1450

Special Cases

Equations (7.22) and (7.23) are simplified for two limiting cases corresponding to very small and very large Prandtl numbers. The local Nusselt number for these cases is given by [3]

$$Nu_x = 0.600\,(PrRa_x)^{1/4}, \quad Pr \to 0, \quad (7.29a)$$

and

$$Nu_x = 0.503\,(PrGr_x)^{1/4}, \quad Pr \to \infty. \quad (7.29b)$$

Example 7.1 Vertical Plate at Uniform Surface Temperature

A square plate measuring $8\,cm \times 8\,cm$ is suspended vertically in air. The plate is maintained at uniform surface temperature of $70\,{}^{\circ}C$. The ambient air is at $10\,{}^{\circ}C$. Of interest is the determination of the flow and heat transfer conditions at the trailing end $x = L$. Specifically, determine:

(1) *Axial velocity u at y = 0.2 cm*
(2) *Air temperature T at y = 0.2 cm*
(3) *Viscous boundary layer thickness δ*
(4) *Thermal boundary layer thickness δ_t*
(5) *Nusselt number*
(6) *Heat transfer coefficient*
(7) *Heat flux*
(8) *Total heat transfer rate from the plate.*

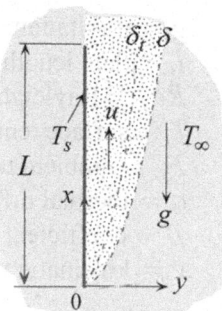

(1) Observations. (i) This is an external free convection problem over a vertical flat plate. (ii) The plate is maintained at uniform surface temperature. (iii) The Rayleigh number should be computed to determine if the flow is laminar. (iv) If the flow is laminar Fig. 7.2 gives the axial velocity u and viscous boundary layer thickness δ. Similarly, Fig. 7.3 gives temperature distribution, and thermal boundary layer thickness δ_t. (v) Attention is focused on the trailing edge of the plate. This means that local values of Nusselt number and heat transfer coefficient should be determined.

(2) Problem Definition. Determine the flow and heat transfer characteristics for free convection over a vertical flat plate at uniform surface temperature.

(3) Solution Plan. Compute the Rayleigh number to determine if the flow is laminar. If it is, use the similarity solution results using Figs. 7.2 and 7.3.

(4) Plan Execution.

   (i) Assumptions. (1) Continuum, (2) Newtonian fluid, (3) steady state, (4) Boussinesq approximations, (5) two-dimensional, (6) laminar flow ($Ra_x < 10^9$), (7) flat plate (8) uniform surface temperature, (9) no dissipation, and (10) no radiation.

   (ii) Analysis and Computation. The Rayleigh number is computed at the trailing edge to establish if the flow is laminar. The Rayleigh number is defined in equation (7.2) as

$$Ra_L = \frac{\beta g (T_s - T_\infty) L^3}{\nu \alpha}, \qquad (7.2)$$

where

  g = gravitational acceleration = 9.81 m/s^2
  L = plate length = 0.08 m
  Ra_L = Rayleigh number at the trailing end $x = L$
  T_s = surface temperature = 70°C
  T_∞ = ambient temperature = 10°C
  α = thermal diffusivity, m^2/s
  β = coefficient of thermal expansion = $1/T_f$ K^{-1}
  ν = kinematic viscosity, m^2/s

Properties are evaluated at the film temperature T_f

$$T_f = \frac{T_s + T_\infty}{2} = \frac{(70+10)(^\circ C)}{2} = 40^\circ C$$

Properties of air at this temperature are

$k = $ thermal conductivity $= 0.0271$ W/m$-^\circ$C

$Pr = 0.71$

$\nu = 16.96 \times 10^{-6} m^2/s$

$$\alpha = \frac{\nu}{Pr} = \frac{16.96 \times 10^{-6} m^2/s}{0.71} = 23.89 \times 10^{-6} m^2/s$$

$$\beta = \frac{1}{40^\circ C + 273.15} = 0.0031936 \ K^{-1}$$

Substituting into (7.2)

$$Ra_L = \frac{0.0031936(K^{-1})9.81(m/s^2)(70-10)(K)(0.08^3)}{16.96(K^{-1})(m^2/s)23.89 \times 10^{-6}(m^2/s)} = 2.3753 \times 10^6$$

Thus the flow is laminar. Axial velocity u is given by (7.20)

$$u = 2\nu \frac{\sqrt{Gr_x}}{x} \frac{d\xi}{d\eta}, \tag{7.20}$$

where $d\xi/d\eta$ is given in Fig. 7.2 as a function of η, given by

$$\eta = \left(\frac{Gr_x}{4}\right)^{1/4} \frac{y}{x}, \tag{a}$$

where, according to (7.2)

$$Gr_x = Gr_L = \frac{Ra_L}{Pr} = \frac{2.3792 \times 10^6}{0.71} = 3.3455 \times 10^6$$

Using (a) to evaluate η at $x = 0.08$ m and $y = 0.002$ m

$$\eta = \left(\frac{3.3455 \times 10^6}{4}\right)^{1/4} \frac{0.0032(m)}{0.08(m)} = 1.21$$

At $\eta = 1.21$ and $Pr = 0.71$, Fig. 7.2 gives

$$\frac{d\xi}{d\eta} = u\frac{x}{2v\sqrt{Gr_x}} \approx 0.27$$

Solving for u at $x = L = 0.08$ m

$$u = 0.27\frac{2v\sqrt{Gr_L}}{L} = 0.27\frac{2(16.96\times10^{-6})(m^2/s)\sqrt{3.3455\times10^6}}{0.08(m)} = 0.2094\,m/s$$

At $\eta = 1.21$ and $Pr = 0.71$, Fig. 7.3 gives the temperature at $x = 0.8$ m and $y = 0.002$ m

$$\theta = \frac{T - T_\infty}{T_s - T_\infty} \approx 0.43$$

$$T \approx T_\infty + 0.43(T_s - T_\infty) = 10(^\circ C) + 0.43(70 - 10)(^\circ C) = 35.8^\circ C$$

At the edge of the viscous boundary layer, $y = \delta$, the axial velocity vanishes. Fig. 7.2 gives the corresponding value of η as

$$\eta(x,\delta) \approx 5 = \left(\frac{Gr_L}{4}\right)^{1/4}\frac{\delta}{L}$$

Solving for δ

$$\delta = 5(0.08)(m)\left(\frac{4}{3.3455\times10^6}\right)^{1/4} = 0.0132\,m = 1.32\,cm$$

At the edge of the thermal boundary layer, $y = \delta_t$, the temperature reaches ambient value and thus $\theta \approx 0$. Fig. 7.3 gives the corresponding value of η as

$$\eta(x,\delta_t) \approx 4.5 = \left(\frac{Gr_L}{4}\right)^{1/4}\frac{\delta}{L}$$

Solving for δ_t

$$\delta_t = 4.5(0.08)(m)\left(\frac{4}{3.3455\times10^6}\right)^{1/4} = 0.0119\,m = 1.19\,cm$$

The local Nusselt number is given by equation (7.26)

$$Nu_x = \frac{hx}{k} = -\left[\frac{Gr_x}{4}\right]^{1/4} \frac{d\theta(0)}{d\eta}. \qquad (7.26)$$

Table 7.1 gives the temperature gradient at the surface $d\theta(0)/d\eta$ for various Prandtl numbers. Interpolation to $Pr = 0.71$ gives

$$\frac{d\theta(0)}{d\eta} = -0.5018$$

Using (7.26), the local Nusselt number is evaluated at the trailing end, $x = L = 0.08$ m

$$Nu_L = \frac{hL}{k} = -\left[\frac{Gr_L}{4}\right]^{1/4} \frac{d\theta(0)}{d\eta} = 0.5018 \left[\frac{3.3455 \times 10^6}{4}\right]^{1/4} = 15.18$$

The local heat transfer coefficient at the trailing end is obtained from the definition of the Nusselt number above

$$h(L) = \frac{k}{L} Nu_L = \frac{0.0271(\text{W/m}-^\circ\text{C})}{0.08(\text{m})} 15.18 = 5.14 \text{ W/m}^2-^\circ\text{C}$$

Newton's law of cooling gives surface heat flux q_s''

$$q_s'' = h(T_s - T_\infty) = 5.14(\text{W/m}-^\circ\text{C})(70-10)(^\circ\text{C}) = 308.4 \text{ W/m}^2$$

Total heat transfer from the plate is determined using the average heat transfer coefficient $\overline{h}$

$$q_T = \overline{h}A(T_s - T_\infty), \qquad (b)$$

where A is surface area and $\overline{h}$ is given by (7.27)

$$\overline{h} = -\frac{4}{3} \frac{k}{L} \left(\frac{Gr_L}{4}\right)^{1/4} \frac{d\theta(0)}{d\eta}, \qquad (7.27)$$

$$\overline{h} = -\frac{4}{3} \frac{(0.0271)(\text{W/m}-^\circ\text{C})}{0.08(\text{m})} \left[\frac{3.3455 \times 10^6}{4}\right]^{1/4} (-0.5018) = 6.85 \text{ W/m}^2-^\circ\text{C}$$

Substituting into (b)

$$q_T = 6.86(\text{W/m}^2 - ^\circ\text{C})0.08(\text{m})0.08(\text{m})(70-10)(^\circ\text{C}) = 2.63 \text{ W}$$

(iii) Checking. *Dimensional check*: Computations showed that units for $u, T, \delta, \delta_t, Nu, h, q''$ and q_T are consistent.

Quantitative check: (i) The heat transfer coefficient is within the range given in Table 1.1 for free convection of gases.

(ii) Computations showed that $\delta > \delta_t$. This must be the case since the thermal boundary layer thickness cannot be greater than the velocity boundary layer thickness. Why?

(5) Comments. (i) As expected, fluid velocity and heat transfer coefficient are relatively small in free convection.

(ii) The local heat transfer coefficient at location x is smaller than the average for a plate of length x. This is due to the fact that the heat transfer coefficient decreases as the distance from the leading edge is increased.

7.5 Laminar Free Convection over a Vertical Plate: Uniform Surface Heat Flux

Fig. 7.4 shows a vertical plate with uniform surface heat flux. The plate is submerged in an infinite fluid at temperature T_∞. Analytical determination of the velocity and temperature distribution follows the procedure used in Section 7.4 replacing uniform surface temperature with uniform surface heat flux. The two problems are based on the same assumptions and governing equations. They differ by the thermal boundary condition at the surface, which takes the form

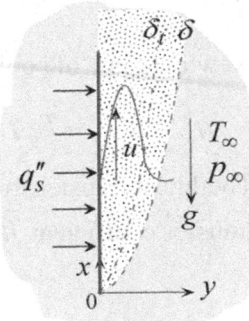

Fig. 7.4

$$-k\frac{\partial T(x,0)}{\partial y} = q_s'', \qquad (7.30)$$

where q_s'' is surface heat flux. It is important to recognize that for uniform surface heat flux, surface temperature varies along the plate. Thus, of interest is the determination of surface temperature variation $T_s(x)$ and the local Nusselt number Nu_x. This problem was solved by similarity transformation [4]. The solution for surface temperature variation is given by

$$T_s(x) - T_\infty = -\left[5\frac{\nu^2(q_s'')^4}{\beta g k^4} x\right]^{1/5} \theta(0), \qquad (7.31)$$

where $\theta(0)$ is a dimensionless parameter which depends on the Prandtl number and is given in Table 7.2 [4]. The local Nusselt number is given by

$$Nu_x = -\left[\frac{\beta g q_s''}{5\nu^2 k} x^4\right]^{1/5} \frac{1}{\theta(0)}. \qquad (7.32)$$

Table 7.2 [4]

Pr	$\theta(0)$
0.1	- 2.7507
1.0	- 1.3574
10	- 0.76746
100	- 0.46566

For a wide range of Prandtl numbers, the parameter $\theta(0)$ may also be determined using the following correlation equation [5]

$$\theta(0) = -\left[\frac{4 + 9Pr^{1/2} + 10Pr}{5Pr^2}\right]^{1/5}, \quad 0.001 < Pr < 1000. \qquad (7.33)$$

Properties are determined at the film temperature T_f, defined as

$$T_f = [T_\infty + T_s(L/2)]/2. \qquad (7.34)$$

Example 7.2: Vertical Plate at Uniform Surface Flux

An 8 cm high plate is suspended vertically in air. The plate is heated at uniform surface flux of 308.4 W/m^2. The ambient temperature is $10°C$. Determine surface temperature, Nusselt number and heat transfer coefficient at x = 2, 4, 6 and 8 cm from the leading end.

(1) Observations. (i) This is an external free convection problem over a vertical flat plate. (ii) The plate is heated at uniform surface flux. (iii) The Rayleigh number should be computed to determine if the flow is laminar. (iv) Surface temperature is given by equation (7.31) and the local Nusselt number is given by equation (7.32).

(2) Problem Definition. Determine the distribution of surface temperature, Nusselt number and heat transfer coefficient along a uniformly heated vertical plate under free convection.

(3) Solution Plan. Compute the Rayleigh number to determine if the flow is laminar. Use (7.31) and (7.32) to determine surface temperature and Nusselt number at the trailing end. Use Newton's law to determine the heat transfer coefficient.

(4) Plan Execution.

(i) **Assumptions.** (1) Continuum, (2) Newtonian fluid, (3) steady state, (4) Boussinesq approximations, (5) two-dimensional, (6) laminar flow ($Ra_x < 10^9$), (7) flat plate, (8) uniform surface heat flux, (9) no dissipation, and (10) no radiation.

(ii) **Analysis and Computation.** The Rayleigh number is computed at the trailing edge to establish if the flow is laminar. The Rayleigh number is defined in equation (7.2) as

$$Ra_L = \frac{\beta g (T_s - T_\infty) L^3}{\nu \alpha}. \tag{7.2}$$

Since surface temperature is unknown, the Rayleigh number cannot be computed. To proceed, the flow is assumed laminar and subsequently verified once surface temperature is computed. For laminar flow, surface temperature and Nusselt number are given by

$$T_s(x) - T_\infty = -\left[5 \frac{\nu^2 (q_s'')^4}{\beta g k^4} x \right]^{1/5} \theta(0), \tag{7.31}$$

$$Nu_x = -\left[\frac{\beta g q_s''}{5\nu^2 k} x^4 \right]^{1/5} \frac{1}{\theta(0)}, \tag{7.32}$$

where $\theta(0)$ is given in Table 7.2. It can also be determined using (7.33). Properties are determined at the film temperature defined in equation (7.34).

(iii) **Computations.** Assume $T_s(L/2) = 70°C$. Compute film temperature using (7.34)

$$T_f = \frac{T_\infty + T_s(L/2)}{2} = \frac{(10 + 70)(°C)}{2} = 40°C$$

Properties of air at this temperature are

$$k = 0.0271 \text{ W/m}-^\circ\text{C}$$

$$Pr = 0.71$$

$$v = 16.96 \times 10^{-6} \text{ m}^2/\text{s}$$

$$\alpha = \frac{v}{Pr} = \frac{16.96 \times 10^{-6} \text{ m}^2/\text{s}}{0.71} = 23.89 \times 10^{-6} \text{ m}^2/\text{s}$$

$$\beta = \frac{1}{40^\circ\text{C} + 273.15} = 0.0031936 \text{ K}^{-1}$$

Equations (7.33) is used to evaluate $\theta(0)$

$$\theta(0) = -\left[\frac{4 + 9(0.71)^{1/2} + 10(0.71)}{5(0.71)^2}\right]^{1/5} = -1.4928$$

The assumed surface temperature at $x = L/2$ is verified first using (7.31)

$$T_s(L/2) = 10(^\circ\text{C}) - \left[5\frac{(16.96 \times 10^{-6})^2(\text{m}^4/\text{s}^2)(308.4)^4(\text{W}^4/\text{m}^8)}{0.0031934(1/\text{K})9.81(\text{m/s}^2)(0.0271)^4(\text{W}^4/\text{m}^4-\text{K}^4)}(0.04)(\text{m})\right]^{1/5}(-1.4928)$$

$$= 56.96 \,^\circ\text{C}$$

This is lower than the assumed temperature. This procedure is repeated until a satisfactory agreement is obtained between the assumed and computed surface temperature at $x = L/2$. Following this iterative procedure, surface temperature at $x = L/2$ is found to be

$$T_s(L/2) = 56.7^\circ\text{C}$$

The corresponding film temperature is

$$T_f = \frac{(10 + 56.7)(^\circ\text{C})}{2} = 33.35^\circ\text{C}$$

Properties of air at this temperature are

$$k = 0.02662 \text{ W/m}-^\circ\text{C}$$

$$Pr = 0.71$$

$$v = 16.3283 \times 10^{-6} \text{ m}^2/\text{s}$$

$$\beta = 0.002662 \text{ K}^{-1}$$

Equation (7.33) gives

$$\theta(0) = -1.4922$$

Equation (7.32) is used to determine the Nusselt number at the trailing end

$$
Nu_x = -\left[\frac{0.0032626(1/K)9.81(m/s^2)308.3(W/m^2)}{5(16.328\times10^{-6})^2(m^4/s^2)0.02662(W/m-^\circ C)}(0.08)^4(m^4)\right]^{1/5}\frac{1}{-1.4922}
$$

$$
= 17.28
$$

The local heat transfer coefficient at the trailing end is obtained from the definition of Nusselt number

$$
h(L) = \frac{k}{L}Nu_L = \frac{0.02662(W/m-^\circ C)}{0.08(m)}17.28 = 5.75\ W/m^2-^\circ C
$$

Surface temperature, Nusselt number, and heat transfer coefficient at various locations along the plate are determined following the above procedure. The result is tabulated below.

x(m)	$T_s(x)(^\circ C)$	Nu_x	$h(x)(W/m^2-^\circ C)$	$q_s''(W/m^2)$
0.02	50.6	5.70	7.59	308.2
0.04	56.7	9.92	6.60	308.2
0.06	60.6	13.73	6.09	308.2
0.08	63.6	17.28	5.75	308.2

To verify that the flow is laminar throughout, equation (7.2) is used to compute the Rayleigh number at $x = L$

$$
Ra_L = \frac{0.0032626\ (K^{-1})9.81\ (m/s^2)(63.6-10)(K)(0.08)^3(m^3)}{(16.3283\times10^{-6})^2(m^2/s)^2}0.711
$$

$$
= 2.3424 \times 10^6
$$

Since $Ra_L < 10^9$, the flow is laminar.

(iii) Checking. *Dimensional check*: Computations showed that units for T_s, Nu and h are consistent.

Quantitative check: (i) The heat transfer coefficient is within the range given in Table 1.1 for free convection of gases. (ii) The tabulated results show that application of Newton's law at the four locations along the plate gives uniform surface flux.

(5) Comments. (i) Since surface temperature is unknown, the problem is solved by an iterative procedure. (ii) As with forced convection over a flat plate, the heat transfer coefficient decreases as the distance from the leading edge is increased.

7.6 Inclined Plates

We consider a plate which is inclined at an angle θ from the vertical. In Fig. 7.5a the heated side of the plate is facing downward while Fig. 7.5b the cooled side is facing upward. Note that the flow field is identical for both cases and consequently the same solution holds for both. Note further that gravity component for the inclined plate is $g\cos\theta$ while for the vertical

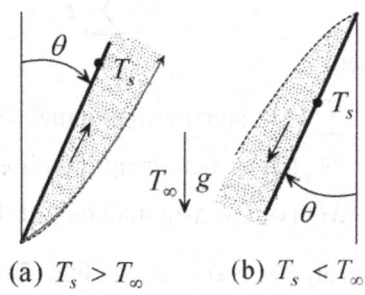

(a) $T_s > T_\infty$ (b) $T_s < T_\infty$

Fig. 7.5

plate it is g. Studies have shown that the vertical plate solutions of Sections 7.4 and 7.5 apply to inclined plates, with g replaced by $g\cos\theta$ [6-8]. However, this approximation deteriorates at large values of θ. Thus, this approach is recommended for $\theta \le 60°$.

7.7 Integral Method

The integral method can be applied to obtain approximate solutions to free convection boundary layer flows problems. As an example, consider the problem of a vertical plate at uniform surface temperature shown in Fig. 7.6. An exact analytic solution to this problem is presented in Section 7.4.

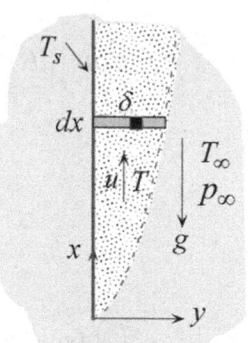

Fig 7.6

7.7.1 Integral Formulation of Conservation of Momentum

The starting point in integral solutions is the integral formulation of conservation of momentum. To simplify the analysis we assume that the viscous and thermal boundary layers are the same. That is

$$\delta = \delta_t. \tag{a}$$

This approximation is valid for Prandtl numbers in the neighborhood of unity. Application of the momentum theorem in the x-direction to the element $\delta \times dx$ shown in Fig. 7.6, gives

$$\sum F_x = M_x(\text{out}) - M_x(\text{in}), \tag{b}$$

where

$\sum F_x$ = sum of all external forces acting on element in the x-direction

$M_x(\text{in})$ = x-momentum of the fluid entering element

$M_x(\text{out})$ = x-momentum of the fluid leaving element

The element $\delta \times dx$ of Fig. 7.6 is enlarged in Fig. 7.7 showing the x-momentum and x-forces. The forces acting on the element are due to shearing stress τ_o at the wall, pressure forces p and gravity force (weight) dW. Applying equation (b) and using the notations in Fig. 7.7, we obtain

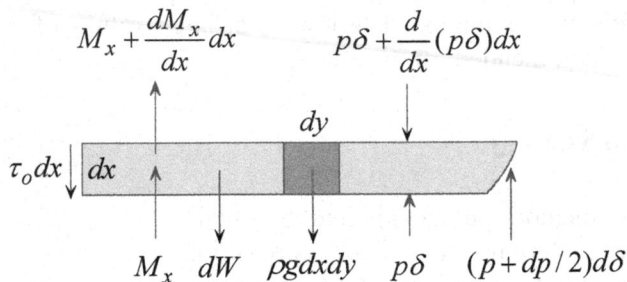

Fig. 7.7

$$p\delta + \left(p + \frac{dp}{2}\right)d\delta - p\delta - \frac{d}{dx}(p\delta)dx - \tau_o dx - dW = \left(M_x + \frac{dM_x}{dx}dx\right) - M_x, \tag{c}$$

Simplifying equation (c) and neglecting higher order terms

$$-\delta dp - \tau_o dx - dW = \frac{dM_x}{dx}dx. \tag{d}$$

Wall shearing stress is given by

$$\tau_o = \mu \frac{\partial u(x,0)}{\partial y}.$$ (e)

In determining the weight of the element $\delta \times dx$, one must take into consideration the variation of density in the y-direction. This requires integration of the weight of a differential element $dx \times dy$ along the thickness δ of the boundary layer. Thus

$$dW = dx \int_0^{\delta} \rho g \, dy.$$ (f)

The x-momentum of the fluid entering the element is

$$M_x = \rho \int_0^{\delta(x)} u^2 \, dy,$$ (g)

where $u = u(x, y)$ is axial velocity. Note that the density ρ is assumed constant in evaluating momentum. This is consistent with the Boussinesq approximation used in obtaining similarity solution to this problem. Substituting (e), (f) and (g) into (d) and rearranging

$$-\mu \frac{\partial u(x,0)}{\partial y} - \delta \frac{dp}{dx} - \int_0^{\delta} \rho g \, dy = \rho \frac{d}{dx} \int_0^{\delta} u^2 \, dy.$$ (h)

Pressure and gravity terms in (h) will now be combined. Pressure gradient in boundary layer flow is given by

$$\frac{dp}{dx} \cong \frac{dp_{\infty}}{dx} = -\rho_{\infty} g.$$ (i)

Thus the pressure gradient term in (h) can be rewritten as

$$\delta \frac{dp}{dx} = -\rho_{\infty} g \delta = -\int_0^{\delta} \rho_{\infty} g \, dy.$$ (j)

Substituting (j) into (h) and rearranging

$$-\mu \frac{\partial u(x,0)}{\partial y} + g \int_0^\delta (\rho_\infty - \rho)dy = \rho \frac{d}{dx} \int_0^\delta u^2 dy. \qquad \text{(k)}$$

However, density change can be expressed in terms of temperature change

$$\rho_\infty - \rho = \rho\beta(T - T_\infty). \qquad (2.28)$$

Introducing (2.28) into (k) and treating β and ρ as constants, gives

$$-v \frac{\partial u(x,0)}{\partial y} + \beta g \int_0^\delta (T - T_\infty)dy = \frac{d}{dx} \int_0^\delta u^2 dy, \qquad (7.35)$$

where $v = \mu / \rho$. Note the following:

(1) There is no shearing force on the slanted surface since the velocity gradient at the edge of the boundary layer vanishes, i.e. $\partial u(x,\delta)/\partial y \approx 0$.

(2) Equation (7.35) applies to laminar as well as turbulent flow.

(3) Although u and T are functions of x and y, once the integrals in (7.35) are evaluated one obtains a first order ordinary differential equation with x as the independent variable.

7.7.2 Integral Formulation of Conservation of Energy

The following assumptions are made in the integral formulation of conservation of energy:

(1) No changes in kinetic and potential energy
(2) Negligible axial conduction
(3) Negligible dissipation
(4) Properties are constant

Based on the above assumptions, integral formulation of conservation of energy for free convection boundary layer flow is the same as that for forced convection. Thus equation (5.7) is applicable to free convection

$$-\alpha \frac{\partial T(x,0)}{\partial y} = \frac{d}{dx} \int_0^{\delta(x)} u(T - T_\infty)dy, \qquad (7.36)$$

where α is thermal diffusivity.

7.7.3 Integral Solution

Consider the vertical plate shown in Fig. 7.6. The plate is maintained at uniform temperature T_s and the quiescent fluid is at uniform temperature T_∞. Following the procedure used in integral solution of forced convection, velocity and temperature profiles are assumed. Recall that in forced convection a velocity profile is assumed in terms of a single unknown function $\delta(x)$. Application of the integral form of momentum is used to determine $\delta(x)$. Similarly, a temperature profile is assumed in terms of a single unknown function $\delta_t(x)$. Application of the integral form of energy is used to determine $\delta_t(x)$. However, in the integral formulation of conservation of momentum and energy for free convection we assumed that $\delta \approx \delta_t$. Thus we have two equations, (7.35) and (7.36) for the determination of a single unknown δ. This presents a quandary which must be resolved so that both (7.35) and (7.36) are used to insure that conservation of momentum and energy are satisfied. The problem is resolved by introducing a second unknown function in the assumed velocity profile.

Assumed Velocity Profile. To proceed, we assume laminar boundary layer flow. Thus a polynomial is an appropriate velocity profile. Assume a fourth degree polynomial for the axial velocity $u(x,y)$

$$u(x, y) = a_0(x) + a_1(x)y + a_2(x)y^2 + a_3(x)y^3. \tag{a}$$

The coefficients $a_n(x)$ are determined using the following known exact and approximate boundary conditions on the velocity

(1) $u(x,0) = 0$,

(2) $u(x,\delta) \approx 0$,

(3) $\dfrac{\partial u(x,\delta)}{\partial y} \approx 0$,

(4) $\dfrac{\partial^2 u(x,0)}{\partial y^2} = -\dfrac{\beta g}{\nu}(T_s - T_\infty)$.

Note that the second and third conditions are approximate since the edge of the boundary layer is not uniquely defined. The fourth condition is obtained by setting $y = 0$ in the x-component of the equations of motion, (7.5).

Equation (a) and the four boundary conditions give the coefficients $a_n(x)$

$$a_0 = 0, \quad a_1 = \frac{\beta g(T_s - T_\infty)}{4v} \delta, \quad a_2 = -\frac{\beta g(T_s - T_\infty)}{2v},$$

$$a_3 = -\frac{\beta g(T_s - T_\infty)}{4v} \frac{1}{\delta}.$$

Substituting the above into (a) and rearranging

$$u = \frac{\beta g(T_s - T_\infty)}{4v} \delta y \left[1 - 2\frac{y}{\delta} + \frac{y^2}{\delta^2} \right].$$

This can be written as

$$u = \left[\frac{\beta g(T_s - T_\infty)}{4v} \delta^2 \right] \frac{y}{\delta} \left[1 - \frac{y}{\delta} \right]^2. \tag{b}$$

To introduce a second unknown function in the assumed velocity profile (b), we define

$$u_o(x) = \left[\frac{\beta g(T_s - T_\infty)}{4v} \delta^2 \right]. \tag{c}$$

Equation (b) becomes

$$u = u_o(x) \frac{y}{\delta} \left[1 - \frac{y}{\delta} \right]^2. \tag{7.37}$$

Note that replacing the term in bracket in (c) with $u_o(x)$ implies that $u_o(x)$ is independent of $\delta(x)$. Thus in (7.32) both $\delta(x)$ and $u_o(x)$ are unknown. This means that both conservation of momentum and energy are needed to solve the problem.

Assumed Temperature Profile. A third degree polynomial is assumed for the temperature profile

$$T(x,y) = b_0(x) + b_1(x)y + b_2(x)y^2. \tag{d}$$

The boundary conditions are

(1) $T(x,0) = T_s$,

(2) $T(x,\delta) \approx T_\infty$,

(3) $\dfrac{\partial T(x,\delta)}{\partial y} \approx 0$.

Equation (d) and the three boundary conditions give the coefficients $b_n(x)$

$$b_0 = T_s, \quad b_1 = -2(T_s - T_\infty)\frac{1}{\delta}, \quad b_2 = (T_s - T_\infty)\frac{1}{\delta^2}.$$

Substituting the above into (d) and rearranging

$$T(x,y) = T_\infty + (T_s - T_\infty)\left[1 - \frac{y}{\delta}\right]^2. \tag{7.38}$$

Heat Transfer Coefficient and Nusselt Number. Equation (7.24) gives the heat transfer coefficient h

$$h = \frac{-k\dfrac{\partial T(x,0)}{\partial y}}{T_s - T_\infty}. \tag{7.24}$$

Substituting (7.38) into (7.24)

$$h = \frac{2k}{\delta(x)}. \tag{7.39}$$

Thus the local Nusselt number is

$$Nu_x = \frac{hx}{k} = 2\frac{x}{\delta(x)}. \tag{7.40}$$

The problem becomes one of determining the boundary layer thickness $\delta(x)$.

Solution. To determine the functions $\delta(x)$ and $u_o(x)$ we substitute (7.37) and (7.38) into (7.35)

$$-v\frac{u_o}{\delta} + \beta g(T_s - T_\infty) \int_0^\delta \left[1 - \frac{y}{\delta}\right]^2 dy = \frac{d}{dx}\left\{\frac{u_o^2}{\delta^2} \int_0^\delta y^2 \left[1 - \frac{y}{\delta}\right]^4 dy\right\}.$$

(e)

Evaluating the integrals in (e) and rearranging

$$\frac{1}{105}\frac{d}{dx}\left[u_o^2\delta\right] = \frac{1}{3}\beta g(T_s - T_\infty)\delta - v\frac{u_o}{\delta}.$$

(7.41)

Similarly, substituting (7.37) and (7.38) into (7.36), gives

$$2\alpha(T_s - T_\infty)\frac{1}{\delta} = (T_s - T_\infty)\frac{d}{dx}\left\{\frac{u_o}{\delta} \int_0^{\delta(x)} y\left[1 - \frac{y}{\delta}\right]^4 dy\right\}.$$

(f)

Evaluating the integrals and rearranging

$$\frac{1}{60}\frac{d}{dx}\left[u_o\delta\right] = \alpha\frac{1}{\delta}.$$

(7.42)

Equations (7.41) and (7.42) are two simultaneous fir-st order ordinary differential equations. The two dependent variables are $\delta(x)$ and $u_o(x)$. We assume a solution of the form

$$u_o(x) = Ax^m,$$

(7.43)

$$\delta(x) = Bx^n.$$

(7.44)

where A, B, m and n are constants. To determine these constants we substitute (7.43) and (7.44) into (7.41) and (7.42) to obtain

$$\frac{2m+n}{105}A^2Bx^{2m+n-1} = \frac{1}{3}\beta g(T_o - T_\infty)Bx^n - \frac{A}{B}v x^{m-n},$$

(7.45)

$$\frac{m+n}{210}ABx^{m+n-1} = \alpha\frac{1}{B}x^{-n}.$$

(7.46)

To satisfy (7.45) and (7.46) at all values of x, the exponents of x in each term must be identical. Thus, (7.45) requires that

$$2m + n - 1 = n = m - n.$$ \hfill (g)

Similarly, (7.46) requires that

$$m + n - 1 = -n.$$ \hfill (h)

Solving (g) and (h) for m and n gives

$$m = \frac{1}{2}, \quad n = \frac{1}{4}.$$ \hfill (i)

Introducing (i) into (7.45) and (7.46) gives two simultaneous algebraic equations for A and B

$$\frac{1}{85} A^2 B = \frac{1}{3} \beta g (T_o - T_\infty) B - \frac{A}{B} v,$$ \hfill (j)

$$\frac{1}{280} AB = \alpha \frac{1}{B}.$$ \hfill (k)

Solving equations (j) and (k) for A and B, gives

$$A = 5.17 v \left[Pr + \frac{20}{21} \right]^{-1/2} \left[\frac{\beta g (T_s - T_\infty)}{v^2} \right]^{1/2},$$ \hfill (l)

and

$$B = 3.93 \, Pr^{-1/2} \left(Pr + \frac{20}{21} \right)^{1/4} \left[\frac{\beta g (T_s - T_\infty)}{v^2} \right]^{-1/4}.$$ \hfill (m)

Substituting (i) and (m) into (7.44), rearranging and introducing the definition of Rayleigh number, gives the solution to $\delta(x)$

$$\frac{\delta}{x} = 3.93 \left[\frac{20}{21} \frac{1}{Pr} + 1 \right]^{1/4} (Ra_x)^{-1/4}.$$ \hfill (7.47)

Introducing (7.47) into (7.40) gives the local Nusselt number

$$Nu_x = 0.508 \left[\frac{20}{21} \frac{1}{Pr} + 1 \right]^{-1/4} (Ra_x)^{1/4}. \qquad (7.48)$$

7.7.4 Comparison with Exact Solution for Nusselt Number

Equation (7.26) gives the exact solution to the local Nusselt number for free convection over a vertical plate at uniform temperature

$$Nu_x = -\left[\frac{Gr_x}{4} \right]^{1/4} \frac{d\theta(0)}{d\eta}. \qquad (7.26)$$

To compare this result with the integral solution (7.48), equation (7.26) is rewritten as

$$\left[\frac{Gr_x}{4} \right]^{-1/4} Nu_x = -\frac{d\theta(0)}{d\eta}. \qquad (7.49)$$

To facilitate comparison, integral solution (7.48) is rewritten as

$$\left[\frac{Gr_x}{4} \right]^{-1/4} Nu_x = 0.508 \left[\frac{20}{21} \frac{1}{Pr} + 1 \right]^{-1/4} (4Pr)^{1/4}. \qquad (7.50)$$

The accuracy of the integral solution depends on the agreement of the right hand side of (7.50) with $-d\theta(0)/d\eta$ of exact solution (7.49). Temperature gradient $d\theta(0)/d\eta$ depends on the Prandtl number and is given in Table 7.2. The two solutions are compared in Table 7.3. The exact solution for the limiting case of $Pr \rightarrow 0$ is given in (7.29a)

Table 7.3

Pr	$-\dfrac{d\theta(0)}{d\eta}$	$0.508\left[\dfrac{20}{21}\dfrac{1}{Pr}+1\right]^{-1/4}(4Pr)^{1/4}$
0.01	0.0806	0.0725
0.09	0.219	0.213
0.5	0.442	0.4627
0.72	0.5045	0.5361
1.0	0.5671	0.6078
2.0	0.7165	0.7751
5.0	0.954	1.0285
10	1.1649	1.2488
100	2.191	2.2665
1000	3.9660	4.0390

$$Nu_x|_{exact} = 0.600 (PrRa_x)^{1/4}, \quad Pr \rightarrow 0. \qquad (7.29a)$$

Applying integral solution (7.47) to $Pr \to 0$ gives

$$Nu_x\big|_{\text{integral}} = 0.514(PrRa_x)^{1/4}, \quad Pr \to 0. \qquad (7.51a)$$

Similarly, exact and integral solutions for the limiting case of $Pr \to \infty$ are given by

$$Nu_x\big|_{\text{exact}} = 0.503(Ra_x)^{1/4}, \quad Pr \to \infty, \qquad (7.29b)$$

and

$$Nu_x\big|_{\text{integral}} = 0.508(Ra_x)^{1/4}, \quad Pr \to \infty. \qquad (7.51b)$$

The following observations are made regarding the above comparisons:

(1) The error ranges from 1% for $Pr \to \infty$ to 14% for $Pr \to 0$.

(2) Although the integral solution is based on the assumption that $\delta = \delta_t$ ($Pr \approx 1$), the solution is reasonably accurate for a wide range of Prandtl numbers.

REFERENCES

[1] Ostrach, S., "An Analysis of Laminar Free Convection Flow and Heat Transfer About a Flat Plate Parallel to the Direction of the Generating Body Force," NACA, Report 1111, 1953.

[2] Edie, J. A., "Advances in Free Convection," in Advances in Heat Transfer, Vol. 4, 1067, pp.1-64.

[3] LeFevre, E. J., "Laminar Free Convection from a Vertical Plane Surface," Ninth International Congress of Applied Mechanics, Brussels, Vol. 4, 1956, pp. 168-174.

[4] Sparrow, E. M. and J. L. Gregg, "Laminar Free Convection from a Vertical Plate with Uniform Surface Heat Flux," Trans. ASME, J. Heat Transfer, Vol. 78, 1956, pp. 435-440.

[5] Fuji, T. and M. Fuji, " The Dependence of Local Nusselt Number on Prandtl Number in the Case of Free Convection Along a Vertical Surface with Uniform Heat Flux," Int. J. Heat Mass Transfer, Vol. 19, 1976, pp. 121-122.

[6] Rich, B. R. "An Investigation of Heat Transfer from an Inclined Flat Plate in Free Convection," Trans. ASME, Vol. 75, 1953, pp. 489-499.

[7] Rich, B. R. "Natural Convection Local Heat Transfer on Constant Heat Flux Inclined Surfaces," . J. Heat Transfer, Vol. 9, 1969, pp. 511-516.

[8] Fuji, T. and H. Imura, "Natural Convection from a Plate with Arbitrary Inclination," Int. J. Heat Mass Transfer, Vol. 15, 1972, pp. 755-767.

PROBLEMS

7.1   Explain why

[a] δ_t can not be larger than δ.

[b] δ can be larger than δ_t.

7.2   A vertical plate 6.5cm high and 30 wide cm is maintained at $82\,^{\circ}C$. The plate is immersed in water at $18\,^{\circ}C$. Determine:

[a] The viscous boundary layer thickness at the trailing end.
[b] The thermal boundary layer thickness at the trailing end.
[c] The average heat transfer coefficient.
[d] Total heat added to water.

7.3   Use Fig. 7.3 to determine $d\theta(0)/d\eta$ for $Pr = 0.01$ and 100. Compare your result with the value given in table 7.1.

7.4   In designing an air conditioning system for a pizza restaurant an estimate of the heat added to the kitchen from the door of the pizza oven is needed. The rectangular door is $50\,cm \times 120\,cm$ with its short side along the vertical direction. Door surface temperature is 110°C. Estimate the heat loss from the door if the ambient air temperature is 20°C.

7.5   To compare the rate of heat transfer by radiation with that by free convection, consider the following test case. A vertical plate measuring 12 cm $\times$ 12 cm is maintained at a uniform surface temperature of 125°C. The ambient air and the surroundings are at 25°C. Compare the two modes of heat transfer for surface emissivities of 0.2 and 0.9. A simplified model for heat loss by radiation q_r is given by

$$q_r = \varepsilon \sigma A (T_s^4 - T_{sur}^4),$$

where A is surface area, ε is emissivity, and $\sigma = 5.67 \times 10^{-8}$ $W/m^2 - K^4$. Surface and surroundings temperatures are measured in degrees kelvin.

7.6 A sealed electronic package is designed to be cooled by free convection. The package consists of components which are mounted on the inside surfaces of two cover plates measuring $7.5\,cm \times 7.5\,cm$ cm each. Because the plates are made of high conductivity material, surface temperature may be assumed uniform. The maximum allowable surface temperature is 70°C. Determine the maximum power that can be dissipated in the package without violating design constraints. Ambient air temperature is 20°C.

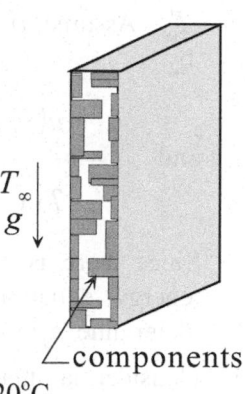

7.7 Assume that the electronic package of Problem 7.6 is to be used in an underwater application. Determine the maximum power that can be dissipated if the ambient water temperature is 20°C.

7.8 Consider laminar free convection from a vertical plate at uniform surface temperature. Two 45° triangles are drawn on the plate as shown. Determine the ratio of the heat transfer rates from the two triangles.

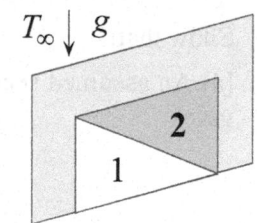

7.9 A vertical plate measuring 21 cm × 21 cm is at a uniform surface temperature of 80°C. The ambient air temperature is 25°C. Determine the heat flux at 1 cm, 10 cm, and 20 cm from the lower edge.

7.10 200 square chips measuring $1\,cm \times 1\,cm$ each are mounted on both sides of a thin vertical board $10\,cm \times 10\,cm$. The chips dissipate 0.035 W each. Assume uniform surface heat flux. Determine the maximum surface temperature in air at 22°C.

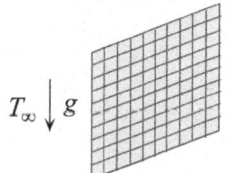

7.11 $12\,\text{cm} \times 12\,\text{cm}$ power boards dissipate 15 watts uniformly. Assume that all energy leaves the board from one side. The maximum allowable surface temperature is 82°C. The ambient fluid is air at 24°C. Would you recommend cooling the board by free convection?

7.12 Use the integral method to obtain solution to the local Nusselt number for laminar flow over a vertical plate at uniform surface temperature T_s. Assume $\delta = \delta_t$ and a velocity and temperature profiles given by

$$u(x,y) = a_0(x) + a_1(x)y + a_2(x)y^2 + a_3(x)y^3,$$

and

$$T(x,y) = b_0(x) + b_1(x)y + b_2(x)y^2 + b_3(x)y^3.$$

Since there is a single unknown $\delta_t(x)$, either the momentum or energy equation may be used. Select the energy equation to determine $\delta_t(x)$.

7.13 Consider laminar free convection over a vertical plate at uniform surface flux q''_s. Assume $\delta = \delta_t$ and a third degree polynomial velocity profile given by

$$u(x,y) = u_o(x)\frac{y}{\delta}\left[1 - \frac{y}{\delta}\right]^2.$$

Show that:

[a] An assumed second degree polynomial for the temperature profile gives

$$T(x,y) = T_\infty + \frac{1}{2}\frac{q''_s}{k}\left[\delta - 2y + \frac{y^2}{\delta}\right].$$

[b] The local Nusselt number is given by

$$Nu_x = \left[\frac{4(Pr)^2}{36 + 45Pr}\frac{\beta g q''_s}{k\nu^2}x^4\right]^{1/5}.$$

8

CONVECTION IN EXTERNAL TURBULENT FLOW

Glen E. Thorncroft
California Polytechnic State University
San Luis Obispo, California

8.1 Introduction

Turbulent flow is a complicated physical phenomenon, and a daunting subject for students of engineering. Turbulent flow is disordered, with random and unsteady velocity fluctuations. This description by itself suggests that exact predictions cannot be determined, and that analysis of turbulent flows will not be easy. However intimidating turbulence may be, its importance is clear: turbulent flows are found in many industrial and natural processes, and affect quantities of great practical importance to engineers, like the local velocity distribution, drag force, and heat transfer. Moreover, a better understanding of this flow allows engineers to make design decisions that could either reduce the effects of turbulence or enhance them, in order to improve the performance of their devices.

In spite of its complexity (and perhaps *because* of its complexity), turbulent flow is an exciting subject to study. The phenomenon has been known for centuries (drawings by Leonardo da Vinci contain turbulent-looking flows [1]), but the serious study of turbulence did not begin until the late 1800s. Much progress in the understanding of turbulence has been made in the last century, and yet our understanding still relies on empirical data and rudimentary conceptual drawings. An exact solution to the governing equations is not possible. However, with advanced measurement techniques and high-powered computers, the last 30 years has seen tremendous growth in research. New concepts and pictures of turbulence are emerging, and although a turbulent flow may never be fully understood, major breakthroughs could come in your lifetime.

8.1.1 Examples of Turbulent Flows

(i) Mixing Processes.   Perhaps the most straightforward example of turbulence is in *mixing processes*.   Turbulent mixing is essential in combustion processes, for example, where the proper mixing of fuel and air is one of the "three T's" of combustion efficiency: time, temperature, and turbulence.   Turbulence is also an advantage in chemical processing, such as the production of chemical emulsions or polymers.   Unfortunately, *turbulent mixing* is not always possible, and manufacturers have to contend with *laminar mixing*, especially when the liquids are highly viscous and/or are mixed slowly.   Laminar mixing is not always undesirable; an example of this can be found in your grocery, in the form of vanilla/fudge swirl ice cream.

(ii) Free Shear Flows.   Another example of turbulent flows is *free shear* flows, like those illustrated in Fig. 8.1.   In jet flows (Fig. 8.1a), the presence of turbulence helps dissipate the energy of the jet to the surrounding fluid.   The turbulent wake behind an object in a flow (Fig. 8.1b) serves to transfer energy between the ambient flow and the object, and contributes to the object's form drag.   The exhaust from a smokestack (Fig. 8.1c) is dispersed by its turbulent wake.

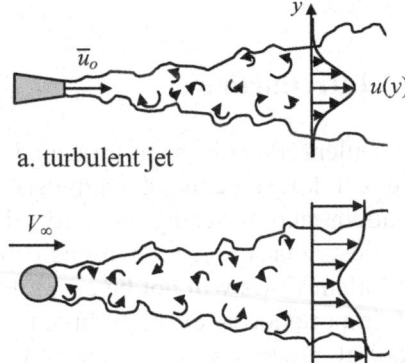

a. turbulent jet

b. turbulent wake behind body

(iii) Wall-Bounded Flows.   Within the category of shear flows are *wall-bounded* flows, like the flow of air over a flat plate or airfoil, or the flow of fluid inside a pipe.   We will focus on wall-bounded turbulent flows in this text. One of the simplest pictures of turbulent flow near a wall is illustrated in Fig. 8.2.   Irregular or random motions cause the shape of

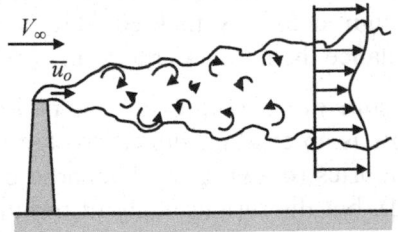

c. turbulent wake behind smokestack

Fig. 8.1

the velocity profile, as well as the location of the boundary layer edge, to change with time (Fig 8.2a).   For simplicity, the instantaneous velocity profiles can be time-averaged, with the mean profile appearing as a dashed line in Fig. 8.2a.   In this way, we could visualize the turbulent velocity as

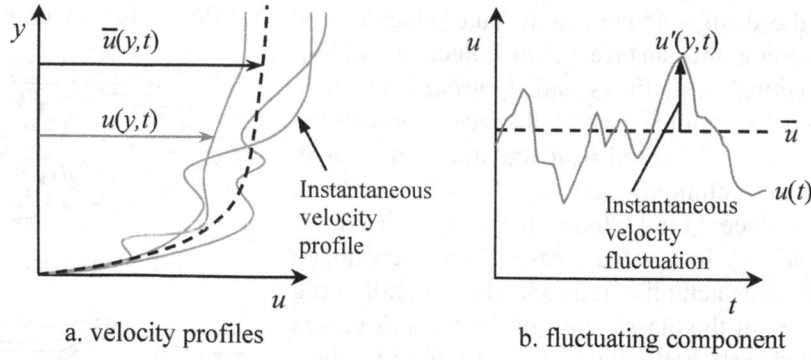

a. velocity profiles   b. fluctuating component

Fig. 8.2

being a combination of a mean velocity, $\bar{u}$, and a fluctuating component, u'. It is the fluctuating component u' that varies with time, as illustrated in Fig. 8.2b. We will later find that this decomposition of velocity into steady (mean) and unsteady (fluctuating) components is a useful technique in the analysis of turbulent flows.

The irregular, random fluctuations result in a turbulent velocity profile that is different from that of laminar flow. Figure 8.3 compares the boundary layer velocity profiles for turbulent and laminar flow. Two features are apparent: First, the velocity fluctuations that cause mixing in a turbulent flow sweep higher velocity fluid closer to the wall, resulting in a steeper velocity profile than that of laminar flow. Conversely, the same mixing behavior disperses the lower-velocity fluid particles farther from the surface, and thus the boundary layer is larger. The shape of the turbulent velocity profile suggests something about the drag on the surface: the steeper velocity profile means that the velocity gradient is larger, which in turn tells us that the shear stress at the wall is

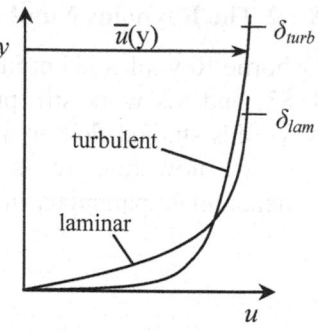

Fig. 8.3

higher in turbulent flow. In short, turbulent fluctuations enhance momentum transfer between the surface and the flow, and therefore we expect the surface (skin) friction to be higher in turbulent flow. A similar argument could be applied to the temperature profile near the wall, suggesting that turbulent fluctuations enhance heat transfer between the surface and the fluid.

Armed with even this rudimentary understanding of turbulence, we can alter the design of devices to take advantage of this flow. For example, roughening the surface of an object, or adding protrusions (sometimes called *turbulators*) to a surface promotes or enhances turbulence, resulting in increased heat transfer. Of course, the same technique increases the shear stress at the surface (and thus the skin friction). Ironically, in some cases the enhanced turbulence actually reduces the overall drag force, as in the classic case of the smooth versus dimpled golf ball (Fig. 8.4). The dimples in a golf ball "trip" the flow around the ball to become turbulent, and the increased velocity in the boundary layer delays the onset of separation. As a result, the lower-pressure turbulent wake region is smaller, and the form drag is drastically reduced. The reduction in form drag outweighs the increase in skin friction,

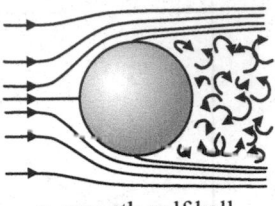

a. smooth golf ball

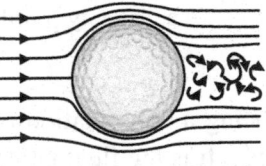

b. dimpled golf ball

Fig. 8.4

and so the overall drag is lower. Recently, the same technique was used in the design of Olympic swimsuits, leading in 2007 to the ban of protrusions or tripwires on the surface of swimsuits [2].

8.1.2 The Reynolds Number and the Onset of Turbulence

Osborne Reynolds [3] produced one of the earliest studies of turbulence in 1883, and his work still provides insight into how turbulence originates. Reynolds studied flow in a circular tube, and observed the two regimes of flow we now refer to as laminar and turbulent. He also introduced a dimensionless parameter that now bears his name:

$$Re_D = \frac{\bar{u}D}{\nu}. \tag{8.1}$$

As we mentioned in Section 2.3, the Reynolds number is a predictor of the onset of turbulence: for flow through tubes, $Re_t = \bar{u}D/\nu \approx 2300$ is the approximate transition value, while for uniform flow over a semi-infinite flat plate, $Re_t = V_\infty x_t/\nu \approx 500,000$.

Why does the Reynolds number play a role in predicting turbulence? To answer this question, recall that the Reynolds number represents the ratio of inertial to viscous forces (This can be shown by examining the acceleration and viscous force terms in the Navier-Stokes equations and performing scale analysis; see Problem 8.2). Inertial forces are those forces associated with accelerating a fluid particle, while viscous forces are those associated with slowing, or damping, the motion of the particle. The reason this ratio is important is that instabilities and disturbances are present in nearly all flows[*]. For a fluid flow at sufficiently low velocity, the viscous forces dominate the inertial forces acting on the fluid particle. Thus any infinitesimal disturbance to the particle is quickly damped out, the particle is forced into a particular "lane," and the flow remains laminar. At sufficiently high fluid velocity, however, inertial forces dominate, and viscous forces cannot prevent a wayward particle from its motion. In fact, the flow amplifies the disturbance. Disturbances create more disturbances, and ultimately, chaotic flow ensues.

For wall-bounded flows, turbulence initiates near the wall. Consider the development of turbulent flow on a semi-infinite flat plate (Fig. 8.5). Within the boundary layer, where velocity profile is steep, disturbances are especially important. A particle of fluid very close to the wall has a low velocity; if it is forced by some disturbance to move away from the wall, it will collide with particles that are much faster. Those particles are forced out of their paths, and collide with other particles moving at different velocities. If viscous forces are not sufficient to damp these motions, these collisions generate still more collisions, eventually causing the random disturbances and chaotic motion identified with the turbulent boundary layer.

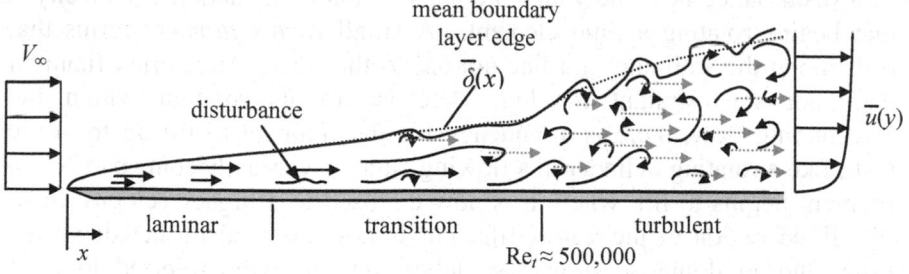

Fig. 8.5

----

[*]Chaos Theory shows that the non-linear convective derivative in the Navier-Stokes equations (see equation 1.24) is the source of instability and chaos. See Reference [1] for an introduction to this topic.

8.1.3 Eddies and Vorticity

We know that turbulence is characterized by random fluctuations in velocity and the chaotic motion of fluid particles. But within this chaotic motion is a localized identifiable structure: regions of intermittent, swirling patches of fluid called *eddies*. To put it in the language of fluid mechanics, an eddy is particle of vorticity, ω, defined as

$$\vec{\omega} = \nabla \times \vec{V} .$$ (8.2)

You may recall from your earlier coursework that vorticity can be interpreted as representing twice the angular velocity of a fluid element.

For most turbulent flows, eddies form in regions of velocity gradient. Consider again, for example, flow over a flat plate. Assuming for the moment that the flow field is two-dimensional, the vorticity can be found from Eqn. (8.2) to be

$$\omega_z = \frac{\partial v}{\partial x} - \frac{\partial u}{\partial y} .$$

Within the boundary layer, and at the wall in particular, the high shear (steep velocity profile) results in high vorticity. It seems plausible, then, that the most likely place for eddies to form is at the wall. In fact, the formation and behavior of eddies within the boundary layer is complicated, and not fully understood.

A common view of eddy formation is depicted in Fig. 8.6. An eddy begins as a disturbance near the wall – perhaps a small fluctuation in velocity – that begins rotating a fluid element. A small *vortex filament* forms that rolls along the wall and in a line normal to the flow. The vortex filament does not stay straight for long: because of its rotation within the surrounding flow, there is a tendency for the filament to lift up from the wall, like a rotating cylinder in a flowing fluid. Eventually some part of the filament begins to lift, where it is now exposed to a higher velocity flow. The lifted region of the vortex filament is now dragged further down the flow, and in doing so it gets stretched into a shape referred to as a *horseshoe* or *hairpin vortex*. Vortex stretching increases the kinetic energy of the vortex, and is thought to be a major mechanism for the main flow to transfer energy to the turbulence. Weaker, secondary vortices may form next to the hairpin vortex as well. Eventually the vortex becomes unstable, and breaks up from its own oscillations or by interaction with some other eddy. *Streamwise rolls*, which may be the remnants of prior hairpin vortices, may also be responsible for lifting vortex filaments from the wall.

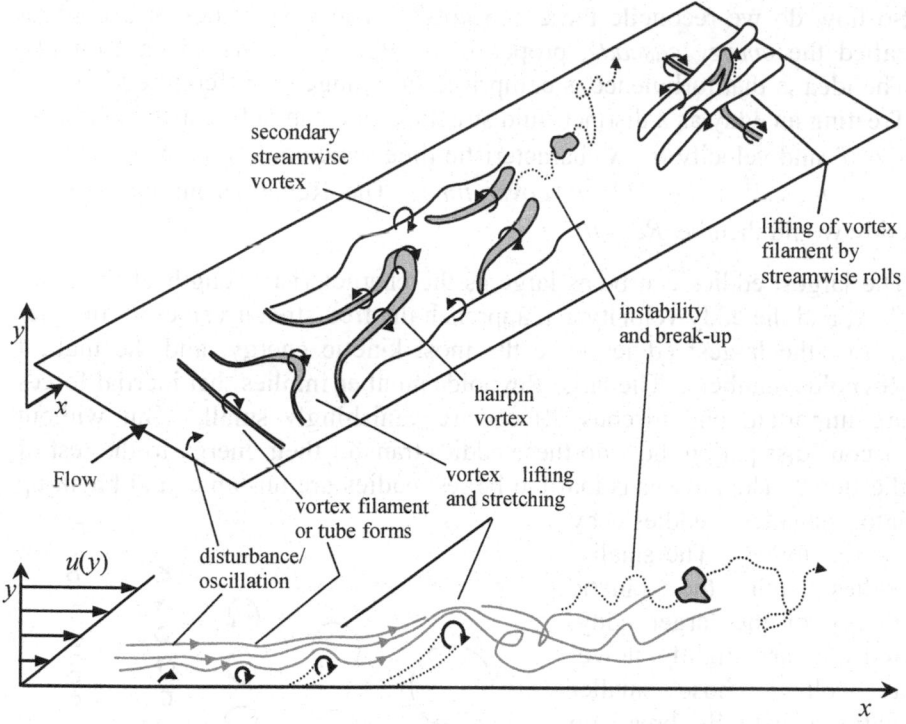

Fig. 8.6

However complex the origination and propagation of turbulence may be, it is clear that turbulence is a highly rotational phenomenon, with eddies that provide bulk motion (advection) and mixing within the boundary layer. It is the advection, via eddies, that differentiates turbulent from laminar flow. Without bulk motion, laminar flow must rely solely on viscous diffusion to transfer momentum, as well as to grow the boundary layer.

8.1.4 Scales of Turbulence

One of the more common misconceptions about turbulent flow is that viscous effects are not important – after all, turbulent flows have high Reynolds numbers, which implies that inertial effects dominate the flow. On the contrary, we have seen that most turbulence originates from shear flow and the rotation of a fluid element occurs under the action of viscous shear. Indeed, turbulent flow would not exist in a truly inviscid fluid. Viscous effects, in fact, play a vital role in turbulence.

So how do we reconcile these concepts?   The answer lies in a concept called the *energy cascade*, proposed in 1922 by Lewis Richardson [4]. The idea is that turbulence is comprised of a range of different eddy sizes. Treating an eddy as a distinct fluid structure, one can define a characteristic size ℓ and velocity u. A characteristic time scale could then be defined as $t = \ell / u$, called the eddy *turnover time*.   The Reynolds number for the eddy would then be $Re = u\ell / v$.

The largest eddies can be as large as the characteristic length of the main flow, and the eddy velocity can approach the free-stream velocity.   In other words, the largest eddies have the most kinetic energy, and the highest Reynolds numbers. The large Reynolds number implies that inertial forces are important and viscous effects are vanishingly small.   But without viscous dissipation, how do these eddies transfer their energy to the rest of the flow?   The answer is that the largest eddies are unstable, and break-up into smaller eddies by inertial forces. The smaller eddies split the kinetic energy of the larger eddy, and thus are slightly slower as well.   Those smaller eddies eventually break up due to inertial forces, and so on, until eventually the eddies reach a small enough size that their Reynolds number approaches the order of unity. Viscosity then becomes important, and viscous forces quickly dissipate the energy of the smallest eddies into heat. This process is illustrated in Figure 8.7.

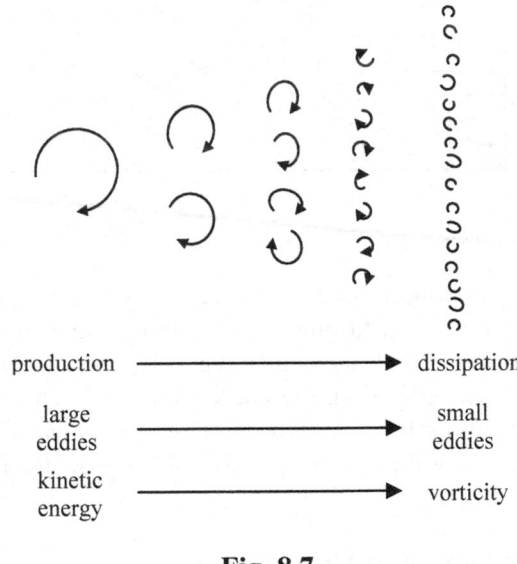

production	⟶	dissipation
large eddies	⟶	small eddies
kinetic energy	⟶	vorticity

Fig. 8.7

It is important to point out that, although the smallest eddies have much lower kinetic energy than the largest eddies, they have very high vorticity. This is because eddies decrease in size much faster than in velocity.   The smallest eddies therefore experience very large gradients in velocity, and so have very high vorticity. It makes sense, then, that the smallest eddies are dissipated by viscosity, since they also experience such high shear rates. To summarize, while the largest eddies contain the bulk of the kinetic energy of a turbulent flow, the smallest eddies contain the bulk of the vorticity, and therefore the mechanism of viscous dissipation.

We can observe that the largest eddies are as large as the characteristic length of the main flow; say, the width of a turbulent wake. But can we predict the size of the smallest eddies? Andrey Kolmogorov [5] proposed a model in 1942 based on the idea that the largest eddies contain all the kinetic energy of the turbulence, and that the smallest scales reach a Reynolds number of unity prior to dissipating into heat. Kolmogorov found

$$\eta / \ell \sim Re^{-3/4}, \tag{8.3a}$$

$$v / u \sim Re^{-1/4}, \tag{8.3b}$$

and
$$\tau / t \sim Re^{-1/2}, \tag{8.3c}$$

where η / ℓ is the ratio of length scales of the larges and smallest eddies, v / u is the ratio of their velocities, and τ / t is the ratio of their turnover times. The Reynolds number in these expressions is based on the largest eddies, $Re = u\ell / v$. The variables η, v, and τ are referred to as the *Kolmogorov microscales* (the variables ℓ, v, and t are called the *integral scales*).

With equations (8.3), we can get a feel for the relative scale of the smallest eddies. For example, consider a 10 m/s airflow over a 10-cm-diameter cylinder. The largest eddies can be as large as the wake, which is on the order of the cylinder diameter, 10 cm; their velocity could be as high as the free stream flow, or 10 m/s; thus their turnover time is on the order of $t = \ell / u \sim 0.01\text{s}$. For standard air, the large-eddy Reynolds number is approximately 69,000, yielding the following estimates for the Kolmogorov microscales:

$$\eta \sim \ell \, Re^{-3/4} = 2 \times 10^{-5} \text{ m}$$

$$v \sim u \, Re^{-1/4} = 0.6 \text{ m/s}$$

$$\tau \sim t \, Re^{-1/2} = 4 \times 10^{-5} \text{ s}$$

There are three important points to draw from Kolmogorov's model. First, there is a vast range of eddy sizes, velocities, and time scales in a turbulent flow. This could make modeling difficult, as we shall discuss later. Second, the smallest eddies are not infinitesimally small; this is because viscosity dissipates them into heat before they can become too small. Finally, note that the scale of the smallest eddies is determined by the scale of the largest eddies through the Reynolds number. If the free stream velocity over the cylinder were doubled, the size of the wake wouldn't change significantly. The size of the largest eddies would therefore not change appreciably. However, their velocities would increase. Performing the same analysis as above, we would predict that the smallest scales are even smaller than before. In other words, a turbulent flow responds to an increase in velocity by generating smaller eddies. Generating smaller eddies is how the viscous dissipation is increased to compensate for the increased production of turbulence. This is in fact what is observed in the real world: faster turbulent flows tend to have finer turbulent structure.

8.1.5 Characteristics of Turbulence

At this point, it is useful to summarize some of the main features and characteristics of turbulence:

- Turbulence is comprised of irregular, chaotic, three-dimensional fluid motion, but containing coherent structures.
- Turbulence occurs at high Reynolds numbers, where instabilities give way to chaotic motion.
- Turbulence is comprised of many scales of eddies, which dissipate energy and momentum through a series of scale ranges. The largest eddies contain the bulk of the kinetic energy, and break up by inertial forces. The smallest eddies contain the bulk of the vorticity, and dissipate by viscosity into heat.
- Turbulent flows are not only dissipative, but also dispersive through the advection mechanism.

8.1.6 Analytical Approaches

Given the complicated picture of turbulence that we have just described, an analytical solution does not seem promising, to say the least. The good news is that the governing equations of fluid mechanics still apply, whether the flow is laminar or turbulent. The challenge, then, is to find a way to *apply* the equations.

We noted in Section 8.1.5 that the smallest scales of turbulence can be very small – orders of magnitude smaller than the largest eddies. The first question we need to ask is: can we still apply the continuum hypothesis to the governing equations? If not, we will have to analyze the motion of the individual molecules of fluid. Consider the estimates of the Kolmogorov microscale in the example of Section 8.1.5. The smallest eddies are small indeed, approximately 2×10^{-5} m. To put this in perspective, the mean free path of air at atmospheric pressure is on the order of 10^{-8} m, three orders of magnitude smaller. It would appear, then, that a differential control volume that is small enough to capture the behavior of the smallest eddies will still be large enough that the fluid can be considered a continuum.[†]

It would seem that, with the development of more powerful computers, Computational Fluid Dynamics (CFD) would be the easiest way to solve the governing equations. Indeed, the use of what is called *direct numerical simulation* (DNS) is a widespread topic of research, and is yielding a better understanding of the structure of turbulent flows. However, there is a practical limitation to DNS, and it has to do with the scales of turbulence. Consider again the example in Section 8.1.3. A computer simulation would have to not only predict the largest scales of fluid motion, but also resolve the smallest eddy scales, down to 20 microns in the example. The grid size alone makes solving for the entire flow field virtually impossible. Now consider the time scale, which for the smallest eddies is on the order of 40 microseconds. We would have to solve the spatial grid at such a minute time step that the computing time would be enormous. At present, direct numerical simulation is useful in studying fundamental motions in the flow, but for practical results like drag and heat transfer, we will have to rely on more traditional analytical techniques.

Our approach in this text will be to pursue the more common analytical approaches to turbulent flow. Traditionally, the focus has been on predicting the macroscale properties and parameters of practical interest to engineering such as velocity, drag force, and heat transfer. In doing so, attempts at revealing the microscale structure of the flow are abandoned. Instead, the chaotic and random nature of the turbulent fluctuations lend themselves to statistical analysis. In fact, powerful statistical analysis has been applied in turbulent theory, although this text will only scratch the surface.

Before we begin developing models for turbulent flow, it is useful to note two idealizations that are commonly made to simplify analysis:

(i) *Homogeneous Turbulence*: turbulence whose microscale motion, on average, does not change from location to location and time to time.

(ii) *Isotropic Turbulence*: turbulence whose microscale motion, on average, does not change as the coordinate axes are rotated.

[†] It is important to remember that for rarified gas flows, the continuum hypothesis breaks down, whether or not the flow is turbulent.

As you might expect, these idealizations simplify the statistical tools, but are they realistic? Not necessarily. Certainly the flow behaves differently at the edge of a boundary layer, for example, than near the wall, so the homogeneous assumption is not likely appropriate. Furthermore, near any solid boundary, you might expect that turbulent fluctuations normal to the boundary will diminish relative to those in the direction of the flow; this is counter to the isotropic assumption. Idealizations are sometimes necessary in order to obtain a solution, of course. However, a direct benefit of these idealizations is that they can be approximated in the laboratory: grid-induced turbulence, for example, approaches homogeneous and isotropic behavior. This means that we have a ready source of experimental data with which to compare and test our simplified analytical models.

8.2 Conservation Equations for Turbulent Flow

8.2.1 Reynolds Decomposition

The simplest approach to solve the governing equations for turbulent flow is to recognize that, despite the fluctuations in velocity and other properties, the behavior of the flow seems well-behaved *on average*. For example, the time-averaged velocity profile appears to be a well-behaved function – almost predictable. We noted in Section 8.1.1 that the velocity in a turbulent flow could be considered as the superposition of time-averaged and fluctuating components (see Fig. 8.2). In fact, it was Osborne Reynolds who proposed this technique, now referred to as *Reynolds decomposition* [6]. The approach is as follows: First, each fluctuating property appearing in the governing equations is decomposed into time-averaged and fluctuating components. Then, the entire equation is itself time-averaged.

Consider some scalar property g. The Reynolds decomposition of g is

$$g = \overline{g} + g' , \tag{8.4}$$

where the time-averaged component is determined by

$$\overline{g} = \frac{1}{\tau} \int_0^\tau g(t)dt . \tag{8.5}$$

By definition, the time average of the fluctuating component is zero:

$$\overline{g'} = \frac{1}{\tau} \int_0^\tau g'(t)dt = 0 . \tag{8.6}$$

In applying time averaging to the governing equations, it will be necessary to make use of some identities. The following averaging identities will be useful, and the reader is encouraged to verify their validity. For two variables $a = \bar{a} + a'$ and $b = \bar{b} + b'$:

$$\bar{\bar{a}} = \bar{a} \qquad (8.7a)$$

$$\overline{\bar{a}\bar{b}} = \bar{a}\bar{b} \qquad (8.7b)$$

$$\overline{(\bar{a})^2} = (\bar{a})^2 \qquad (8.7c)$$

$$\overline{\bar{a}a'} = 0 \qquad (8.7d)$$

$$\overline{ab} = \bar{a}\bar{b} + \overline{a'b'} \qquad (8.7e)$$

$$\overline{a^2} = (\bar{a})^2 + \overline{(a')^2} \qquad (8.7f)$$

$$\overline{a+b} = \bar{a} + \bar{b} \qquad (8.7g)$$

$$\frac{\overline{\partial a}}{\partial x} = \frac{\partial \bar{a}}{\partial x} \qquad (8.7h)$$

$$\frac{\partial \bar{a}}{\partial t} = 0 \qquad (8.7i)$$

$$\overline{\frac{\partial a}{\partial t}} = 0 \qquad (8.7j)$$

Example 8.1: Proofs of Averaging Identities

Prove the averaging identities (8.7a) and (8.7g).

(1) Observations. This is an application of Reynolds decomposition.

(2) Problem Definition. Using Reynolds decomposition (Eqn. 8.4) and the definition of time average (Eqn. 8.5), prove Equations (8.7a) and (8.7b).

(3) Solution Plan. Apply either (8.4) or (8.5) to the left side of the given equation, and demonstrate that it simplifies to the expression on the right side.

(4) Plan Execution.

(i) **Assumptions.** (1) The functions a and b can be decomposed into mean and fluctuating components.

(ii) **Analysis.** (a) To prove identity (8.7a), the definition of time average (8.5) is applied to the left-hand side:

$$\bar{\bar{a}} = \frac{1}{\tau} \int_0^\tau \bar{a}\,dt. \qquad (a)$$

Since $\bar{a}$ is a constant with respect to time, (a) can be written as

$$\overline{\overline{a}} = \frac{1}{\tau} \overline{a} \int\limits_0^\tau dt \,, \tag{b}$$

which reduces to the desired identity:

$$\overline{\overline{a}} = \overline{a} \,. \tag{8.7a}$$

(b) To prove identity (8.7g), the definition of the time average is first substituted into the left-hand side:

$$\overline{a+b} = \frac{1}{\tau} \int\limits_0^\tau (a+b)dt \,. \tag{c}$$

The integral is a linear operator, meaning that (d) can be written as

$$\overline{a+b} = \frac{1}{\tau} \int\limits_0^\tau a\,dt + \frac{1}{\tau} \int\limits_0^\tau b\,dt \,. \tag{d}$$

Finally, the two terms on the right-hand side are the definition of time average, and so the above can be written as

$$\overline{a+b} = \overline{a} + \overline{b} \,,$$

which is (8.7g).

(5) Checking. Both analyses yielded the proper form of the equations.

(6) Comments. (i) The same approach to proving (8.7a) can be used to prove (8.7b) and (8.7c).

(ii) Identity (8.7g) shows that averaging is a linear operation, like integration or differentiation.

(iii) Identity (8.7g) could also be proven using Reynolds decomposition. Substituting (8.4) for a and b, the left side of (8.7g) becomes

$$\overline{a+b} = \overline{\overline{a} + a' + \overline{b} + b'} \,. \tag{e}$$

Recognizing that the time average (8.5) is a linear operation, the above becomes

$$\overline{a+b} = \overline{\overline{a}} + \overline{a'} + \overline{\overline{b}} + \overline{b'} \,. \tag{f}$$

The first and third terms on the right-hand side of (f) can be replaced by (8.7a), and the second and fourth terms are zero by (8.6).   Thus (f) reduces to the original identity, $\overline{a+b} = \overline{a} + \overline{b}$.

8.2.2 Conservation of Mass

In Cartesian coordinates, conservation of mass is given by:

$$\frac{\partial \rho}{\partial t} + \frac{\partial (\rho u)}{\partial x} + \frac{\partial (\rho v)}{\partial y} + \frac{\partial (\rho w)}{\partial z} = 0. \qquad (2.2a)$$

We will limit the analysis to incompressible flow, and for convenience, two-dimensional flow. Then, substituting the Reynolds-decomposed velocities $u = \bar{u} + u'$ and $v = \bar{v} + v'$,

$$\frac{\partial (\bar{u} + u')}{\partial x} + \frac{\partial (\bar{v} + v')}{\partial y} = 0. \qquad (a)$$

The derivative operator is linear, so (a) is expanded as

$$\frac{\partial \bar{u}}{\partial x} + \frac{\partial u'}{\partial x} + \frac{\partial \bar{v}}{\partial y} + \frac{\partial v'}{\partial y} = 0. \qquad (b)$$

Now the entire equation can be time-averaged. Noting that time-averaging is also a linear operation (by equation 8.7g),

$$\overline{\frac{\partial \bar{u}}{\partial x}} + \overline{\frac{\partial u'}{\partial x}} + \overline{\frac{\partial \bar{v}}{\partial y}} + \overline{\frac{\partial v'}{\partial y}} = 0. \qquad (c)$$

Then, we can simplify each term by invoking identity (8.7h):

$$\frac{\partial \overline{\bar{u}}}{\partial x} + \frac{\partial \overline{u'}}{\partial x} + \frac{\partial \overline{\bar{v}}}{\partial y} + \frac{\partial \overline{v'}}{\partial y} = 0. \qquad (d)$$

By identity (8.7a), $\overline{\bar{u}} = \bar{u}$ and $\overline{\bar{v}} = \bar{v}$, and by (8.6), $\overline{u'} = \overline{v'} = 0$, so the above reduces to

$$\frac{\partial \bar{u}}{\partial x} + \frac{\partial \bar{v}}{\partial y} = 0. \qquad (8.8)$$

Equation (8.8) is the time-averaged, turbulent flow continuity equation. Note that (8.8) is identical in form to the original equation, so no new terms appear in the equation due to the presence of fluctuating velocities. Also, subtracting (8.8) from (b) gives

$$\frac{\partial u'}{\partial x} + \frac{\partial v'}{\partial y} = 0, \qquad (8.9)$$

which demonstrates that the divergence of the fluctuating terms is also zero. The usefulness of this equation is not obvious, since we are generally more interested in the average velocities. However, this equation will be useful in the derivations that follow.

8.2.3 Momentum Equations

The x- and y-momentum equations are given by

$$\rho\left(\frac{\partial u}{\partial t} + u\frac{\partial u}{\partial x} + v\frac{\partial u}{\partial y} + w\frac{\partial u}{\partial z}\right) = \rho g_x - \frac{\partial p}{\partial x} + \mu\left(\frac{\partial^2 u}{\partial x^2} + \frac{\partial^2 u}{\partial y^2} + \frac{\partial^2 u}{\partial z^2}\right),$$

$$(2.10x)$$

and $$\rho\left(\frac{\partial v}{\partial t} + u\frac{\partial v}{\partial x} + v\frac{\partial v}{\partial y} + w\frac{\partial v}{\partial z}\right) = \rho g_y - \frac{\partial p}{\partial y} + \mu\left(\frac{\partial^2 v}{\partial x^2} + \frac{\partial^2 v}{\partial y^2} + \frac{\partial^2 v}{\partial z^2}\right),$$

$$(2.10y)$$

where we have assumed incompressible flow and constant properties. Again, we will limit ourselves to two-dimensional flow for convenience, and neglect body forces. We will also assume the flow is, on average, steady state; that is, that $\bar{u}$ and $\bar{v}$ do not change with time (the overall velocity, however, contains a fluctuating component, and so u and v still vary with time). Thus equations (2.10) become

$$\rho\left(\frac{\partial u}{\partial t} + u\frac{\partial u}{\partial x} + v\frac{\partial u}{\partial y}\right) = -\frac{\partial p}{\partial x} + \mu\left(\frac{\partial^2 u}{\partial x^2} + \frac{\partial^2 u}{\partial y^2}\right), \qquad (8.10x)$$

and $$\rho\left(\frac{\partial v}{\partial t} + u\frac{\partial v}{\partial x} + v\frac{\partial v}{\partial y}\right) = -\frac{\partial p}{\partial y} + \mu\left(\frac{\partial^2 v}{\partial x^2} + \frac{\partial^2 v}{\partial y^2}\right). \qquad (8.10y)$$

At this point, we could perform Reynolds decomposition on equations (8.10) and time-average the result. This is left as homework Problem 8.6 for the case of the y-momentum equation. However, a slightly simpler analysis results if we first manipulate the left-hand sides of the equations. For the x-momentum equation, the terms $u(\partial u/\partial x)$ and $v(\partial u/\partial y)$ can be replaced by the following relations, derived from the product rule of differentiation:

$$u\frac{\partial u}{\partial x} = \frac{\partial u^2}{\partial x} - u\frac{\partial u}{\partial x}, \qquad (a)$$

and
$$v\frac{\partial u}{\partial y} = \frac{\partial(uv)}{\partial y} - u\frac{\partial v}{\partial y}. \tag{b}$$

Substituting (a) into the x-momentum equation (8.10x) yields

$$\rho\left(\frac{\partial u}{\partial t} + \frac{\partial u^2}{\partial x} - u\frac{\partial u}{\partial x} + \frac{\partial(uv)}{\partial y} - u\frac{\partial v}{\partial y}\right) = -\frac{\partial p}{\partial x} + \mu\left(\frac{\partial^2 u}{\partial x^2} + \frac{\partial^2 u}{\partial y^2}\right). \tag{c}$$

$$\underbrace{\quad}_{\textcircled{a}} \qquad \underbrace{\quad}_{\textcircled{b}}$$

Note that terms marked $\textcircled{a}$ and $\textcircled{b}$ in the above can be combined as

$$-u\left(\frac{\partial u}{\partial x} + \frac{\partial v}{\partial y}\right), \tag{d}$$

which is zero by conservation of mass. Thus the x-momentum equation reduces to

$$\rho\left(\frac{\partial u}{\partial t} + \frac{\partial u^2}{\partial x} + \frac{\partial(uv)}{\partial y}\right) = -\frac{\partial p}{\partial x} + \mu\left(\frac{\partial^2 u}{\partial x^2} + \frac{\partial^2 u}{\partial y^2}\right). \tag{8.11}$$

We are now ready to perform Reynolds decomposition and time-averaging. It is left as homework exercises (Probs. 8.5 and 8.6) to show that the x- and y-momentum equations for turbulent flow are

$$\rho\left(\bar{u}\frac{\partial\bar{u}}{\partial x} + \bar{v}\frac{\partial\bar{u}}{\partial y}\right) = -\frac{\partial\bar{p}}{\partial x} + \mu\left(\frac{\partial^2\bar{u}}{\partial x^2} + \frac{\partial^2\bar{u}}{\partial y^2}\right) - \rho\frac{\partial\overline{(u')^2}}{\partial x} - \rho\frac{\partial\overline{u'v'}}{\partial y}$$

$$\tag{8.12x}$$

$$\rho\left(\bar{u}\frac{\partial\bar{v}}{\partial x} + \bar{v}\frac{\partial\bar{v}}{\partial y}\right) = -\frac{\partial\bar{p}}{\partial y} + \mu\left(\frac{\partial^2\bar{v}}{\partial x^2} + \frac{\partial^2\bar{v}}{\partial y^2}\right) - \rho\frac{\partial\overline{u'v'}}{\partial x} - \rho\frac{\partial\overline{(v')^2}}{\partial y}$$

$$\tag{8.12y}$$

Equations (8.12) are the turbulent x- and y-momentum equations for turbulent flow. The above equations are identical to the original equations (8.10) except in two ways. First, the transient terms $\partial u/\partial t$ and $\partial v/\partial t$ disappear. Recall that we are assuming that the flow is steady on average, but regardless, the terms disappear by identity (8.7j). The second, and more important, difference is the introduction of new terms containing fluctuating velocity components. The terms originate from the convective derivatives on the left side of the momentum equations. The significance of these new terms will be discussed in Section 8.3.

8.2.4 Energy Equation

For incompressible flow, the energy equation is given by

$$\rho c_p \left(\frac{\partial T}{\partial t} + u \frac{\partial T}{\partial x} + v \frac{\partial T}{\partial y} + w \frac{\partial T}{\partial z} \right) = k \left(\frac{\partial^2 T}{\partial x^2} + \frac{\partial^2 T}{\partial y^2} + \frac{\partial^2 T}{\partial z^2} \right) + \mu \, \Phi,$$

(2.19b)

where we have also neglected heat generation, and assumed constant thermal properties. We assume steady-on-average flow again, and for simplicity, we limit the derivation to two-dimensional flow. Furthermore, we will neglect the dissipation function $\mu \, \Phi$; this assumption is appropriate as long as the flow is not highly viscous or compressible. The Energy Equation reduces to

$$\rho c_p \left(\frac{\partial T}{\partial t} + u \frac{\partial T}{\partial x} + v \frac{\partial T}{\partial y} \right) = k \left(\frac{\partial^2 T}{\partial x^2} + \frac{\partial^2 T}{\partial y^2} \right).$$

(8.13)

Again, it is left as an exercise (Prob. 8.7) that following Reynolds decomposition and time averaging, Eqn. (8.13) becomes

$$\rho c_p \left(\overline{u} \frac{\partial \overline{T}}{\partial x} + \overline{v} \frac{\partial \overline{T}}{\partial y} \right) = k \left(\frac{\partial^2 \overline{T}}{\partial x^2} + \frac{\partial^2 \overline{T}}{\partial y^2} \right) - \rho c_p \frac{\partial \left(\overline{u'T'} \right)}{\partial x} - \rho c_p \frac{\partial \left(\overline{v'T'} \right)}{\partial y}.$$

(8.14)

Eqn. (8.14) is almost identical to the steady-state Energy Equation (8.13), except for the disappearance of the transient term $\rho c_p \partial T / \partial t$, and the appearance of two terms containing fluctuating velocities and temperature. Like in the momentum equation, these new terms come out of the convective terms on the left side of the equation, and their significance will be discussed further in Section 8.3.

8.2.5 Summary of Governing Equations for Turbulent Flow

Continuity:
$$\frac{\partial \overline{u}}{\partial x} + \frac{\partial \overline{v}}{\partial y} = 0 \; .$$

(8.8)

x-momentum:

$$\rho \left(\overline{u} \frac{\partial \overline{u}}{\partial x} + \overline{v} \frac{\partial \overline{u}}{\partial y} \right) = -\frac{\partial \overline{p}}{\partial x} + \mu \left(\frac{\partial^2 \overline{u}}{\partial x^2} + \frac{\partial^2 \overline{u}}{\partial y^2} \right) - \rho \frac{\partial \overline{(u')^2}}{\partial x} - \rho \frac{\partial \overline{u'v'}}{\partial y} \; .$$

(8.12x)

y-momentum:

$$\rho\left(\bar{u}\frac{\partial\bar{v}}{\partial x}+\bar{v}\frac{\partial\bar{v}}{\partial y}\right)=-\frac{\partial\bar{p}}{\partial y}+\mu\left(\frac{\partial^{2}\bar{v}}{\partial x^{2}}+\frac{\partial^{2}\bar{v}}{\partial y^{2}}\right)-\rho\frac{\partial\overline{u'v'}}{\partial x}-\rho\frac{\partial\overline{(v')^{2}}}{\partial y}.\quad(8.12y)$$

Energy:

$$\rho c_{p}\left(\bar{u}\frac{\partial\bar{T}}{\partial x}+\bar{v}\frac{\partial\bar{T}}{\partial y}\right)=k\left(\frac{\partial^{2}\bar{T}}{\partial x^{2}}+\frac{\partial^{2}\bar{T}}{\partial y^{2}}\right)-\rho c_{p}\frac{\partial\overline{\left(u'T'\right)}}{\partial x}-\rho c_{p}\frac{\partial\overline{\left(v'T'\right)}}{\partial y}.\quad(8.14)$$

8.3 Analysis of External Turbulent Flow

Having established the governing equations for turbulent flow, our goal now is to solve these equations to determine quantities of practical interest, like drag force and heat transfer. For many engineering applications, drag and heat transfer involve flow along a surface. Therefore, just as for the case of laminar flow, we will invoke the boundary layer concept.

The development that follows is almost identical that of the laminar flow boundary layer equations (Chapter 4). However, this development differs in two ways: First, the governing equations for turbulent flow contain time-averaged quantities like $\bar{u}$ and $\bar{T}$. These quantities are handled exactly as u and T were in the original development. However, the second difference is the presence of terms containing fluctuating quantities like u' and T'. These terms will require additional consideration.

8.3.1 Turbulent Boundary Layer Equations

(i) Turbulent Momentum Boundary Layer Equation

Let's consider a flat plate in turbulent flow, as depicted in Fig. 8.8. We will assume steady-state, incompressible flow with constant properties. In addition, we will assume that the boundary layer is thin; that is, $\delta \ll L$, or

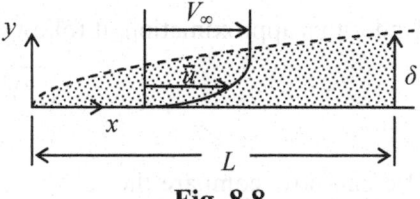

Fig. 8.8

$$\frac{\delta}{L}\ll 1.\quad(8.15)$$

Following the same arguments as for the laminar boundary layer, the following scalar arguments are made:

$$\bar{u} \sim V_\infty,$$ (8.16a)

$$x \sim L,$$ (8.16b)

and $$y \sim \delta.$$ (8.16c)

Then, following an analysis identical to Section 4.2, it can be shown that the viscous dissipation terms in (8.12x) compare as follows:

$$\frac{\partial^2 \bar{u}}{\partial x^2} \ll \frac{\partial^2 \bar{u}}{\partial y^2}.$$ (8.17)

Also, the pressure gradient in the y-direction is negligible,

$$\frac{\partial \bar{p}}{\partial y} \approx 0,$$ (8.18)

and the pressure gradient in the x-direction can be expressed as

$$\frac{\partial \bar{p}}{\partial x} = \frac{d\bar{p}}{dx} = \frac{dp_\infty}{dx}.$$ (8.19)

The fluctuation terms $\dfrac{\partial \overline{(u')^2}}{\partial x}$ and $\dfrac{\partial \overline{u'v'}}{\partial y}$ require us to make additional scaling arguments. If the fluctuation terms are the result of eddies, one could argue that there is no preferred direction to the fluctuations. This approximation is the same as assuming the turbulence is isotropic. Thus,

$$u' \sim v'.$$ (8.20)

Under this approximation, it follows that

$$\overline{(u')^2} \sim \overline{u'v'}.$$ (8.21)

We can now compare the relative magnitudes of $\dfrac{\partial \overline{(u')^2}}{\partial x}$ and $\dfrac{\partial \overline{u'v'}}{\partial y}$ using scale analysis:

First fluctuation term: $$\frac{\partial \overline{(u')^2}}{\partial x} \sim \frac{\overline{(u')^2}}{L}.$$ (a)

Second fluctuation term: $\dfrac{\partial \overline{u'v'}}{\partial y} \sim \dfrac{\overline{u'v'}}{\delta} \sim \dfrac{\overline{(u')^2}}{\delta}$.

$$\hspace{8cm}\text{(b)}$$

Since $\delta \ll L$, we can conclude that

$$\frac{\partial \overline{(u')^2}}{\partial x} \ll \frac{\partial \overline{u'v'}}{\partial y} . \tag{8.22}$$

Finally, using the simplifications (8.17) and (8.22), the x-momentum equation for the turbulent boundary layer reduces to

$$\rho\left(\overline{u}\,\frac{\partial \overline{u}}{\partial x} + \overline{v}\,\frac{\partial \overline{u}}{\partial y}\right) = -\frac{d\overline{p}}{dx} + \mu\frac{\partial^2 \overline{u}}{\partial y^2} - \rho\frac{\partial \overline{u'v'}}{\partial y} . \tag{8.23}$$

(ii) Turbulent Boundary Layer Energy Equation

Again following the derivation for the laminar flow boundary layer equations, we begin with the following scaling arguments for the thermal boundary layer:

$$x \sim L , \tag{8.16b}$$

$$y \sim \delta_t , \tag{8.24}$$

$$\Delta T \sim T_s - T_\infty . \tag{8.25}$$

Just as in Chapter 4, scales for the velocity components depend on the relative sizes of δ_t and δ. But in either case, we find that the second-derivative terms compare as follows:

$$\frac{\partial^2 \overline{T}}{\partial x^2} \ll \frac{\partial^2 \overline{T}}{\partial y^2} . \tag{8.26}$$

The question remains as to the relative importance of the fluctuation terms $\rho c_p \dfrac{\partial}{\partial x}\left(\overline{u'T'}\right)$ and $\rho c_p \dfrac{\partial}{\partial y}\left(\overline{v'T'}\right)$. Again, arguing that there is no preferred direction to the fluctuations,

$$u' \sim v' . \tag{8.20}$$

It can then be argued that the terms $\overline{u'T'}$ and $\overline{v'T'}$ are of the same order of magnitude:

$$\overline{u'T'} \sim \overline{v'T'} . \tag{8.27}$$

It is left as an exercise to show that the fluctuating terms compare as follows:

$$\frac{\partial\overline{\left(u'T'\right)}}{\partial x} << \frac{\partial\overline{\left(v'T'\right)}}{\partial y} \ . \tag{8.28}$$

Applying the simplifications (8.26) and (8.28), the energy equation then reduces to the following form for the turbulent boundary layer:

$$\rho c_p\left(\overline{u}\frac{\partial\overline{T}}{\partial x} + \overline{v}\frac{\partial\overline{T}}{\partial y}\right) = k\frac{\partial^2\overline{T}}{\partial y^2} - \rho c_p\frac{\partial\overline{\left(v'T'\right)}}{\partial y} \ . \tag{8.29}$$

8.3.2 Reynolds Stress and Heat Flux

With a slight modification, we can write the x-momentum and energy boundary layer equations (equations 8.23 and 8.29) as

$$\rho\left(\overline{u}\frac{\partial\overline{u}}{\partial x} + \overline{v}\frac{\partial\overline{u}}{\partial y}\right) = -\frac{d\overline{p}}{dx} + \frac{\partial}{\partial y}\left(\mu\frac{\partial\overline{u}}{\partial y} - \rho\overline{u'v'}\right) \tag{8.30}$$

and

$$\rho c_p\left(\overline{u}\frac{\partial\overline{T}}{\partial x} + \overline{v}\frac{\partial\overline{T}}{\partial y}\right) = \frac{\partial}{\partial y}\left(k\frac{\partial\overline{T}}{\partial y} - \rho c_p\overline{v'T'}\right) \ . \tag{8.31}$$

The equations above are useful because they provide physical insight into the time-averaged fluctuation terms $\overline{u'v'}$ and $\overline{v'T'}$—and ultimately, a way to model them.

Based on their location in the equation, the terms in parentheses on the right-hand side of (8.30) appear to represent shear stress. Certainly the first term represents the molecular shear stress resulting from the time-averaged velocity. Might we view the second term as a shear imposed by the time-averaged turbulent velocity fluctuations? As a matter of fact, this suggestion was first made in 1877 by the physicist Joseph Valentin Boussinesq[‡].

To make sense of his idea, consider a particle "fluctuation" imposed on some average velocity profile, as depicted in Figure 8.9. In Figure 8.9a, a fluid particle is forced by some fluctuation toward the wall, and as a result the particle finds itself in a region of lower local velocity. Relative to the local velocity, the fluid particle has a velocity that is higher by a value of

---

[‡] Boussinesq is also known, among other things, for the *Boussinesq Approximation* (see Section 2.8).

$+u'$. Thus the velocity fluctuation in the y-direction, $-v'$, results in a velocity fluctuation in the x-direction, $+u'$. It appears from Fig. 8.9a that the magnitude of u' depends on the slope of the mean velocity profile, so

$$-\overline{u'v'} \propto \frac{\partial \overline{u}}{\partial y} .$$

Just like for viscous shear stress, the time-averaged fluctuation term $\overline{u'v'}$ is also proportional to the velocity gradient. This suggests that the fluctuation term $\overline{u'v'}$ behaves like a shear in the flow. Note that the left-hand side of the equation above is actually positive, since the fluctuation term $\overline{u'v'}$ is negative. You can verify the sign of $\overline{u'v'}$ is the same in the case where a particle is forced to move away from the wall (Fig. 8.9b).

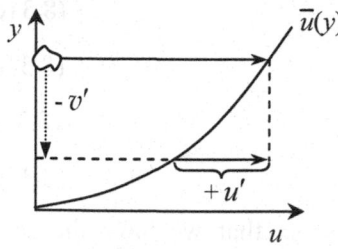
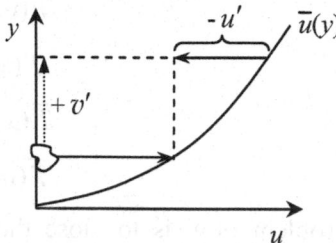

a. Particle motion toward wall   b. Particle motion away from wall

Fig. 8.9

To summarize, the *apparent shear stress* experienced by the flow is made up of two parts: the molecular shear imposed by the time-averaged velocity profile, and a turbulent shear imposed by the time-averaged velocity fluctuations. The latter term, $\rho \overline{u'v'}$, is usually called the *turbulent shear stress* or the *Reynolds stress*. Similarly, the *apparent heat flux* in the energy equation includes a turbulence-induced heat flux, $\rho c_p \overline{v'T'}$, which is sometimes referred to as the *turbulent heat flux* or the *Reynolds heat flux*.

8.3.3 The Closure Problem of Turbulence

We can summarize the turbulent boundary layer equations as follows:

Continuity:
$$\frac{\partial \overline{u}}{\partial x} + \frac{\partial \overline{v}}{\partial y} = 0 . \tag{8.8}$$

x-Momentum:

$$\rho\left(\bar{u}\frac{\partial \bar{u}}{\partial x} + \bar{v}\frac{\partial \bar{u}}{\partial y} \right) = -\frac{d\bar{p}}{dx} + \frac{\partial}{\partial y}\left(\mu\frac{\partial \bar{u}}{\partial y} - \rho\overline{u'v'} \right). \tag{8.30}$$

Energy:

$$\rho c_p\left(\bar{u}\frac{\partial \bar{T}}{\partial x} + \bar{v}\frac{\partial \bar{T}}{\partial y} \right) = \frac{\partial}{\partial y}\left(k\frac{\partial \bar{T}}{\partial y} - \rho c_p\overline{v'T'} \right). \tag{8.31}$$

These equations are subject to the boundary conditions:

$$\bar{u}(x,0) = 0, \tag{8.31a}$$

$$\bar{v}(x,0) = 0, \tag{8.31b}$$

$$\bar{u}(x,\infty) = V_\infty, \tag{8.31c}$$

$$\bar{u}(0,y) = V_\infty, \tag{8.31d}$$

$$\bar{T}(x,0) = T_s, \tag{8.31e}$$

$$\bar{T}(x,\infty) = T_\infty, \tag{8.31f}$$

$$\bar{T}(0,y) = T_\infty. \tag{8.31g}$$

The problem now is to close the equations, so that we have the same number of equations as unknowns. Regarding the pressure $\bar{p}(x)$, we can show that the pressure gradient can be expressed as

$$\frac{d\bar{p}}{dx} = \frac{dp_\infty}{dx}, \tag{8.32}$$

and from inviscid flow theory, outside the boundary layer we have

$$V_\infty\frac{dV_\infty}{dx} = -\frac{1}{\rho}\frac{dp_\infty}{dx}. \tag{8.33}$$

In other words, if we know the velocity field outside the boundary layer, the pressure can be determined.

This leaves us with the three equations (8.8), (8.30), and (8.31), but five unknowns: $\bar{u}$, $\bar{v}$, $\bar{T}$, $\overline{u'v'}$, and $\overline{v'T'}$. This is *the closure problem of turbulence*. The latter two terms are due to turbulent fluctuations, and the challenge is that we have never seen expressions like these before. In short, if we are to solve the governing equations for the turbulent boundary layer, we need to find some way to model the new unknowns $\overline{u'v'}$ and $\overline{v'T'}$.

Even if we are able to model the two new unknowns, it is important to realize that there is no exact solution to the turbulent boundary layer equations. This wasn't true for the laminar boundary layer equations, for which we developed the Blasius and Pohlhausen solutions (see Chapter 4). The turbulent boundary layer equations will require either numerical solution, or some kind of approximate method. We will keep this in mind as we proceed with our modeling.

8.3.4 Eddy Diffusivity

Based on Boussinesq's hypothesis, it is customary to model the Reynolds stress as follows,

$$- \rho \overline{u'v'} = \rho \varepsilon_M \frac{\partial \overline{u}}{\partial y}, \tag{8.34}$$

where ε_M is called the *momentum eddy diffusivity*. The term $\rho \varepsilon_M$ is often referred to as the *eddy viscosity*. Similarly, we can model the Reynolds heat flux as

$$- \rho c_p \overline{v'T'} = \rho c_p \varepsilon_H \frac{\partial \overline{T}}{\partial y}, \tag{8.35}$$

where ε_H is called the *thermal eddy diffusivity*, and $\rho c_p \varepsilon_H$ is sometimes referred to as the *eddy conductivity*.

We can then write the boundary layer momentum and energy equations as

$$\rho \left(\overline{u} \frac{\partial \overline{u}}{\partial x} + \overline{v} \frac{\partial \overline{u}}{\partial y} \right) = - \frac{d\overline{p}}{dx} + \frac{\partial}{\partial y} \left[(\mu + \rho \varepsilon_M) \frac{\partial \overline{u}}{\partial y} \right], \tag{8.36}$$

and

$$\rho c_p \left(\overline{u} \frac{\partial \overline{T}}{\partial x} + \overline{v} \frac{\partial \overline{T}}{\partial y} \right) = \frac{\partial}{\partial y} \left[(k + \rho c_p \varepsilon_H) \frac{\partial \overline{T}}{\partial y} \right]. \tag{8.37}$$

It is customary to simplify the above equations by dividing (8.36) by ρ and (8.37) by ρc_p:

$$\overline{u} \frac{\partial \overline{u}}{\partial x} + \overline{v} \frac{\partial \overline{u}}{\partial y} = \frac{\partial}{\partial y} \left[(\nu + \varepsilon_M) \frac{\partial \overline{u}}{\partial y} \right] \tag{8.38}$$

and

$$\overline{u} \frac{\partial \overline{T}}{\partial x} + \overline{v} \frac{\partial \overline{T}}{\partial y} = \frac{\partial}{\partial y} \left[(\alpha + \varepsilon_H) \frac{\partial \overline{T}}{\partial y} \right]. \tag{8.39}$$

Again, the terms in brackets in (8.38) and (8.39) represent the *apparent shear stress* and *apparent heat flux*, respectively:

$$\frac{\tau_{app}}{\rho} = \left(v + \varepsilon_M\right)\frac{\partial \overline{u}}{\partial y}, \tag{8.40}$$

and
$$-\frac{q''_{app}}{\rho c_p} = \left(\alpha + \varepsilon_H\right)\frac{\partial \overline{T}}{\partial y}, \tag{8.41}$$

where the negative sign in (8.41) assigns the correct direction to the heat transfer. Note that ε_M and ε_H are properties of the flow, not the fluid. This means that, to some extent, we have to know something about the velocity field in order to evaluate ε_M (and well as temperature field for ε_H). Also, note that we haven't reduced the number of unknowns in the governing equations; we've merely replaced the fluctuation terms with expressions containing different unknowns, ε_M and ε_H. Finding ways to evaluate ε_M and ε_H is one of the main goals of turbulence research, and is the focus of the discussion that follows.

8.4 Momentum Transfer in External Turbulent Flow

You may recall that the momentum and energy equations are decoupled: solving the energy equation requires knowledge of the velocity field, but solving the momentum equation does not require knowledge of the temperature field. Therefore solving for momentum transfer requires the solution of only the continuity and momentum equations. This still leaves us with a closure problem, with two equations and three unknowns. However, the problem is slightly simpler in that we have only one new term, ε_M, that requires further modeling.

8.4.1 Modeling Eddy Diffusivity: Prandtl's Mixing Length Theory

The simplest model for eddy diffusivity was suggested by Boussinesq, who postulated that ε_M was constant. The problem with this model can be seen by examining equation (8.34): a constant ε_M does not allow $\overline{u'v'}$ to approach zero at the wall (except in the trivial case of the separation point). We expect that the turbulent fluctuations diminish, or are damped out, very close to the wall. Therefore a more realistic model for eddy diffusivity would be one where ε_M approaches zero at the wall.

In 1925, Ludwig Prandtl [7] reasoned that particles of fluid acted analogous to molecules in the kinetic theory of gases. We can use the following scaling argument to justify his choice of model. First, consider Fig. 8.10, which depicts a particle of fluid being forced toward the wall by a velocity fluctuation v'. We will

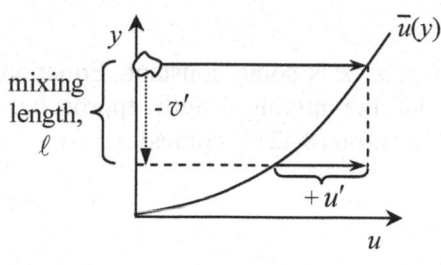

Fig. 8.10

define the mixing length ℓ as the distance the particle travels toward the wall as the result of a fluctuation. The velocity fluctuation u' that results can be approximated from a Taylor series as

$$u_{final} \approx u_{initial} + \frac{\partial \overline{u}}{\partial y} dy ,\qquad \text{(a)}$$

or, with $u' = u_{final} - u_{initial}$,

$$u' \sim \ell \frac{\partial \overline{u}}{\partial y} .\qquad \text{(b)}$$

If we assume, as we have before, that fluctuations have no preferred direction, then $u' \sim v'$, and so

$$v' \sim \ell \frac{\partial \overline{u}}{\partial y} .\qquad \text{(c)}$$

One could argue, then, that the turbulent stress term $-\overline{u'v'}$ is of the following scale:

$$-\overline{u'v'} \sim (u')(v') \sim \ell^2 \left(\frac{\partial \overline{u}}{\partial y} \right)^2 .\qquad \text{(d)}$$

Finally, we can solve Equation (8.34), for the eddy viscosity:

$$\varepsilon_M = \frac{-\overline{u'v'}}{\partial \overline{u} / \partial y} \sim \ell^2 \left| \frac{\partial \overline{u}}{\partial y} \right| ,\qquad \text{(8.42)}$$

where the absolute value is imposed on the derivative to ensure that the eddy diffusivity remains positive.

What remains is to be determined is a model for the mixing length itself. The model depends on the type of flow; for flow over a flat plate, Prandtl proposed the following model,

$$\ell = \kappa y, \tag{8.43}$$

where κ is some constant. Equation (8.43) implies, as we might expect, that the mixing length approaches zero as y approaches zero. Then Equation (8.42) becomes

$$\varepsilon_M = \kappa^2 y^2 \left| \frac{\partial \overline{u}}{\partial y} \right|. \tag{8.44}$$

Equation (8.44) is *Prandtl's mixing-length model*. We could now substitute this model into the momentum equation, (8.38), and attempt solve the boundary layer equations (8.8) and (8.38) either numerically or using an approximate solution technique. However, the variable κ is now the unknown, and it turns out that no one single value for κ is effective throughout the entire boundary layer. We need to further understand the behavior of the boundary layer before we can develop a suitable model for κ.

8.4.2 Universal Turbulent Velocity Profile

When we studied laminar flow, we discovered that we could find an approximate solution to the integral form of the boundary layer equations (Chapter 5). To do this, all we need is an approximate solution to the velocity profile. But is there a function $u(y)$ that we can confidently apply to a turbulent flow? The pursuit of such a *universal velocity profile* will also provide physical insight that we can apply to other solution techniques as well.

(i) Large-Scale Velocity Distribution: "Velocity Defect Law"

If we are looking for a universal boundary layer velocity profile, the first step might be to normalize the variables that make up the velocity distribution. As a first guess, we choose y / δ for one axis, and $\overline{u} / V_\infty$ for the other, where V_∞ is the velocity outside the boundary layer.

The result of such a plot is illustrated in Fig. 8.11 (based on data presented by Clauser [8]). The different curves correspond to different values of wall friction. If the plot were truly universal, the velocity curves would collapse into a single curve. Clearly, more work is needed.

Since wall friction is important, it might make sense to normalize the data by the friction factor,

$$C_f = \frac{\tau_o}{(1/2)\rho V_\infty^2},$$

where τ_o is the shear stress at the wall. In fact, after some manipulation, invoking the friction factor turns out to be a good idea. First, for convenience, we cast the data relative to the velocity outside the boundary layer,

$$(\bar{u} - V_\infty). \qquad (8.45)$$

Second, we define a *friction velocity* as

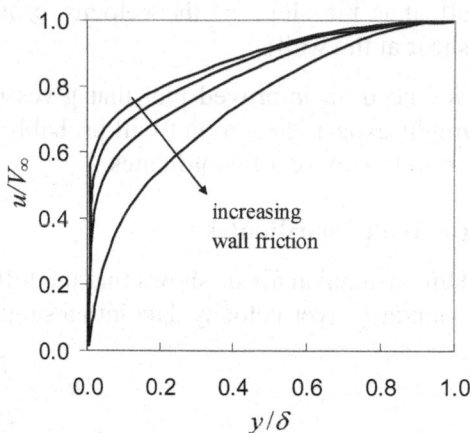

Fig. 8.11

$$u^* \equiv \sqrt{\tau_o / \rho}, \qquad (8.46)$$

You can show that the friction velocity can also be written as

$$u^* = V_\infty \sqrt{C_f / 2}, \qquad (8.47)$$

so u^* has the same dimensions as velocity. Therefore, dividing (8.45) by u^* will normalize the velocity difference. This is called the *velocity defect*:

$$\frac{(\bar{u} - V_\infty)}{u^*}. \qquad (8.48)$$

The velocity defect is plotted against y/δ in Figure 8.12 (adapted from Clauser [8]). The curves indeed collapse into a single curve. The dashed lines represent data from a range of wall roughness.

The defect plot is a good start, but unfortunately it doesn't show enough detail close to the wall, where the shape of the velocity curve is the most important (after

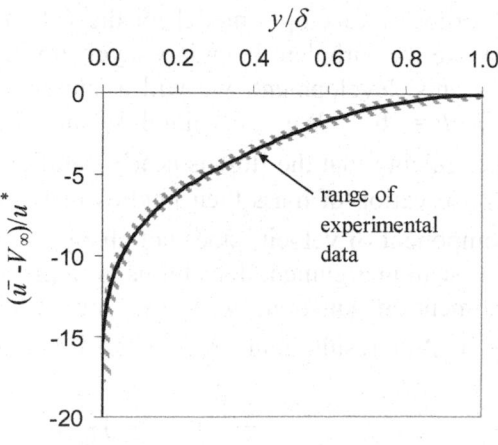

Fig. 8.12

all, it is the slope of the velocity profile at the wall that determines the shear at the wall).

We need an improved plot that gives more detail near the wall. As you might expect, such a plot will probably be logarithmic, and will require an entirely new set of coordinates.

(ii) Wall Coordinates

Dimensional analysis shows that the following coordinates will collapse the boundary layer velocity data into a single curve reasonably well:

$$u^+ \equiv \frac{\overline{u}}{u^*}, \tag{8.49}$$

and

$$y^+ \equiv \frac{yu^*}{v}, \tag{8.50}$$

Similarly, $v^+ = \overline{v}/u^*$, and $x^+ = xu^*/v$. These variables are dimensionless, and are commonly referred to as the *wall coordinates*. The resulting plot is depicted in Figure 8.13, based on data presented by Clauser [8]. To get a sense of scale, the boundary layer extends out to approximately $\delta \approx 2000$ to 5000. The range of experimental data includes data obtained from flow in a pipe as well as flow along a flat plate, so the velocity profile does indeed appear universal. We are now going to make scaling arguments to develop a model for this velocity profile.

(iii) Near-Wall Profile: Couette Flow Assumption

In order to develop a model for the velocity profile near the wall, we can invoke the turbulent boundary layer momentum equation, equation (8.38). For this development, we will assume flow over a flat plate, for which $d\overline{p}/dx = 0$. We can simplify the momentum equation further by recognizing that the flow is nearly parallel close to the wall; that is, $\overline{v} \sim 0$. Conservation of mass then implies that $\partial \overline{u}/\partial x \sim 0$, meaning that the $\overline{u}$ component of velocity does not change significantly along the wall. What this scaling argument does for us is suggest that the convective terms in the momentum equation, $\overline{u}\,\partial \overline{u}/\partial x$ and $\overline{v}\,\partial \overline{u}/\partial y$, are each approximately zero. As a result, in this region (8.38) becomes

$$\frac{\partial}{\partial y}\left[(v + \varepsilon_M)\frac{\partial \overline{u}}{\partial y}\right] \sim 0 \quad \text{near wall},$$

where the bracketed term is the apparent shear stress, τ_{app} / ρ (Eqn. 8.40). The above relationship implies that the apparent stress is approximately constant (with respect to y):

$$\frac{\tau_{app}}{\rho} = \left(v + \varepsilon_M\right)\frac{\partial \overline{u}}{\partial y} \sim \text{constant}. \qquad (8.51)$$

You may recall from Chapter 3 that this result is similar to Couette Flow, where a linear velocity profile implies that the local shear is constant with respect to y. In fact, since the local shear is constant in (8.51), we can replace τ_{app} with its value at the wall, τ_o. A key difference between (8.51) and Couette flow is the presence of the eddy diffusivity, which, as we argued in Section 8.4.1, varies in the y direction. In other words, (8.51) only implies that the shear stress is constant; the velocity is not necessarily linear. Regardless of this distinction, Equation (8.51) is referred to as the *Couette Flow Assumption.*

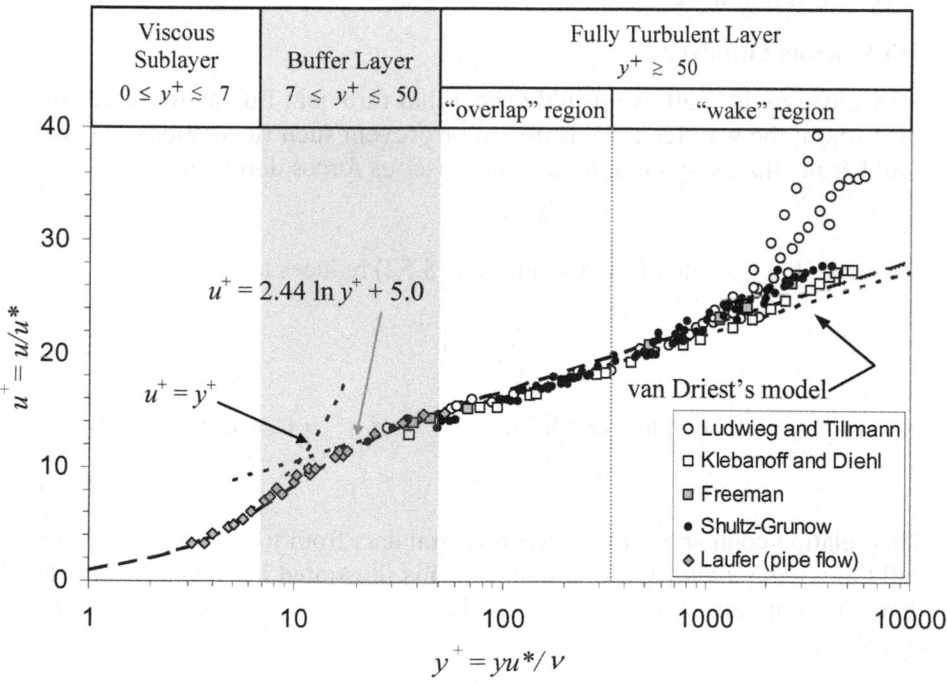

Fig. 8.13

What do we do with this? We can use (8.51) to develop an expression for the velocity profile. First, we need to express (8.51) in terms of the wall coordinates u^+ and y^+. Substituting their definitions, it can be shown that

$$\left(1 + \frac{\varepsilon_M}{\nu}\right)\frac{\partial u^+}{\partial y^+} = 1 ,\qquad (8.52)$$

for which, after rearranging and integrating,

$$u^+ = \int_0^{y^+} \frac{dy^+}{(1 + \varepsilon_M / \nu)} . \qquad (8.53)$$

This is a general expression for the universal velocity profile in wall coordinates. A simpler way to evaluate this integral is to divide the boundary layer into two near-wall regions: (1) a region very close to the wall where viscous forces dominate, and (2) a region where turbulent fluctuations dominate.

(iv) Viscous Sublayer

Very close to the wall, we would expect that turbulent fluctuations become very small; the wall tends to damp out or prevent such fluctuations. So we could argue that, very close to the wall, viscous forces dominate,

$$\nu \gg \varepsilon_M .$$

Therefore the Couette Flow Assumption (8.52) reduces to

$$\frac{\partial u^+}{\partial y^+} = 1 .$$

Integrating, with boundary condition $u^+ = 0$ at $y^+ = 0$, we find

$$u^+ = y^+, \quad (0 \lesssim y^+ \lesssim 7) . \qquad (8.54)$$

This relation compares well to experimental data from $y^+ \approx 0$ to 7, which we call the *viscous sublayer*. Equation (8.54) is illustrated in Figure 8.13 (note that the curvature is a result of being plotted in semi-logarithmic coordinates).

(v) Fully Turbulent Region: "Law of the Wall"

Further away from the wall, we might expect that turbulent fluctuations (i.e., Reynolds stresses) dominate:

$$\varepsilon_M \gg \nu .$$

Under this condition, the Couette Flow Assumption (8.52) becomes

$$\frac{\varepsilon_M}{\nu} \frac{\partial u^+}{\partial y^+} = 1. \tag{8.55}$$

As before, τ is constant, so we have again chosen the same value as in the viscous sublayer (which is the value at the wall), and thereby avoiding any discontinuity between regions. Recall that the eddy diffusivity varies with y. However, we have an expression for ε_M from Prandtl's mixing length theory (8.44). Substituting wall coordinates, Prandtl's mixing length is expressed as

$$\varepsilon_M = \kappa^2 (y^+)^2 \nu \frac{\partial u^+}{\partial y^+}. \tag{8.56}$$

Substituting it into Equation (8.55),

$$\kappa^2 (y^+)^2 \left(\frac{\partial u^+}{\partial y^+} \right)^2 = 1,$$

which, solving for the velocity gradient,

$$\frac{\partial u^+}{\partial y^+} = \frac{1}{\kappa y^+}. \tag{8.57}$$

Finally, integrating the above, we obtain

$$u^+ = \frac{1}{\kappa} \ln y^+ + B, \tag{8.58}$$

This equation is sometimes referred to as the *Law of the Wall*. The constant κ is called *von Kármán's constant*, and experimental measurements show that $\kappa \approx 0.41$. The constant of integration, B, can be estimated by noting that the viscous sublayer and the Law of the Wall region appear to intersect at roughly $y^+ = u^+ \sim 10.8$. Using this as a boundary condition, the integration constant is found to be $B \approx 5.0$. Thus, an approximation for the Law of the Wall region is

$$u^+ = 2.44 \ln y^+ + 5.0, \qquad (50 \lesssim y^+ \lesssim 1500), \tag{8.59}$$

(vi) Other Models

The velocity profile described above is called a *two-layer model*, since the velocity profile is modeled as two regions, the viscous sublayer and the fully turbulent (Law of the Wall) region. It was developed in independent works by Prandtl [9] and Taylor [10], and so is sometimes referred to as the Prantdl-Taylor model. *Three-layer* models exist as well, like that of von Kármán (see Section 8.5.1).

Some investigators have also tried to develop a single equation model for the velocity profile. One of these is van Driest's *continuous law of the wall* [11]. A look at this model will give us some idea of how these methods work.

First, we need a model that makes the eddy diffusivity diminish as y approaches zero. van Driest proposed a mixing length model of this form:

$$\ell = \kappa y \left(1 - e^{-y/A}\right) . \tag{8.60}$$

The term in parentheses is damping factor that makes (8.60) approach zero at the wall, just as we would expect for the turbulent fluctuations. Van Driest used this equation with (8.42) in (8.51) to obtain

$$\frac{\tau_{app}}{\rho} = \left[\nu + \kappa^2 y^2 \left(1 - e^{-y/A}\right)^2\right] \frac{\partial \overline{u}}{\partial y} . \tag{8.61}$$

Again, as y approaches zero, the eddy diffusivity approaches zero, leaving pure viscous shear.

Transforming (8.61) into wall coordinates, and solving for $\partial u^+ / \partial y^+$, one can obtain

$$\frac{\partial u^+}{\partial y^+} = \frac{2}{1 + \sqrt{1 + 4\kappa^2 y^{+2} \left(1 - e^{-y^+/A^+}\right)^2}} . \tag{8.62}$$

For flow over a smooth, flat plate, van Driest used $\kappa = 0.4$ and $A^+ = 26$. The equation above can be integrated numerically, with the resulting curve shown in Figure 8.13.

A model commonly used for both flat plates and pipe flow is by D.B. Spalding [12],

$$y^+ = u^+ + e^{-\kappa B}\left[e^{\kappa u^+} - 1 - \kappa u^+ - \frac{(\kappa u^+)^2}{2} - \frac{(\kappa u^+)^3}{6}\right], \tag{8.63}$$

where Spalding used $\kappa = 0.40$ and $B = 5.5$. One challenge with this equation is that u^+ is implicit in (8.63). The equation can be solved for u^+ numerically, but a closed-form equation is not possible.

Reichardt [13] developed the following profile, which has been applied frequently to pipe flow:

$$u^+ = \frac{1}{\kappa}\ln\left(1 + \kappa y^+\right) + C\left[1 - e^{-y^+/X} - \frac{y^+}{X}e^{-0.33y^+}\right], \qquad (8.64)$$

where $\kappa = 0.40$, $C = 7.8$, and $X = 11$.

(vii) Effect of Pressure Gradient

Until now, the velocity profiles being modeled were for flat plates where the pressure gradient is zero. Figure 8.14 depicts how the velocity profile is affected by pressure gradient. The plot, adapted from White [14] for flow over a flat plate, shows that in the presence of an adverse pressure gradient, the velocity profile beyond $y^+ \approx 350$ deviates from the Law of the Wall model. The deviation is referred to as a "wake," and the region $y^+ > 350$ is commonly referred to as the *wake region* (this region is depicted in Fig. 8.13 as well). The region where the data continue to adhere to the Wall Law is called the *overlap region*. The wake increases with adverse pressure gradient, until separation, where the velocity profile deviates even from the overlap region.

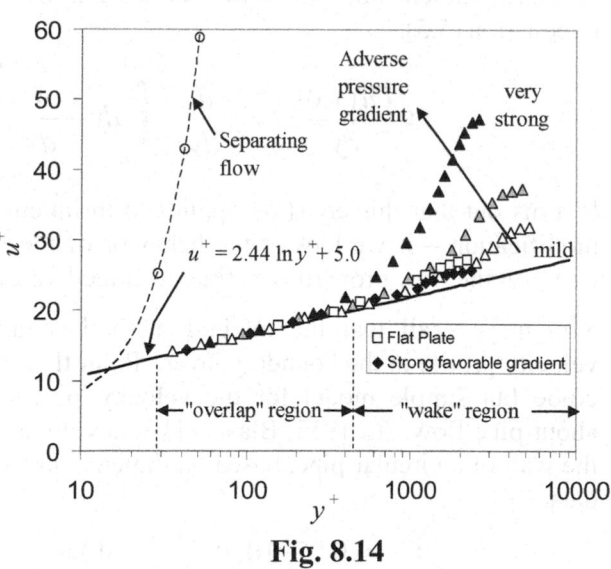

Fig. 8.14

Note that a slight wake exists for zero or even a strong favorable pressure gradient, although the difference between the two sets of data is negligible.

Attempts have been made to model the wake behavior, most notably by Coles [15] (known as *Coles' Law of the Wake*). We will not treat the wake region in this text, but we do note that the Law of the Wall-type models developed earlier model flat plate flow reasonably well in the presence of zero pressure gradient. Another point to realize here is that a favorable pressure gradient is approximately what we encounter in pipe flow, which helps explain why the models developed here apply as well to pipe flow.

8.4.3 Approximate Solution for Momentum Transfer: Momentum Integral Method

Having spent a considerable amount of time developing a model for the velocity profile, let's not lose sight of our primary goal, which is to obtain the drag force on the surface of the body. One way to do this is to invoke the momentum integral equation, just as we did in Chapter 5 for laminar flow. In independent works, Prandtl [7] and von Kármán [16] used precisely this approach to estimate the friction factor on a flat plate.

(i) Prandtl-von Kármán Model

Let's consider a flat, impermeable plate exposed to incompressible, zero-pressure-gradient flow, for which the integral momentum equation reduces to equation (5.5),

$$\nu \frac{\partial \overline{u}(x,0)}{\partial y} = V_\infty \frac{d}{dx} \int_0^{\delta(x)} \overline{u}\, dy - \frac{d}{dx} \int_0^{\delta(x)} \overline{u}^2 dy . \tag{5.5}$$

It turns out that this equation applies to turbulent flows as well – without modification – if we look at the behavior of the flow on average, and we interpret the flow properties as time-averaged values.

You may recall that the integral method requires an estimate for the velocity profile in the boundary layer. Prandtl and von Kármán both used a crude but simple model for the velocity profile using prior knowledge about pipe flow. In 1913, Blasius [17] developed a model for the shear at the wall of a circular pipe, based on dimensional analysis and experimental data,

$$C_f \approx 0.07910\ Re_D^{-1/4} \quad (4000 < Re_D < 10^5), \tag{a}$$

where $C_f = 2\tau_o / \rho u_m^2$, and u_m is the mean velocity over the pipe cross-section. Based solely on this empirical relation, Prandtl [18] and von

Kármán [16] each showed that the velocity profile in the pipe could be modeled as

$$\frac{\bar{u}}{u_{CL}} = \left(\frac{y}{r_o}\right)^{1/7}, \tag{b}$$

where y is the distance from the wall, and u_{CL} is the centerline velocity. This is the well-known *1/7th Law velocity profile*, and is discussed further in Chapter 9. We saw in Fig. 8.13 that the velocity data for pipe flow and flat plate flow (at zero or favorable pressure gradient) have essentially the same shape, so the use of this model to describe flow over a flat plate is not unreasonable.

Of course, in an external flow a mean velocity is not defined, nor is r_o. To make the $1/7^{th}$ power law profile (b) represent conditions in the boundary layer, we'll approximate r_o as the edge of the boundary layer δ, and approximate u_{CL} as representing the free-stream velocity V_∞. Then the velocity profile in the boundary layer is modeled as

$$\frac{\bar{u}}{V_\infty} = \left(\frac{y}{\delta}\right)^{1/7}, \tag{8.65}$$

There is one fundamental problem with using this velocity profile in the momentum integral equation: the gradient of the velocity profile goes to infinity as y approaches zero. This means that, while (8.65) may be appropriate for most of the boundary layer, it cannot be used directly to estimate the wall shear on the left-hand side of (5.5). To avoid this dilemma, Prandtl and von Kármán attempted to model the wall shear differently. Arguing again that the characteristics of the flow near the surface of the plate are similar to that of pipe flow, they adapted the Blasius correlation to find an expression for the wall shear on a flat plate. Recasting (a) in terms of the wall shear and the tube radius, we have

$$\tau_o = 0.03326\rho u_m^2\left(\frac{\nu}{r_o u_m}\right)^{1/4}.$$

It can be shown that for the $1/7^{th}$ velocity profile, $u_m = 0.8167 u_{CL}$, and recall that we are modeling u_{CL} as V_∞. Thus, the above can be written as

$$\frac{\tau_o}{\rho} = \nu\frac{\partial \bar{u}(x,0)}{\partial y} = 0.02333 V_\infty^2\left(\frac{\nu}{V_\infty \delta}\right)^{1/4}. \tag{8.66}$$

Or, in terms of friction factor, we could write

$$\frac{C_f}{2} = \frac{\tau_o}{\rho V_\infty^2} = 0.02333 \left(\frac{V_\infty \delta}{\nu} \right)^{-1/4} . \tag{8.67}$$

This expression can now be used in the integral momentum equation.

Example 8.2: Integral Solution for Turbulent Boundary Layer Flow over a Flat Plate

Consider turbulent flow over a flat plate, depicted in Fig. 8.8. Using the $1/7^{th}$ law velocity profile (8.65) and the expression for friction factor (8.67), obtain expressions for the boundary layer thickness and friction factor along the plate.

(1) Observations. The solution parallels that of Chapter 5 for laminar flow over a flat plate.

(2) Problem Definition. Determine expressions for the boundary layer thickness and friction factor as a function of x.

(3) Solution Plan. Start with the integral Energy Equation (5.5), substitute the power law velocity profile (8.65) and friction factor (8.67), and solve.

(4) Plan Execution.

  (i) Assumptions. (1) Boundary layer simplifications hold, (2) constant properties, (3) incompressible flow, (4) impermeable flat plate.

  (ii) Analysis. Substituting (8.65) into (5.5), and noting that the left side is equal to τ_o / ρ :

$$\frac{\tau_o}{\rho} = V_\infty \frac{d}{dx} \int_0^{\delta(x)} V_\infty \left(\frac{y}{\delta} \right)^{1/7} dy - \frac{d}{dx} \int_0^{\delta(x)} V_\infty^2 \left(\frac{y}{\delta} \right)^{2/7} dy . \tag{a}$$

Dividing the expression by V_∞^2, and collecting terms,

$$\frac{\tau_o}{\rho V_\infty^2} = \frac{d}{dx} \int_0^{\delta(x)} \left[\left(\frac{y}{\delta} \right)^{1/7} - \left(\frac{y}{\delta} \right)^{2/7} \right] dy . \tag{b}$$

After integrating,

$$\frac{\tau_o}{\rho V_\infty^2} = \frac{7}{72} \frac{d\delta}{dx} . \tag{8.68}$$

Now substituting (a) into (8.67),

$$0.02333\left(\frac{V_\infty \delta}{\nu}\right)^{-1/4} = \frac{7}{72}\frac{d\delta}{dx}. \tag{c}$$

Then, separating variables and integrating,

$$\frac{4}{5}\delta^{5/4} = 0.02333\left(\frac{72}{7}\right)\left(\frac{V_\infty}{\nu}\right)^{-1/4} x + C. \tag{8.69}$$

To complete the solution, a boundary condition is needed. A simple one is to assume that $\delta(x)$ is zero at $x = 0$. This is certainly true, but we are ignoring the fact that, unless the boundary layer is tripped, there will be some initial region that is laminar, as demonstrated in Fig. 8.15. Effectively, we are treating the entire flow along the plate as being turbulent, beginning from the leading edge. This assumption was first suggested by Prandtl. However crude the assumption, we find that the results of this analysis compare well to experimental data.

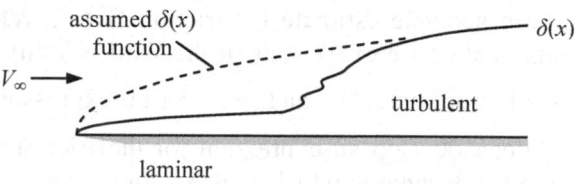

Fig. 8.15

With the boundary condition established, the integration constant C equals zero. Then solving (8.69) for $\delta(x)$,

$$\delta(x) = 0.3816\left(\frac{V_\infty x}{\nu}\right)^{-1/5} x, \tag{d}$$

or

$$\frac{\delta}{x} = \frac{0.3816}{Re_x^{1/5}}, \tag{8.70}$$

Finally, with an expression for $\delta(x)$ in hand, we can solve for friction factor. Substituting (8.70) into (8.67),

$$\frac{C_f}{2} = 0.02333\left(\frac{V_\infty (0.3816)\left(\frac{V_\infty x}{\nu}\right)^{-1/5} x}{\nu}\right)^{-1/4}, \tag{e}$$

which reduces to

$$\frac{C_f}{2} = \frac{0.02968}{Re_x^{1/5}}.$$ (8.71)

(5) Checking. Equations (8.70) and (8.71) are both dimensionless, as expected.

(6) Comments. Note that, according to this model, the turbulent boundary layer δ/x varies as $Re_x^{-1/5}$, as does the friction factor C_f. This is contrast to laminar flow, in which δ/x and C_f vary as $Re_x^{-1/2}$.

(ii) Newer Models

One limitation of the Prandtl-von Kármán model is that the approximation for the wall shear, Eqn. (8.66), is based on limited experimental data, and considered to be of limited applicability even for pipe flow. To obtain a more accurate estimate for friction factor, White [14] presents a method that makes use of the Law of the Wall velocity profile (8.59). Because the wall coordinates u^+ and y^+ can be expressed as functions of $\sqrt{C_f/2}$, we can develop an expression for the friction factor from the Law of the Wall (a technique which is also seen in analysis of pipe flow – see Section 9.5). First, substituting the definitions of u^+ and y^+, as well as u^*, into the Law of the Wall expression (8.59),

$$\frac{\bar{u}}{V_\infty}\sqrt{\frac{2}{C_f}} = 2.44\ln\left(\frac{yV_\infty}{\nu}\sqrt{\frac{C_f}{2}}\right) + 5.0.$$

In theory, any y value within the wall law layer would satisfy this expression, but a useful value to choose is the edge of the boundary layer, where $\bar{u}(y=\delta) = V_\infty$. Then, the above can be expressed as

$$\frac{1}{\sqrt{C_f/2}} = 2.44\ln\left(Re_\delta\sqrt{\frac{C_f}{2}}\right) + 5.0,$$ (8.72)

where $Re_\delta = V_\infty\delta/\nu$. Equation (8.72) relates the skin friction to the boundary layer thickness, which we can use in the momentum integral equation. However, the above equation is cumbersome; a simpler expression can be found by curve-fitting values obtained from (8.72) over a range of values from $Re_\delta \approx 10^4$ to 10^7. Doing this yields the approximate relation

$$C_f \approx 0.02 Re_\delta^{-1/6}. \tag{8.73}$$

We can now use this expression to estimate the wall shear in the integral method. For the velocity profile, the $1/7^{th}$ power law, (8.65), is used as before. It can be shown that the solution to the momentum integral equation in this case becomes

$$\frac{\delta}{x} = \frac{0.16}{Re_x^{1/7}} \tag{8.74}$$

and

$$\frac{C_f}{2} = \frac{0.0135}{Re_x^{1/7}}. \tag{8.75}$$

Equations (8.74) and (8.75) replace the less accurate Prandtl-von Kármán correlations, and White recommends these expressions for general use.

Perhaps a more accurate correlation would result if we use one of the more advanced velocity profiles to estimate the wall shear, as well as to replace the crude $1/7^{th}$ power law profile. Kestin and Persen [19] developed such a model, using Spalding's law of the wall (8.63). The resulting model is extremely accurate, but cumbersome. White [20] modified the result to obtain the simpler relation

$$C_f = \frac{0.455}{\ln^2 (0.06 Re_x)}. \tag{8.76}$$

White reports that this expression is accurate to within 1% of Kestin and Persen's model.

(iii) Total Drag

The total drag is found by integrating the wall shear along the entire plate. Assuming that laminar flow exists along the initial portion of the plate, the total drag on a plate of width w would be

$$F_D = \int_0^{x_{crit}} (\tau_o)_{lam} \, w dx + \int_{x_{crit}}^{L} (\tau_o)_{turb} \, w dx. \tag{8.77}$$

Dividing by $\frac{1}{2}\rho V_\infty^2 A = \frac{1}{2}\rho V_\infty^2 wL$, the *drag coefficient* C_D is

$$C_D = \frac{1}{L}\left[\int_0^{x_{crit}} C_{f,lam}dx + \int_{x_{crit}}^L C_{f,turb}dx\right].$$ (8.78)

Substituting Eqn. (4.48) for laminar flow and using White's model (8.75) for turbulent flow, we obtain with some manipulation,

$$C_D = \frac{0.0315}{Re_L^{1/7}} - \frac{1477}{Re_L},$$ (8.79)

assuming $x_{crit} = 5 \times 10^5$.

8.4.4 Effect of Surface Roughness on Friction Factor

Thus far we have examined flow only over smooth walls. As you might expect, the presence of a rough wall affects the velocity profile and ultimately the friction factor. It turns out that modeling flow over rough surfaces is not easy, and the interaction between the already complex turbulent flow and the complex, random geometric features of a rough wall is the subject of advanced study and numerical modeling. However, with crude modeling and some experimental study we can gain at least some physical insight.

Let's define k as the average height of roughness elements on the surface. For comparison purposes, we will transform k into wall coordinates as $k^+ = ku^*/v$, as we did to define y^+. Experiments show that for small values of k^+ (less than approximately 5), the velocity profile and friction factor are unaffected by the roughness. This makes sense, because the roughness is contained within the viscous sublayer: As in laminar flow, disturbances in this region are likely to be damped out by the viscosity-dominated flow, and so the surface is essentially smooth.

For $k^+ > 10$ or so, however, the roughness extends beyond the viscous sublayer, and the viscous sublayer begins to disappear, likely due to the enhanced mixing that the roughness provides. Finally, beyond $k^+ > 70$ viscous effects are virtually eliminated, and the flow is referred to as *fully rough*. Beyond this value of roughness, the shape of the velocity profile changes very little. Consequently, we might expect that once the surface is fully rough, increasing the roughness would not change the friction factor.

Figure 8.16, adapted from Arpaci and Larsen [21], illustrates how the near-wall velocity profile is affected by roughness. In short, roughness tends to shift the Law of the Wall down and to the right. Practically, the shift in velocity profile means that the velocity gradient at the wall is greater, and therefore the friction factor increases as we expect. The slope of the wall law curve, however, is not affected by roughness.

Few correlations are available for friction factor on rough plates, and those that do exist are highly dependent on the geometry of the roughness. One common way to simulate surface roughness is by gluing sand of uniform particle size to the surface of the plate. For fully rough flow over sand-rough plates, White [14] suggests the following correlation, based on a wall law velocity profile developed from experimental data:

$$ C_f = \left[1.4 + 3.7 \log_{10}\left(\frac{x}{k}\right) \right]^{-2}, \qquad \frac{x}{k} > \frac{Re_x}{1000}. \qquad (8.80) $$

Note that this correlation does not include Reynolds number. For fully rough flow, the fact that viscous effects are eliminated suggests that the friction factor in this case is not dependent on Re_x.

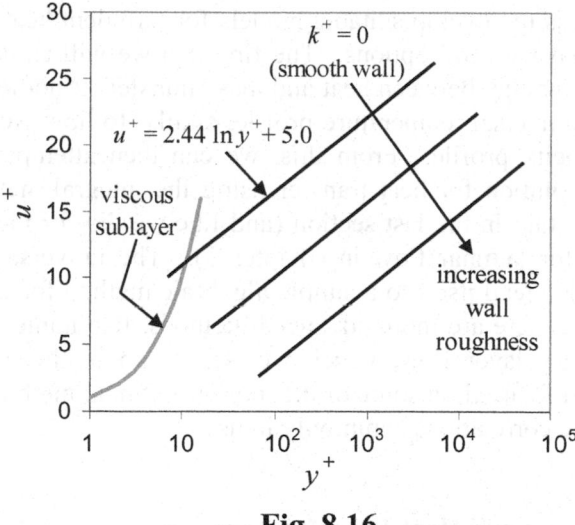

Fig. 8.16

8.5 Energy Transfer in External Turbulent Flow

We have just seen how the presence of turbulence greatly complicates the flow along a body. Turbulence also complicates our analysis of the momentum transfer. Not surprisingly, energy transfer is also greatly complicated under turbulent flow. We found in Chapter 2 that the heat transfer for flow over a geometrically similar body like a flat plate (neglecting both buoyancy and viscous dissipation) could be correlated through dimensionless analysis by

$$Nu_x = f(x^*, Re, Pr),$$ (2.52)

where x^* is the dimensionless location along the body. Looking at the turbulent boundary layer equations, (8.38) and (8.39), we see that turbulence introduces two new variables into the analysis: the momentum and thermal eddy diffusivities, ε_M and ε_H. One way to deal with these new terms is to introduce a new dimensionless parameter,

$$Pr_t = \frac{\varepsilon_M}{\varepsilon_H},$$ (8.81)

called the *turbulent Prandtl number*. Practically speaking, if we were planning experiments to obtain an empirical correlation for turbulent heat transfer, we would have an additional parameter with which to contend.

Our goal here is to develop suitable models for turbulent heat transfer, and to do so we have several options. The first that we will explore is to find some kind of analogy between heat and mass transfer. Another approach is to develop a universal temperature profile, similar to how we developed a universal velocity profile. From this, we can then attempt to obtain an approximate solution for heat transfer using the integral method, like we did for momentum in the last section (and like we did for momentum and heat transfer for laminar flow in Chapter 5). The universal temperature profile may also lend itself to a simple algebraic method for evaluating the heat transfer. There are more advanced methods, like numerical solutions to the boundary layer flow, which we will forgo in this text. We will instead remain focused on some of the more traditional methods, which are the basis of the correlations commonly in use.

8.5.1 Momentum and Heat Transfer Analogies

Osborne Reynolds was the first to discover a link between momentum and heat transfer, in a study on the behavior of turbulent flow inside steam boilers. In his work, published in 1874, he theorized that the heat transfer and the frictional resistance in a pipe are proportional to each other [22]. The implication of his work seems incredible even today: if we can measure or predict the friction along a wall or pipe, we can determine the heat transfer simply by using a multiplying factor. This approach would allow us to solve for the heat transfer directly, avoiding the difficulty of solving the energy equation.

The simplest of these momentum and heat transfer analogies is named after Reynolds, and we will develop it for external flow as follows.

(i) Reynolds Analogy

Consider parallel flow over a flat plate. The pressure gradient dp/dx is zero, and the boundary layer momentum and energy equations (8.38) and (8.39) reduce to

$$\bar{u}\frac{\partial \bar{u}}{\partial x} + \bar{v}\frac{\partial \bar{u}}{\partial y} = \frac{\partial}{\partial y}\left[(v + \varepsilon_M)\frac{\partial \bar{u}}{\partial y}\right], \tag{8.82a}$$

and

$$\bar{u}\frac{\partial \bar{T}}{\partial x} + \bar{v}\frac{\partial \bar{T}}{\partial y} = \frac{\partial}{\partial y}\left[(\alpha + \varepsilon_H)\frac{\partial \bar{T}}{\partial y}\right]. \tag{8.82b}$$

The boundary conditions are

$$\bar{u}(y=0) = 0, \quad \bar{T}(y=0) = T_s, \tag{8.83a}$$

and

$$\bar{u}(y\to\infty) = V_\infty, \quad \bar{T}(y\to\infty) = T_\infty. \tag{8.83b}$$

Notice that equations (8.82a&b) and their respective boundary conditions are very similar; if they were identical, their solutions – the velocity and temperature profiles – would be the same.

We can get closer to an analogy by normalizing the variables. Selecting, for example, the following variables,

$$U = \frac{\bar{u}}{V_\infty}, \quad V = \frac{\bar{v}}{V_\infty}, \quad \theta = \frac{\bar{T} - T_s}{T_\infty - T_s}, \quad X = \frac{x}{L}, \text{ and } Y = \frac{y}{L},$$

we can transform equations (8.82) and the boundary conditions (8.83) to

$$U\frac{\partial U}{\partial X} + V\frac{\partial U}{\partial Y} = \frac{1}{V_\infty L}\frac{\partial}{\partial Y}\left[(v + \varepsilon_M)\frac{\partial U}{\partial Y}\right], \tag{8.84a}$$

$$U\frac{\partial \theta}{\partial x} + V\frac{\partial \theta}{\partial y} = \frac{1}{V_\infty L}\frac{\partial}{\partial Y}\left[(\alpha + \varepsilon_H)\frac{\partial \theta}{\partial Y}\right], \tag{8.84b}$$

and

$$U(Y=0) = 0, \quad \theta(Y=0) = 0, \tag{8.85a}$$

$$U(Y\to\infty) = 1, \quad \theta(Y\to\infty) = 1. \tag{8.85b}$$

We see that normalizing the variables has made the boundary conditions identical.

Equations (8.84) can then be made identical if $(\nu + \varepsilon_M) = (\alpha + \varepsilon_H)$, which is possible under two conditions. The first is if the kinematic viscosity and thermal diffusivity are equal:

$$\nu = \alpha . \qquad (8.86)$$

This condition limits the analogy to fluids with $Pr = 1$. Note that this also suggests that the velocity and thermal boundary layers are approximately the same thickness, $\delta \approx \delta_t$ (see Section 4.2.5).

The second condition is if the eddy diffusivities are equal:

$$\varepsilon_M = \varepsilon_H . \qquad (8.87)$$

We can provide some justification for this assumption by arguing that the same turbulent mechanism—the motion and interaction of fluid particles— is responsible for both momentum and heat transfer. Reynolds made essentially the same argument, and so Equation (8.87) by itself is sometimes referred to as Reynolds' analogy. This assumption also means that the turbulent Prandtl number Pr_t is equal to 1.

The analogy is now complete, meaning that the normalized velocity and temperature profiles, $U(X,Y)$ and $\theta(X,Y)$ are equal. This is demonstrated in Fig. 8.17.

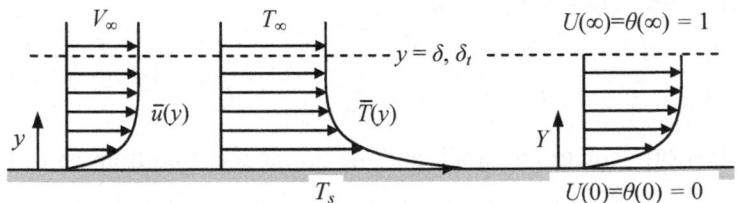

Fig. 8.17

We can now derive a relationship between the shear stress and heat flux at the wall. We begin by writing the ratio of the apparent heat flux and shear stress (equations. 8.40 and 8.41),

$$\frac{q''_{app} / \rho c_p}{\tau_{app} / \rho} = -\frac{(\alpha + \varepsilon_H)\partial \overline{T} / \partial y}{(\nu + \varepsilon_M)\partial \overline{u} / \partial y} . \qquad (8.88)$$

Imposing the two conditions (8.86) and (8.87), the terms in parentheses cancel. Substituting the dimensionless variables into the above and rearranging yields

$$\frac{q''_{app}}{\tau_{app}} = \frac{c_p(T_s - T_\infty)\partial\theta/\partial Y}{V_\infty \partial U/\partial Y}. \tag{8.89}$$

Since the dimensionless velocity and temperature profiles are identical, their derivatives cancel. Another important implication of (8.89) is that the ratio q''_{app}/τ_{app} is constant throughout the boundary layer. This means we can represent this ratio by the same ratio at the wall. Equation (8.89) then becomes

$$\frac{q''_o}{\tau_o} = \frac{c_p(T_s - T_\infty)}{V_\infty}.$$

We now have the result we were seeking, which is an expression relating the shear stress and heat flux at the wall. We can recast this into a more convenient form by substituting $q''_o = h(T_s - T_\infty)$ and $\tau_o = \frac{1}{2}C_f\rho V_\infty^2$ into the above, and rearranging,

$$\frac{h}{\rho V_\infty c_p} = \frac{C_f}{2}.$$

The terms on the left side can also be written in terms of the Reynolds, Nusselt, and Prandtl numbers,

$$St_x \equiv \frac{Nu_x}{Re_x Pr} = \frac{C_f}{2}, \quad (Pr = 1), \tag{8.90}$$

where St_x is called the *Stanton number*. Equation (8.90) is commonly referred to as the *Reynolds analogy*. The same analogy can also be derived for laminar flow over a flat plate (for $Pr = 1$).

A major limitation of (8.90) is that it is limited to $Pr = 1$ fluids. This may be a reasonable approximation for many gases, but for most liquids the Prandtl numbers are much greater than unity—values of up to 700 are possible. Therefore the Reynolds analogy is not appropriate for liquids.

(ii) Prandtl-Taylor Analogy

Another flaw with the Reynolds analogy is that it doesn't account for the varying intensity of molecular and turbulent diffusion in the boundary layer. From the development of the universal turbulent velocity profile

(Section 8.4), we expect that very close to the wall (the viscous sublayer), molecular effects dominate:

$$v \gg \varepsilon_M \text{ and } \alpha \gg \varepsilon_H, \tag{8.91}$$

while turbulent effects dominate further away from the wall,

$$\varepsilon_M \gg v \text{ and } \varepsilon_H \gg \alpha. \tag{8.92}$$

Notice that neither of these conditions restricts us to $Pr = 1$ fluids.

In independent works, Prandtl [9] and Taylor [10] modified the Reynolds analogy by dividing the boundary layer into two regions: a viscous sublayer where molecular effects dominate (equation 8.91), and a turbulent outer layer, where (8.92) is assumed to hold. In order for an analogy to exist, the momentum and boundary layer equations, and their boundary conditions, must be identical in both regions.

We will define the viscous sublayer as the portion of the boundary layer beneath $y = y_1$, where y_1 is some threshold value. The boundary conditions for this region are

$$\overline{u}(0) = 0 , \ \overline{T}(0) = T_s,$$

and
$$\overline{u}(y_1) = \overline{u}_1, \ \overline{T}(y_1) = \overline{T}_1.$$

It can be shown that the boundary conditions above, as well as equations (8.82), can be made identical by defining the following normalized variables:

$$U = \frac{\overline{u}}{\overline{u}_1}, \ V = \frac{\overline{v}}{\overline{u}_1}, \ \theta = \frac{\overline{T}-T_s}{\overline{T}_1-T_s}, \ X = \frac{x}{y_1} \text{ and } Y = \frac{y}{y_1}.$$

Then, for the viscous sublayer, the ratio of the apparent heat flux and apparent shear stress (Eqn. 8.86) leads to the following:

$$T_s - \overline{T}_1 = \frac{q_o''}{\tau_o c_p} Pr \, \overline{u}_1, \tag{8.93}$$

where we have again noted that $q_{app}'' / \tau_{app} = q_o'' / \tau_o = $ constant.

The outer layer closely resembles the Reynolds analogy, with $\varepsilon_M = \varepsilon_H$ (or $Pr_t = 1$), but this time we assume that the turbulent effects outweigh the molecular effects, equation (8.92). This region has boundary conditions

$$\overline{u}(y_1) = \overline{u}_1 , \ \overline{T}(y_1) = \overline{T}_1,$$

and $$\overline{u}(y \to \infty) = V_\infty, \ \overline{T}(y \to \infty) = T_\infty$$

The following normalized variables will make the analogy valid in this region:

$$U = \frac{\overline{u} - \overline{u}_1}{V_\infty - \overline{u}_1}, \ V = \frac{\overline{v} - \overline{u}_1}{V_\infty - \overline{u}_1}, \ \theta = \frac{\overline{T} - \overline{T}_1}{T_\infty - T_1}, \ X = \frac{x}{L}, \text{ and } Y = \frac{y}{L}.$$

Then, for the outer region, the ratio of the apparent heat flux and apparent shear stress (equation 8.86) leads to

$$\overline{T}_1 - T_\infty = \frac{q_o''}{\tau_o c_p}(V_\infty - \overline{u}_1). \tag{8.94}$$

As before, the ratio q_{app}'' / τ_{app} is constant, so we have chosen the value at $y = y_1$ (which, as we found for the viscous sublayer, can be represented by q_o'' / τ_o). Adding (8.93) and (8.94) gives

$$T_s - T_\infty = \frac{q_o''}{\tau_o c_p}V_\infty\left[\frac{\overline{u}_1}{V_\infty}(Pr - 1) + 1\right],$$

and substituting $\tau_o = \frac{1}{2}C_f \rho V_\infty^2$ into the above yields

$$St = \frac{q_o''}{\rho V_\infty c_p (T_s - T_\infty)} = \frac{C_f / 2}{\left[\dfrac{\overline{u}_1}{V_\infty}(Pr - 1) + 1\right]}.$$

The velocity at the edge of the viscous sublayer, $\overline{u}_1$, is still unknown, but we can estimate this value using the universal velocity profile (Fig. 8.13). In this analogy, a value of $u^+ = y^+ \approx 5$ is chosen to approximate the edge of the viscous sublayer. From the definition of u^+,

$$u^+ = 5 = \frac{\overline{u}_1}{V_\infty}\sqrt{\frac{2}{C_f}},$$

or $$\frac{\overline{u}_1}{V_\infty} = 5\sqrt{\frac{C_f}{2}}. \tag{8.95}$$

Thus the *Prandtl-Taylor analogy* is

$$St_x \equiv \frac{Nu_x}{Re_x Pr} = \frac{C_f/2}{\left[5\sqrt{\frac{C_f}{2}}(Pr-1)+1\right]}. \tag{8.96}$$

(iii) von Kármán Analogy

Theodore von Kármán [23] extended the Reynolds analogy even further to include a third layer – a buffer layer – between the viscous sublayer and outer layer. The result, developed in Appendix D, is

$$St_x \equiv \frac{Nu_x}{Re_x Pr} = \frac{C_f/2}{1+5\sqrt{\frac{C_f}{2}}\left\{(Pr-1)+\ln\left[\frac{5Pr+1}{6}\right]\right\}}. \tag{8.97}$$

(iv) Colburn Analogy

Colburn [24] proposed a purely empirical modification to the Reynolds analogy that accounts for fluids with varying Prandtl number. He proposed the following correlation through an empirical fit of available experimental data:

$$St_x Pr^{2/3} = \frac{C_f}{2}. \tag{8.98}$$

The exponent (2/3) on the Prandtl number is entirely empirical, and does not contain any theoretical basis. The *Colburn analogy* is considered to yield acceptable results for $Re_x < 10^7$ (including the laminar flow regime) and Prandtl number ranging from about 0.5 to 60.

Example 8.3: Average Nusselt Number on Flat Plate

Determine the average Nusselt number for heat transfer along a flat plate of length L with constant surface temperature. Use White's model (8.75) for turbulent friction factor, and assume a laminar region exists along the initial portion of the plate.

(1) Observations. This is a mixed-flow type problem, with the initial portion of the plate experiencing laminar flow.

(2) Problem Definition. Determine an expression for the average Nusselt number for a flat plate of length L.

(3) Solution Plan. Start with an expression for average heat transfer coefficient, equation (2.50), and split the integral up between laminar and turbulent regions.

(4) Plan Execution.

   (i) Assumptions. (1) Boundary layer assumptions apply, (2) mixed (laminar and turbulent) flow, (3) constant properties, (4) incompressible flow, (5) impermeable flat plate, (6) uniform surface temperature. (7) transition occurs at $x_c = 5 \times 10^5$.

   (ii) Analysis. The average heat transfer coefficient is found from (2.50),

$$\overline{h}_L = \frac{1}{L} \int_0^L h(x)\,dx, \tag{2.50}$$

which can be split into laminar and turbulent regions as follows:

$$\overline{h}_L = \frac{1}{L} \left[\int_0^{x_c} h_{lam}(x)\,dx + \int_{x_c}^L h_{turb}(x)\,dx \right]. \tag{8.99}$$

From the definition of Nusselt number, we can write the above as

$$\overline{Nu}_L \equiv \frac{\overline{h}_L L}{k} = \int_0^{x_c} \frac{1}{x} Nu_{x,lam}\,dx + \int_{x_c}^L \frac{1}{x} Nu_{x,turb}\,dx \tag{a}$$

To find expressions for local Nusselt number, we will use the friction factors for laminar flow, equation 4.48), and White's model for turbulent flow (8.75), and apply them to Colburn's analogy (8.98). The results are

$$Nu_{x,lam} = 0.332 Pr^{1/3} Re_x^{1/2}, \tag{b}$$

and

$$Nu_{x,turb} = 0.0135 Pr^{1/3} Re_x^{6/7}. \tag{c}$$

Substituting these expressions into (a) gives

$$\overline{Nu}_L = \int_0^{x_c} (0.332) Pr^{1/3} \left(\frac{V_\infty}{\nu}\right)^{1/2} \frac{dx}{\sqrt{x}} + \int_{x_c}^L (0.0135) Pr^{1/3} \left(\frac{V_\infty}{\nu}\right)^{6/7} \frac{dx}{x^{1/7}},$$

which yields

$$\overline{Nu}_L = 0.664 Pr^{1/3} Re_{x_c}^{1/2} + \frac{7}{6}(0.0135) Pr^{1/3} \left(Re_L^{6/7} - Re_{x_c}^{6/7} \right). \quad \text{(d)}$$

Finally, since Re_{x_c} is assumed as 5×10^5, (d) reduces to

$$\overline{Nu}_L = \left(0.0158 Re_L^{6/7} - 739 \right) Pr^{1/3}. \quad (8.100)$$

(iii) Checking. The resulting Nusselt number correlation is dimensionless.

(5) Comments. If the laminar length had been neglected, the resulting correlation would be

$$\overline{Nu}_L = 0.0158 Re_L^{6/7} Pr^{1/3}. \quad (8.101)$$

This result also makes sense when examining the mixed-flow correlation (8.100). If the plate is very long, such that the majority of the plate is in turbulent flow, the second term in the parentheses becomes negligible, leading to (8.101).

8.5.2 Validity of Analogies

Momentum-heat transfer analogies like the ones presented in this text are frequently used to develop heat transfer models for many types of flows and geometries. Therefore it makes sense to pause for a moment to discuss the applicability and validity of this technique. Although derived for a flat plate, these analogies are considered generally valid for slender bodies, where the pressure gradient does not vary greatly from zero. They are approximately valid for internal flows in circular pipes as well, although other analogies have been developed specifically for internal flow (See Chapter 9 and Reference [25]). Also, although they are derived assuming constant wall temperature, the above correlations work reasonably well even for constant heat flux [26].

The large temperature variation near the wall means that the assumption of uniform properties, particularly for Pr, becomes a weakness. One way to address this problem is to evaluate properties at the film temperature:

$$T_f = \frac{T_s + T_\infty}{2}. \quad (8.102)$$

The analogies were also derived assuming that the turbulent Prandtl number is equal to unity. Is this accurate? A thorough review and survey of the state of knowledge regarding the turbulent Prandtl number is given by Kays [27]. Experimentally-measured values of Pr_t are as high as 3 very near the wall, though outside the viscous sublayer the values range from around 1 to 0.7. The turbulent Prandtl number seems to be affected slightly by pressure gradient, though largely unaffected by surface roughness or the presence of boundary layer suction or blowing. A value of $Pr_t \approx 0.85$ is considered reasonable for most flows. This suggests that the analogies should be approximately valid for real flows.

Arguably, the most popular analogy is that of Colburn. It bears repeating, though, that the analogy is as primitive as the Reynolds analogy, adds no new theoretical insight, and is in fact merely a curve-fit of experimental data. So why has this method maintained its usefulness over the decades? It's not just because the equation is easy to use. More advanced models are based on theoretical assumptions that are, at best, approximations. For example, looking back at the development of the Prandtl-Taylor and von Kármán analogies, we note that both assume that the viscous sublayer and conduction sublayer are the same thickness. However, we would expect the "conduction sublayer" to extend further into the flow field when Pr is small, and for large Pr the turbulent effects would extend into the viscous sublayer [28]. Though only an empirical correlation, the Colburn analogy was shown to represent experimental data well over a variety of fluids. Empiricism is sometimes better than pure theoretical arguments; the test is the experimental data.

Despite its widespread use, the Colburn analogy is not without its detractors. Churchill and Zajic [29] compared the Colburn analogy, along with other common correlations, including a new correlation they introduced, to the results of numerical modeling for flow inside round tubes. They demonstrated that that the Colburn analogy under-predicts the Nusselt number by 30-40% for fluids with Prandtl numbers greater than 7.

Whatever their shortcomings, analogies are fairly straightforward, and facilitate the development of empirical correlations that are often reasonably accurate and easy to use. Numerical solutions, on the other hand, are still difficult to obtain and are limited in applicability. For these reasons, heat and mass transfer analogies remain in widespread use, and new correlations are still being developed often based on this technique.

8.5.3 Universal Turbulent Temperature Profile

As was the case with the universal velocity profile, much physical insight can be gained if we try to develop a universal temperature profile in turbulent flow. One direct effect of this analysis is that we can use an approximate temperature profile in an integral approach to solve for the heat transfer.

(i) Near-Wall Profile

We begin with the turbulent energy equation, (8.39). Akin to the Couette flow assumption (Section 8.4.2), we assume that, near the wall, the velocity component $\overline{v} \sim 0$, as is the temperature gradient $\partial \overline{T}/\partial x$. Thus the left-hand-side of (8.39) approaches zero. Then,

$$\frac{\partial}{\partial y}\left[(\alpha + \varepsilon_H)\frac{\partial \overline{T}}{\partial y}\right] \sim 0 \text{ near wall,}$$

implying that the apparent heat flux is approximately constant with respect to y,

$$\frac{q''_{app}}{\rho c_p} = -(\alpha + \varepsilon_H)\frac{\partial \overline{T}}{\partial y} \sim \text{constant.} \tag{8.103}$$

The idea here is the same as we developed for the universal velocity profile: we can solve the above relation for the temperature profile. First, recognize that, since $q''_{app}/\rho c_p$ is constant throughout this region, we can replace q''_{app} with q''_o. Then, substituting wall coordinates u^+ and y^+, (8.103) can be rearranged to

$$-\frac{\partial \overline{T}}{\partial y^+}\frac{\rho c_p u^*}{q''_o} = \frac{\nu}{(\alpha + \varepsilon_H)}, \tag{8.104}$$

which was arranged in this form such that both sides are dimensionless. In fact, (8.104) suggests a definition for a temperature wall coordinate, which is

$$T^+ \equiv (T_s - \overline{T})\frac{\rho c_p u^*}{q''_o}. \tag{8.105}$$

With the wall coordinate T^+ defined as above, (8.104) can be cast in a simpler form:

$$\frac{\partial T^+}{\partial y^+} = \frac{\nu}{(\alpha + \varepsilon_H)}. \tag{8.106}$$

We can now integrate the above expression. Doing this, we obtain

$$T^+ = \int\limits_{0}^{y^+} \frac{v \, dy^+}{\alpha + \varepsilon_H}.$$ (8.107)

This is a general expression for the temperature profile in wall coordinates. Just as with the universal velocity profile, we will divide the boundary layer into two regions in order to evaluate this expression.

(ii) Conduction Sublayer

Very close to the wall, we expect molecular effects to dominate the heat transfer; that is,

$$\alpha \gg \varepsilon_H.$$

Invoking this approximation, (8.107) reduces to

$$T^+ = \int Pr \, dy^+ = Pr \, y^+ + C.$$

The constant of integration, C, can be found by applying the boundary condition that $T^+(y^+ = 0) = 0$. This condition yields $C = 0$, so the temperature profile in the conduction sublayer is

$$T^+ = Pr \, y^+, \quad (y^+ < y_1^+).$$ (8.108)

In the above, y_1^+ is the dividing point between the conduction and outer layers.

(iii) Fully Turbulent Region

Outside the conduction-dominated region close to the wall, we expect that turbulent effects dominate:

$$\varepsilon_H \gg \alpha,$$

for which the temperature profile (8.104) becomes

$$T^+ = T_1^+ + \int\limits_{y_1^+}^{y^+} \frac{v}{\varepsilon_H} dy^+,$$ (8.109)

In order to evaluate (8.109), we need to evaluate ε_H. We might try to develop a new "mixing length" type of model for eddy thermal diffusivity, but instead we can simply relate it to the momentum eddy diffusivity as follows:

$$\varepsilon_H = \frac{Pr_t}{\varepsilon_M} .$$

In other words, by invoking the definition of the turbulent Prandtl number (which has been measured empirically), we can express the thermal eddy diffusivity in terms of the momentum eddy diffusivity. Why is this useful? Because we already have a model for the eddy diffusivity from Prandtl's mixing length theory, Equation (8.44):

$$\varepsilon_M = \kappa^2 y^2 \left| \frac{\partial \bar{u}}{\partial y} \right|. \tag{8.44}$$

In terms of wall coordinates, we can write (8.44) as

$$\varepsilon_M = \kappa^2 (y^+)^2 \nu \frac{\partial u^+}{\partial y^+}, \tag{8.110}$$

and the partial derivative $\partial u^+ / \partial y^+$ can be found from the Law of the Wall, Equation (8.58). Substituting the above and (8.58) into (8.109), we obtain

$$T^+ = \int_{y_1^+}^{y^+} \frac{Pr_t}{\kappa y^+} dy^+ . \tag{8.111}$$

Finally, if we assume Pr_t and κ are constants, then (8.110) becomes

$$T^+ = \frac{Pr_t}{\kappa} \ln\left(\frac{y^+}{y_1^+} \right), \quad \left(y^+ > y_1^+ \right). \tag{8.112}$$

Clearly, the temperature profile defined by Equations (8.108) and (8.111) depends on the fluid (Pr), as well as the parameters Pr_t and κ. Kays et al. [30] assumed $Pr_t = 0.85$ and $\kappa = 0.41$, but found that the thickness of the conduction sublayer (y_1^+) varies by fluid. White [14] reports a correlation that can be used for any fluid with $Pr \geq 0.7$:

$$T^+ = \frac{Pr_t}{\kappa} \ln y^+ + 13 \, Pr^{2/3} - 7, \tag{8.113}$$

for which Pr_t is assumed to be approximately 0.9 or 1.0. Figure 8.18 is a plot of this model, along with viscous sublayer, for various values of Pr. For comparison, the plot includes data reported by Kays et al. for water and air. Note that, as we expect, the temperature profile increases with increasing Prandtl number.

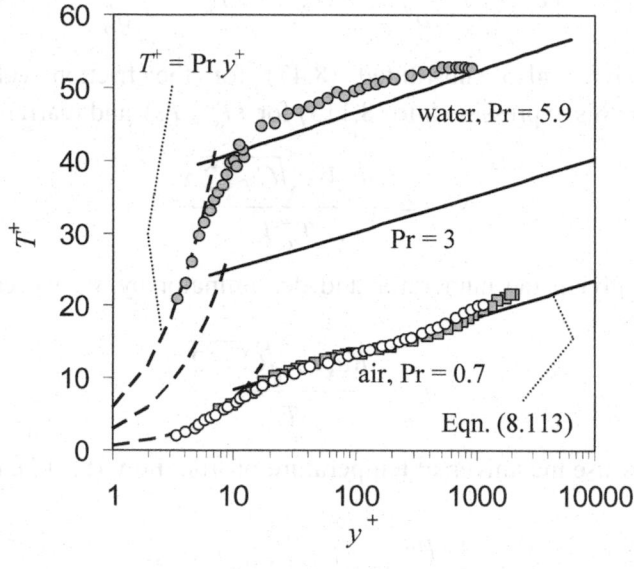

Fig. 8.18

(iv) A 1/7th Law for Temperature

As with the velocity profile, a simpler $1/7^{\text{th}}$ power law relation is sometimes used for the temperature profile:

$$\frac{\overline{T} - T_s}{T_\infty - T_s} = \left(\frac{y}{\delta_t}\right)^{1/7} .$$

(8.114)

8.5.4 Algebraic Method for Heat Transfer Coefficient

The existence of a universal temperature and velocity profile makes for a fairly simple method to estimate the heat transfer. We begin, logically enough, with the definition of the Nusselt number, which can be expressed using Newton's law of cooling as

$$Nu_x \equiv \frac{hx}{k} = \frac{q_o'' x}{(T_s - T_\infty)k} . \qquad (8.115)$$

We want to somehow invoke the universal temperature profile, T^+. Using the definition of T^+, equation 8.105, we can define the free stream temperature as follows,

$$T_\infty^+ = (T_s - T_\infty)\frac{\rho c_p u^*}{q_o''} = (T_s - T_\infty)\frac{\rho c_p V_\infty \sqrt{C_f / 2}}{q_o''} , \qquad (8.116)$$

where we have also substituted (8.47) for the friction velocity u^*. Substituting this expression into (8.115) for $(T_s - T_\infty)$ and rearranging,

$$Nu_x = \frac{\rho c_p V_\infty \sqrt{C_f / 2}\, x}{T_\infty^+ k} .$$

Then, multiplying the numerator and denominator by ν, we can express the above as

$$Nu_x = \frac{Re_x Pr \sqrt{C_f / 2}}{T_\infty^+} . \qquad (8.117)$$

We can now use the universal temperature profile, Eqn. (8.113), to evaluate T_∞^+:

$$T_\infty^+ = \frac{Pr_t}{\kappa}\ln y_\infty^+ + 13 Pr^{2/3} - 7 . \qquad (8.118)$$

Unfortunately, a precise value for y_∞^+ is not easy to determine. However, we can make a clever substitution using the Law of the Wall velocity profile, Eqn. (8.58). In the free stream, we can evaluate (8.58) as

$$u_\infty^+ = \frac{1}{\kappa}\ln y_\infty^+ + B . \qquad (8.119)$$

Substituting (8.119) into (8.118) for $\ln y_\infty^+$, the Nusselt number relation then becomes

$$Nu_x = \frac{Re_x Pr \sqrt{C_f / 2}}{Pr_t (u_\infty^+ - B) + 13 Pr^{2/3} - 7} .$$

We can simplify this expression further. Using the definition of Stanton number, $St = Nu_x /(Re_x Pr)$, selecting $B = 5.0$ and $Pr_t = 0.9$, and noting that the definition of u^+ leads to $u_\infty^+ = \sqrt{2/C_f}$, we can rearrange the relation above to arrive at the final result:

$$St_x = \frac{C_f/2}{0.9 + 13\left(Pr^{2/3} - 0.88\right)\sqrt{C_f/2}}.$$ (8.120)

It is interesting to note how similar this result is to the more advanced momentum-heat transfer analogies, particularly those by Prandtl and Taylor (8.96) and von Kármán (8.97). The similarity is not entirely by accident. All three methods invoke the universal velocity profile, and the universal temperature profile used in this development was produced by dividing the thermal boundary layer into sublayers that are similar to those in the two analogies.

8.5.5 Integral Methods for Heat Transfer Coefficient

One use for the universal temperature profile is that it allows us to model the heat transfer using the integral energy equation. Consider turbulent flow over a flat plate, where a portion of the leading surface is unheated (Fig. 8.19).

The simplest solution is to assume a $1/7^{\text{th}}$ power law profile for both the velocity and temperature (equations 8.65 and 8.114). Even with the simplest of assumed profiles, the development is mathematically cumbersome.

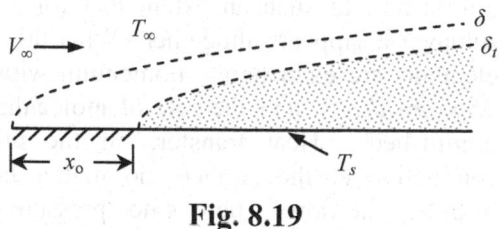

Fig. 8.19

A detailed development appears in Appendix E; the result of the analysis is

$$St_x \equiv \frac{Nu_x}{Re_x Pr} = \frac{C_f}{2}\left[1 - \left(\frac{x_o}{x}\right)^{9/10}\right]^{-1/9},$$ (8.121)

which applies to turbulent flow over a flat plate with unheated starting length x_0.

Note that (8.121) reduces to the Reynolds analogy when $x_o = 0$. This is because the Prandtl number was assumed to be 1 as part of the derivation. However, the model has been used to approximate heat transfer for other fluids as follows. Equation (8.121) can be expressed as

$$Nu_x = \frac{Nu_{x_o=0}}{\left[1-\left(x_o/x\right)^{9/10}\right]^{1/9}} \, , \qquad (8.122)$$

where $Nu_{x_o=0}$ represents the heat transfer in the limit of zero insulated starting length. In this form, other models for heat transfer, like von Kármán's analogy, could be used to approximate $Nu_{x_o=0}$ for $Pr \neq 1$ fluids.

8.5.6 Effect of Surface Roughness on Heat Transfer

While smooth surfaces are common (or at least a common assumption) in engineering practice, the effect of roughness is an important topic. This is an area of great interest, for example, in turbomachinery, where rough surfaces can be used to enhance heat transfer in the cooling of turbine blades.

How does roughness influence heat transfer? In general, we would expect roughness to increase the heat transfer, like it did for the friction factor. However, the mechanisms for momentum and heat transfer are different. We saw in Fig. 8.16 that as roughness increases, the viscous sublayer diminishes, to such an extent that for a fully rough surface the viscous sublayer disappears altogether. What this implies is that the turbulent fluid elements are exchanging momentum with surface directly (like *profile* or *pressure drag*), and the role of molecular diffusion (i.e., *skin friction*) is diminished.   Heat transfer, on the other hand, relies on molecular conduction at the surface, no matter how rough the surface, or how turbulent the flow. There is no "pressure drag" equivalent in heat transfer. Moreover, fluid in the spaces between roughness elements is largely stagnant, and transfers heat entirely by molecular conduction.   The conduction sublayer, then, can be viewed as the average height of the roughness elements.   The stagnant regions between roughness elements effectively create a resistance to heat transfer, and is the major source of resistance to heat transfer [27].

The above discussion tells us that we can not expect roughness to improve heat transfer as much as it increases friction. This is unfortunate, because it also means that we can not predict the heat transfer by simply using a friction factor for rough plates along with one of the momentum-heat transfer analogies.

So if we were to develop a model for the heat transfer from a rough surface, what would influence it? Clearly the roughness size k is a factor.

We would expect that roughness size has no influence until it extends beyond the viscous and conduction sublayers, and its influence reaches a maximum beyond some roughness size (the *fully rough* limit). We would also expect the Prandtl number to appear in any model, since molecular conduction is important. Furthermore, fluids with higher Prandtl number (lower conductivity) would be affected more by roughness. Why? The lower-conductivity fluid trapped between the roughness elements will have a higher resistance to heat transfer. Also, the conduction sublayer is shorter for these fluids, so roughness elements penetrate relatively further into the thermal boundary layer. In contrast, for a liquid metal, the conduction sublayer may fully engulf the roughness elements, virtually eliminating their influence on the heat transfer.

Kays et al. [30] develop a correlation for rough plate, which is equivalent to

$$St = \frac{C_f}{2}\left[Pr_t + C(k_s^+)^{0.2}Pr^{0.44}\sqrt{C_f/2}\right]^{-1}, \qquad (8.123)$$

where $k_s^+ = k_s u^*/v$ is based on the equivalent sand-grain roughness, k_s, and C is a constant that depends on roughness geometry. Indeed, (8.123) displays behavior that we expected in our earlier discussion. In particular, for high-Prandtl-number fluids the second term in the parentheses dominates, but for sufficiently low-Pr fluids the term diminishes, in spite of roughness size. Bogard et al. [31] showed that this model compared well with experimental data from roughened turbine blades. Their results showed a 50% increase in heat transfer over smooth plates. They also demonstrated that increasing roughness beyond some value showed little increase in the heat transfer. This is consistent with (8.123), and our intuition: once the surface is *fully rough*, there is no added advantage to increasing the roughness.

REFERENCES

[1] Davidson, P.A., *Turbulence: An Introduction for Scientists and Engineers*, Oxford University Press, 2007.

[2] Goodgame, C., "High-Tech Swimsuits: Winning Medals Too," *Time*, Aug. 13, 2008.

[3] Reynolds, O., "An Experimental Investigation of the Circumstances Which Determine Whether the Motion of Water Shall be Direct or Sinuous; And of the Law of Resistance in Parallel Channels,"

Philosophical Transactions of the Royal Society of London, Vol. 174, Pt. 3, 1883, pp. 935-982.

[4] Richardson, L.F., *Weather Prediction by Numerical Process*, Cambridge University Press, 1922.

[5] Kolmogorov, A.N., "The Local Structure of Turbulence in Incompressible Viscous Fluid for Very Large Reynolds Numbers," *Dokl. Akad. Nauk SSSR*, Vol. 30, 1941, pp. 301-305.

[6] Reynolds, O., "On the Dynamical Theory of Incompressible Viscous Flows and the Determination of the Criterion," *Philosophical Transactions of the Royal Society of London*, Vol. 186, 1894, pp. 123-161.

[7] Prandtl, L., "On the Frictional Resistance of Air," *Göttinger Ergebinisse*, Vol. 3, p. 1, 1927.

[8] Clauser, F.H., "The Turbulent Boundary Layer," *Advances in Applied Mechanics*, Vol. IV, pp. 91-108, 1954.

[9] Prandtl, L., "Eine Beziehung Zwischen Warmeaustausch und Strömungswiderstand der Flüssigkeit (A Relation between Heat Convection and Flow Resistance in Fluids)" *Phys. Zeitschr.*, Vol. 11, pp. 1072-1078, 1910.

[10] Taylor, G.I., "Conditions at the Surface of a Hot Body Exposed to the Wind," *British Advisory Committee for Aeronautics,* Report and Memorandum No. 272, Vol. 2, 1916, pp. 423-429.

[11] van Driest, E.R., "On Turbulent Flow near a Wall," *J. Aeronautical Sciences*, Vol. 23, No. 11, 1956, pp. 1007-1011.

[12] Spalding, D.B., "A Single Formula for the Law of the Wall," *J. Appl. Mech.*, Vol. 28, 1961, pp. 455-457.

[13] Reichardt, H., "Die Grundlagen des Turbulenten Wärmeübergange (Fundamentals of Turbulent Heat Transfer)," *Arch. Gesamte Waermetech.* Vol. 2, 1951, pp. 129-142.

[14] White, F.M., *Viscous Fluid Flow*, 3rd Ed., McGraw-Hill, Boston, 2006.

[15] Coles, D.E., "The Law of the Wake in the Turbulent Boundary Layer," *J. Fluid Mech.*, Vol. 1, 1956, pp. 191-226.

[16] von Kármán, T., "Über laminare und turbulente Reibung (On laminar and Turbulent Friction) *Z. Angew. Math. Mech.*, Vol. 1, 1921, pp. 233-252.

[17] Blasius, H., *Forschungsarbeiten auf dem Gebiete des Ingenieurwesens,* No. 131, 1913.

[18] Prandtl, L., "Über die Ausgebildete Turbulenz (Investigations on Turbulent Flow)" *Z. Angew. Math. Mech.,* Vol. 5, pp. 136-139 (1925).

[19] Kestin J. and Persen, L.N., "The Transfer of Heat Across a Turbulent Boundary Layer at Very High Prandtl Numbers," *Int. J. Heat Mass Transfer,* Vol. 5, 1962, pp. 355-371.

[20] White, F.M., "A New Integral Method for Analyzing the Turbulent Boundary Layer with Arbitrary Pressure Gradient," *J. Basic Eng.,* Vol. 91, 1969, pp. 371-378.

[21] Arpaci, V.S. and Larsen, P.S., *Convection Heat Transfer,* Prentice-Hall, Inc., Englewood Cliffs, New Jersey, 1984.

[22] Reynolds, O., "On the Extent and Action of the Heating Surface for Steam Boilers," *Proc. Manchester Lit. Philos. Soc.,* Vol. 14, 1874, pp. 7-12.

[23] von Kármán, T., "The Analogy between Fluid Friction and Heat Transfer," *Trans. ASME,* Vol. 61, 1939, pp. 705-710.

[24] Colburn, A.P., "A Method of Correlating Forced Convection Heat Transfer Data and a Comparison with Fluid Friction," *Trans. Am. Inst. Chem. Eng.,* Vol. 29, 1933, pp. 174-210.

[25] Schlichting, H., *Boundary-Layer Theory,* 7th Ed., McGraw-Hill, New York, 1987.

[26] Bejan, A., *Convection Heat Transfer,* 3rd Ed., John Wiley and Sons, Inc., 2004.

[27] Kays, W.M., "Turbulent Prandtl Number – Where Are We?" *J. Heat Transfer,* Vol. 116, 1994, pp. 284-295.

[28] Kakaç, S. and Yener, Y., *Convective Heat Transfer,* 2nd Ed., CRC Press, Boca Raton, 1995.

[29] Churchill, S.W. and Zajic, S.C., "Prediction of Fully Developed Turbulent Convection with Minimal Explicit Empiricism," *AIChE J. Fluid Mech. And Transport Phenomena,* Vol. 48, No. 5, 2002, pp. 927-940.

[30] Kays, W.M., Crawford, M.E., and Weigand, B., *Convective Heat and Mass Transfer,* 4th Ed., McGraw-Hill, Boston, 2005.

[31] Bogard, D.G., Schmidt, D.L., and Tabbita, M., "Characterization and Laboratory simulation of Turbine Airfoil Surface Roughness and Associated Heat Transfer, " J. Turbomachinery, Vol. 2, 1998, pp. 337-343.

PROBLEMS

8.1   Put a check mark in the appropriate column for each of the following statements.

	Statement	true	false	maybe
(a)	The high Reynolds numbers that are characteristic of turbulent flows indicate that viscous forces are not important.			
(b)	All else being equal, the hydrodynamic boundary layer is larger in a turbulent flow than in a laminar flow.			
(c)	Turbulent flows are chaotic and consist of velocity fluctuations that have no discernable structure.			
(d)	The smallest eddies in a flow carry the bulk of the kinetic energy.			
(e)	Numerical simulation of turbulence is difficult because the grid size required to capture the behavior of the smallest eddies violates the continuum hypothesis.			

8.2   Demonstrate, using scale analysis, that the Reynolds number can be interpreted as the ratio of inertial to viscous forces. Consider the x-momentum equation (2.10x), and assume steady, incompressible flow.

8.3   Using Reynolds decomposition and the definition of time average prove the following identities:

[a] $\overline{\overline{a}a'} = 0$ (equation 8.4d).

[b] $\overline{ab} = \overline{a}\,\overline{b} + \overline{a'b'}$ (equation 8.4e).

[c] $\overline{a^2} = (\overline{a})^2 + \overline{(a')^2}$ (equation 8.4f).

8.4   Using Reynolds decomposition and the definition of time average prove the following identities:

[a] $\dfrac{\overline{\partial a}}{\partial x} = \dfrac{\partial \overline{a}}{\partial x}$ (equation 8.4h).

[b] $\dfrac{\partial \overline{a}}{\partial t} = 0$ (equation 8.4i).

[c] $\dfrac{\overline{\partial a}}{\partial t} = 0$ (equation 8.4j).

8.5   Beginning with equation (8.11), derive the turbulent, time-averaged x-momentum equation, (8.12x).

8.6   Beginning with equation (8.10y), derive the turbulent, time-averaged y-momentum equation, (8.12y).   Do not first simplify equation (8.10y) using the product rule, as was done in the text for the x-momentum equation.   Instead, apply Reynolds decomposition directly to the equation.

8.7   Beginning with equation (8.13), derive the turbulent, time-averaged Energy equation, equation (8.14).

8.8   Water at 20°C flows along a plate with a free stream velocity of 20 m/s.   At a distance of 1 m from the leading edge, calculate or estimate:

[a] the Reynolds number, Re_x
[b] the friction velocity, u^*
[c] the thickness of the viscous sublayer
[d] the distance to the inner and outer edges of the "Law of the Wall" region.

8.9   Put a check mark in the appropriate column for each of the following statements regarding the universal velocity and temperature profiles.

	Statement	true	false	maybe
(a)	Beneath the wake region, the u^+ profile for external flow over a flat			

	plate is almost identical to that for pipe flow.			
(b)	In the buffer region, momentum diffusion by viscous effects is of the same order of magnitude as that due to turbulent fluctuations.			
(c)	The viscous sublayer and the conduction sublayer are the same size.			
(d)	The wake region is where velocity data deviate from the Law of the Wall because of compressibility effects.			
(e)	The $1/7^{th}$ power law velocity profile is valid everywhere in the boundary layer.			
(f)	The shape of the universal temperature profile depends on Prandtl number.			

8.10 Using the results of the momentum integral analysis, show that the $1/7^{th}$ power law velocity profile can be expressed in wall coordinates as

$$u^+ \approx 8.56\left(y^+\right)^{1/7} .$$

8.11 Plot the following universal velocity profiles in wall coordinates (from $y^+ = 1$ to 2,000):

[a] The two-layer model consisting of the viscous sublayer and Law of the Wall.
[b] van Driest's continuous model.
[c] Richardson's continuous model.
[d] The $1/7^{th}$ power-law model (see Problem 8.10).

Plot these profiles in semi-log coordinates (like in Figure 8.13), and in normal coordinates.

8.12 The momentum integral method uses a $1/7^{\text{th}}$ power law velocity profile, but other exponents could be chosen. Experimental data for pipe flow suggest that at a higher flow velocity, an exponent of $1/10$ would better fit experimental data in a pipe, with a friction factor of

$$C_f \approx 0.0363 \, \text{Re}_D^{-2/11} \, .$$

Using a $1/10^{\text{th}}$ power law velocity profile, and adapting the friction factor from pipe flow given above, derive equations for friction factor and boundary layer thickness along a flat plate. HINT: For a $1/10^{\text{th}}$ power law velocity profile, $u_m = 0.8768 \, u_{\text{max}}$.

8.13 Derive equations (8.74) and (8.75) in the text.

8.14 Air at 25 °C flows over a flat plate 6 m in length with a velocity of 10 m/s. Plot the local heat transfer coefficient versus x for the portion of the plate that is turbulent. Plot and compare the results of the von Kármán momentum-heat transfer analogy and the Colburn analogy, both using equation (8.75) for the friction factor.

8.15 Put a check mark in the appropriate column for each of the following statements

	Statement	true	false	maybe
(a)	In the limit as for $Pr = 1$, the von Kármán analogy reduces to the Reynolds analogy.			
(b)	In the limit as $Pr = 1$, the Prandtl-Taylor analogy reduces to the Reynolds analogy.			
(c)	For a *fully rough* plate, the friction factor is not dependent on roughness, but increases with Reynolds number.			
(d)	For *laminar* flow over a rough flat plate, the friction factor is the same as for a smooth plate.			
(e)	A rough surface increases heat transfer as much as it increases friction.			

(f)	Roughness reduces the conduction sublayer, and for a *fully rough* surface, the conduction sublayer vanishes.			

8.16 Air at 40°C flows over flat plate with a free stream velocity of 20 m/s. At a distance of 1 m from the leading edge, compare local friction factor obtained from equations (8.71), (8.75), and (8.76). Compare the local Nusselt number obtained from equations (8.95), (8.96) and (8.117), using equation (8.76) for the friction factor.

8.17 Derive equation (8.79) in the text.

8.18 A thin symmetrical airfoil 2m long by 3m wide is oriented parallel to a flow stream that has a velocity of 25 m/s and a free stream temperature of 20°C. The surfaces of the airfoil (both sides) are maintained at a temperature of 50 °C. If the drag force on the airfoil is known to be 10 N, what heat rate has to be applied to the airfoil to maintain the surface temperature?

8.19 Consider Example 8.3. At what value of Reynolds number, Re_L, can the initial laminar region be neglected with a loss of no more that 2% accuracy in the estimate of average Nusselt number? For a flat plate exposed to 20 °C air at 20 m/s, what minimum length does this correspond to?

9

CONVECTION IN TURBULENT CHANNEL FLOW

Glen E. Thorncroft
California Polytechnic State University
San Luis Obispo, California

9.1 Introduction

We discussed laminar flow in channels in Chapter 6, and many of the features of turbulent flow are similar. We will begin this subject with the criteria for fully developed velocity and temperature profiles, and will focus most of our attention on analyzing fully developed flows. As in Chapter 6, our analysis is limited to general boundary conditions: (i) uniform surface temperature, and (ii) uniform surface heat flux.

9.2 Entry Length

We discussed the criteria for the entry length earlier in Chapter 6. As a common rule of thumb, fully developed velocity and temperature profiles exist for

$$\frac{L_h}{D_e} \approx \frac{L_t}{D_e} \approx 10, \tag{6.7}$$

where D_e is the *hydraulic*, or *equivalent* diameter,

$$D_e = \frac{4A_f}{P},$$

where A_f is the flow area, and P is the wetted perimeter. Bejan [1] recommends (6.7) particularly to $Pr = 1$ fluids. More elaborate correlations exist, particularly for hydrodynamic entry length. White [2] recommends the following approximation:

$$\frac{L_h}{D_e} \approx 4.4 Re_{D_e}^{1/6}, \tag{9.1}$$

while Latzko (see Reference 3) suggests

$$\frac{L_h}{D_e} \approx 0.623 Re_{D_e}^{1/4}. \tag{9.2}$$

Thermal entry length, on the other hand, doesn't lend itself to a simple, universally-applicable equation, since the flow is influenced so much by fluid properties and boundary conditions. Consult Kays et al. [3] for more information.

The hydrodynamic entry length is much shorter for turbulent flow than for laminar. In fact, the hydrodynamic entrance region is sometimes neglected in the analysis of turbulent flow. However, the thermal entry length is often important. Analysis of the heat transfer in the thermal entry region is complicated, and not covered in this text. Kays et al. [3] and Burmeister [4] provide introductions to this topic.

9.3 Governing Equations

Consider a circular pipe, as depicted in Fig. 9.1. For convenience, the velocity in the x-direction is labeled as u (we will find that there are many similarities with flow over a flat plate, so we are using u in place of the usual v_x to represent the axial velocity). We begin by assuming two-dimensional, axisymmetric, incompressible flow.

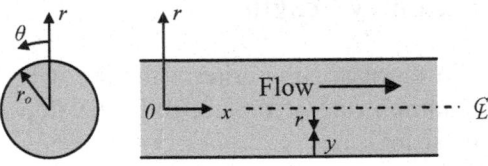

Fig. 9.1

9.3.1 Conservation Equations

Conservation of mass reduces to the following, after Reynolds-averaging:

$$\frac{\partial \overline{u}}{\partial x} + \frac{1}{r}\frac{\partial}{\partial r}(r\overline{v}_r) = 0. \tag{9.3}$$

Using the same conditions, the Reynolds-averaged x-momentum equation reduces to

$$\overline{u}\frac{\partial \overline{u}}{\partial x} + \overline{v}_r \frac{\partial \overline{v}_r}{\partial r} = -\frac{1}{\rho}\frac{\partial p}{\partial x} + \frac{1}{r}\frac{\partial}{\partial r}\left[r(v + \varepsilon_M)\frac{\partial \overline{u}}{\partial r}\right], \qquad (9.4)$$

and conservation of energy becomes:

$$\overline{u}\frac{\partial \overline{T}}{\partial x} + \overline{v}_r \frac{\partial \overline{T}}{\partial r} = \frac{1}{r}\frac{\partial}{\partial r}\left[r(\alpha + \varepsilon_H)\frac{\partial \overline{T}}{\partial r}\right]. \qquad (9.5)$$

9.3.2 Apparent Shear Stress and Heat Flux

We will define the apparent shear stress and heat flux similar to that of the flat plate development:

$$\frac{\tau_{app}}{\rho} = (v + \varepsilon_M)\frac{\partial \overline{u}}{\partial r} \qquad (9.6)$$

and

$$\frac{q''_{app}}{\rho c_p} = -(\alpha + \varepsilon_H)\frac{\partial \overline{T}}{\partial r}. \qquad (9.7)$$

9.3.3 Mean Velocity and Temperature

Correlations for predicting friction and heat transfer in duct flow usually involve the mean velocity and bulk, or mean temperature. The mean velocity is calculated by evaluating the mass flow rate in the duct:

$$m = \rho u_m A = \int_0^{r_o} \rho \overline{u}(2\pi r)dr,$$

or, assuming constant density,

$$u_m = \frac{1}{\pi r_o^2}\int_0^{r_o} \overline{u}(2\pi r)dr = \frac{2}{r_o^2}\int_0^{r_o} \overline{u}r\,dr. \qquad (9.8)$$

The bulk, or mean temperature in the duct is evaluated by integrating the total energy of the flow,

$$mc_p T_m = \int_0^{r_o} c_p \overline{T}\overline{u}(2\pi r)dr.$$

Substituting (9.8) for the mass flow rate, and assuming constant specific heat,

$$T_m \equiv \frac{\displaystyle\int_0^{r_o} \bar{T}\,\bar{u}\,r\,dr}{\displaystyle\int_0^{r_o} \bar{u}\,r\,dr} \,,$$

which can be simplified by substituting the mean velocity, equation (9.8),

$$T_m = \frac{2}{u_m r_o^2} \int_0^{r_o} \bar{T}\,\bar{u}\,r\,dr . \qquad (9.9)$$

9.4 Universal Velocity Profile

9.4.1 Results from Flat Plate Flow

We have already shown that the universal velocity profile in a pipe is very similar to that of flow over a flat plate (see Fig. 8.13), especially when the flat plate is exposed to zero or favorable pressure gradient. We even adapted a pipe flow friction factor model to analyze flow over a flat plate using the momentum integral method (see Section 8.4.3). It is apparent, then, that the characteristics of the flow near the wall of a pipe are not influenced greatly by the curvature of the wall of the radius of the pipe. Therefore a reasonable start to modeling pipe flow is to invoke the two-layer model that we used to model flow over a flat plate:

viscous sublayer: $u^+ = y^+ ,$ (8.54)

Law of the Wall: $u^+ = \dfrac{1}{\kappa}\ln y^+ + B .$ (8.58)

We also have continuous wall law models by Spalding (8.63) and Reichardt (8.64) that have been applied to pipe flow. These profiles were also discussed in Section 8.4.2.

Note that for pipe flow, the wall coordinates are a little different than for flat-plate flow. First, the y-coordinate for pipe flow is

$$y = r_o - r , \qquad (9.10)$$

and so the wall coordinate y^+ is

$$y^+ = r_0^+ - r^+ = \frac{(r_o - r)u^*}{\nu}. \tag{9.11}$$

The velocity wall coordinate is the same as before,

$$u^+ \equiv \frac{\bar{u}}{u^*}, \tag{8.49}$$

and the friction velocity is the same,

$$u^* \equiv \sqrt{\tau_o / \rho}. \tag{8.46}$$

The friction factor, however, is based on the mean flow velocity, instead of the free-stream velocity:

$$C_f = \frac{\tau_o}{(1/2)\rho u_m^2}, \tag{9.12}$$

and so the friction velocity can be expressed as

$$u^* = u_m \sqrt{C_f / 2}. $$

9.4.2 Development in Cylindrical Coordinates

We were fortunate that the velocity profile data for pipe flow matches that of flat plate flow, because it allowed us to develop expressions for universal velocity profiles solely from flat plate (Cartesian) coordinates. Would we have achieved the same results if we had started from the governing equations for pipe flow (i.e., cylindrical coordinates)? Yes, but only after revealing some important issues and insights.

Let's assume fully developed flow. The left side of the x-momentum equation (9.4) goes to zero, and we are left with

$$\frac{1}{r}\frac{\partial}{\partial r}\left(\frac{r\tau}{\rho}\right) = \frac{1}{\rho}\frac{\partial \overline{p}}{\partial x}. \tag{9.13}$$

Rearranging and integrating, we obtain an expression for the shear stress anywhere in the flow:

$$\tau(r) = \frac{r}{2}\frac{\partial \overline{p}}{\partial x} + C. \tag{9.14}$$

The constant C is zero, since we would expect the velocity gradient (and hence the shear stress) to zero at $r = 0$. Evaluating (9.14) at r and r_o and taking the ratio of the two gives

$$\frac{\tau(r)}{\tau_o} = \frac{r}{r_o}.$$ (9.15)

Equation (9.15) shows that the local shear is a linear function of radial location. This result raises an important issue: equation (9.15) is a linear shear profile, as depicted in Fig. 9.2. This expression contradicts how we expect the fluid to behave near the wall: recall that for flat plate flow, the Couette Flow assumption led us to the idea that τ is approximately constant in the direction normal to the wall. And yet, we saw in Section 8.4 that the universal velocity profile that resulted from this assumption works well for flat plate flow as well as pipe flow.

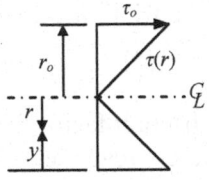

Fig. 9.2

How do we reconcile the Couette flow assumption with (9.15)? Remember that the near-wall region over which we make the Couette flow assumption covers a very small distance. Therefore we could assume that, in that small region vary close to the wall of the pipe, the shear is nearly constant, $\tau \approx \tau_o$. Thus the Couette assumption approximates the behavior near the pipe wall as

$$(v + \varepsilon_M)\frac{\partial \overline{u}}{\partial r} = \frac{\tau_o}{\rho} = \text{constant}.$$ (9.16)

This may not seem like a convincing argument, but remember that we have experimental data that show that near the wall, the velocity profiles for flat plate flow and pipe flow are essentially the same (see Fig. 8.13). The data suggest that the near-wall behavior is not influenced by the outer flow, or even the curvature of the wall.

9.4.3 Velocity Profile for the Entire Pipe

Note from the previous development that we expect the velocity gradient (and the shear stress) to be zero at the centerline of the pipe. Unfortunately, none of the universal velocity profiles we've developed so far behave this way.

Reichardt [5] attempted to account for the entire region of the pipe. He suggested a model for eddy viscosity:

$$\frac{\varepsilon_M}{\nu} = \frac{\kappa y^+}{6}\left(1+\frac{r}{r_o}\right)\left[1+2\left(\frac{r}{r_o}\right)^2\right],\qquad(9.17)$$

which leads to the following expression for the velocity profile:

$$u^+ = \frac{1}{\kappa}\ln\left[y^+\frac{1.5(1+r/r_o)}{1+2(r/r_o)^2}\right]+B,\qquad(9.18)$$

where Reichardt used $\kappa = 0.40$ and $B = 5.5$. It can be shown that the slope of this profile is zero at $r = 0$, so the profile matches the behavior of the flow in the core of the pipe. The profile does not account for the viscous sublayer, but as $r \to r_o$, equation (9.18) does reduce to the original Law of the Wall form, equation (8.58).

9.5 Friction Factor for Pipe Flow

9.5.1 Blasius Correlation for Smooth Pipe

Based on dimensional analysis and experimental data, Blasius [6] developed a purely empirical correlation for flow through a smooth circular pipe:

$$C_f \approx 0.0791\, Re_D^{-1/4}\quad(4000 < Re_D < 10^5),\qquad(9.19)$$

where the friction factor is based on the mean flow velocity, $C_f \equiv \tau_o /(1/2)\rho u_m^2$. Later correlations have proven to be more accurate and versatile. However, equation (9.19) led to the development of the $1/7^{th}$ Power Law velocity profile.

9.5.2 The $1/7^{th}$ Power Law Velocity Profile

The Blasius correlation (equation 9.17) led to a crude but simple approximation for the velocity profile in a circular pipe [9], which was discovered by Prandtl [7] and von Kármán [8]. We begin with the Blasius correlation, equation (9.19), which can be recast in terms of wall shear stress:

$$\frac{\tau_o}{\frac{1}{2}\rho u_m^2} = 0.0791\left(\frac{2r_o u_m}{\nu}\right)^{-1/4},$$

or, rearranging,

$$\tau_o = 0.03326\rho u_m^{7/4} r_o^{-1/4}\nu^{1/4}.\qquad(a)$$

We will assume a power law can be used to approximate the velocity profile,

$$\frac{\bar{u}}{u_{CL}} = \left(\frac{y}{r_o}\right)^q,$$ (b)

with u_{CL} representing the centerline velocity. Furthermore, we assume that the mean velocity in the flow can be related to the centerline velocity as

$$u_{CL} = (const)u_m.$$ (c)

Substituting (b) and (c) for the mean velocity in (a) yields

$$\tau_o = (const)\rho\left[\bar{u}\left(\frac{y}{r_o}\right)^{-1/q}\right]^{7/4} r_o^{-1/4}\nu^{1/4},$$

which simplifies to

$$\tau_o = (const)\rho\bar{u}^{7/4}y^{(-7/4q)}r_o^{(7/4q-1/4)}\nu^{1/4}.$$ (d)

Now, both Prandtl and von Kármán argued that the wall shear stress is not a function of the size of the pipe, a point that we also argued earlier in Section 9.4.1. If this is true, then the exponent on r_o should be equal to zero. Setting the exponent to zero, the value of q must be equal to $1/7$, leading to the classic $1/7^{th}$ power law velocity profile,

$$\frac{\bar{u}}{u_{CL}} = \left(\frac{y}{r_o}\right)^{1/7}.$$ (9.20)

Experimental data show that this profile adequately models the velocity profile through a large portion of the pipe, and is frequently used in models for momentum and heat transfer. (You may recall that the $1/7^{th}$ Law was also adapted by Prandtl [12] and von Kármán [8] for turbulent flow over a flat-plate flow, and was used in the integral method in Section 8.4.3 to develop an estimate for the friction factor).

The model does have its limitations, though. The model is accurate for a narrow range of Reynolds numbers (roughly, 10^4 to 10^6), it yields an infinite velocity gradient at the wall, and does not yield a gradient of zero at the centerline.

Another student of Prandtl, Nikuradse [10] measured velocity profiles in smooth pipe over a wide range of Reynolds numbers, and reported that the exponent varied with Reynolds number,

$$\frac{\bar{u}}{u_{CL}} = \left(\frac{y}{r_o}\right)^n.$$ (9.21)

Table 9.1 lists results of Nikuradse's measurements, including measurement of pipe friction factor of the form

$$C_f = \frac{C}{\text{Re}_D^{1/m}}.$$ (9.22)

As one might expect, Nikuradse's results show that the velocity profile becomes fuller as the mean velocity increases.

Table 9.1

Re_D	n	C	m
4×10^3	6	0.1064	3.5
2.3×10^4	6.6	0.0880	3.8
1.1×10^5	7	0.0804	4
1.1×10^6	8.8	0.0490	4.9
2.0×10^6	10	0.0363	5.5
3.2×10^6	10	0.0366	5.5

9.5.3 Prandtl's Law for Smooth Pipe

Whereas the Blasius correlation (9.19) is purely empirical, we can develop a more theoretical model for friction factor by employing the universal velocity profile. We begin with the Law of the Wall, equation (8.58). Substituting the wall coordinates u^+ and y^+, as well as the friction velocity $u^* = \sqrt{\tau_o / \rho} = u_m \sqrt{C_f / 2}$, (8.58) becomes

$$\frac{\bar{u}}{u_m}\sqrt{\frac{2}{C_f}} = \frac{1}{\kappa}\ln\left(\frac{yu_m}{\nu}\sqrt{\frac{C_f}{2}}\right) + B.$$ (9.23)

If we assume that the equation holds at any value of y, we could evaluate the expression at the centerline of the duct, $y = r_o = D/2$, where $\bar{u} = u_{CL}$. Substituting these and the definition of the Reynolds number, we obtain

$$\frac{u_{CL}}{u_m}\sqrt{\frac{2}{C_f}} = \frac{1}{\kappa}\ln\left(\frac{Re_D}{2}\sqrt{\frac{C_f}{2}}\right) + B. \qquad (9.24)$$

We see that by looking merely at the Law of the Wall profile, we are able to obtain a functional relationship for the friction factor. (Recall that we followed the same process when developing one of the newer models for friction factor on a flat plate, in Section 8.4.3.) Unfortunately, the ratio u_{CL}/u_m is still unknown.

To evaluate the mean velocity u_m, we can simply substitute the Law of the Wall velocity profile, Eqn. (8.58), into the expression for the mean velocity, Equation (9.8). First, we perform a variable substitution on (9.8) to

$$u_m = \frac{1}{\pi r_o^2}\int_0^{r_o}\bar{u}(2\pi r)dr = \frac{2}{r_o^2}\int_0^{r_o}\bar{u}(r_o - y)dy, \qquad (9.25)$$

where $y = r_o - r$. Then, substituting the velocity profile (8.58) and performing the integration, it can be shown that the mean velocity becomes

$$u_m = u^*\left[\frac{1}{\kappa}\ln\left(\frac{r_o u^*}{\nu}\right) + B - \frac{3}{2\kappa}\right]. \qquad (9.26)$$

Or, making substitutions again,

$$u_m = u_m\sqrt{\frac{C_f}{2}}\left[\frac{1}{\kappa}\ln\left(\frac{Re_D}{2}\sqrt{\frac{C_f}{2}}\right) + B - \frac{3}{2\kappa}\right]. \qquad (9.27)$$

Ironically, the term we were trying to evaluate, u_m, cancels out of the expression. However, u_{CL} does not appear either. In other words, we can use the above expression directly to find an expression for C_f. Rearranging, and substituting the values $\kappa = 0.41$ and $B = 5.0$ gives

$$\frac{1}{\sqrt{C_f/2}} = 2.44\ln\left(Re_D\sqrt{C_f/2}\right) - 0.349,$$

This expression is not yet complete. Note that, in this development we assumed that the Law of the Wall is accurate everywhere. This assumption ignores the presence of a viscous sublayer or a wake region. As a result, we find we have to adjust the constants in this equation to better fit experimental data. Doing so, we obtain

$$\frac{1}{\sqrt{C_f/2}} = 2.46 \ln\left(C_f/2_D\sqrt{C_f/2}\right) + 0.29, \quad (Re_D > 4000). \quad (9.28)$$

This is called Prandtl's universal law of friction for smooth pipes [11], although it is sometimes referred to as the Kármán-Nikuradse equation [3]. Note that, despite the empiricism of using a curve fit to obtain the constants in (9.28), using a more theoretical basis to develop the function has given the result a wider range of applicability than Blasius's correlation.

Equation (9.28) must be solved iteratively for C_f. A simpler, empirical relation that closely matches Prandtl's is [3]

$$\frac{C_f}{2} \approx 0.023 Re_D^{-1/5}, \quad (3 \times 10^4 < Re_D < 10^6). \quad (9.29)$$

This correlation is also suitable for non-circular ducts, with the Reynolds number calculated using the hydraulic diameter.

9.5.4 Effect of Surface Roughness

From our discussion of turbulent flow over a rough flat plate, we saw that roughness shifts the universal velocity profile downward (see Section 8.4.4 and Fig 8.16). We could write the velocity profile in the logarithmic layer as

$$u^+ = \frac{1}{\kappa} \ln y^+ + B - \Delta B,$$

where ΔB is the shift in the curve, which increases with wall roughness k^+. The behavior doesn't just depend on the average roughness; it also depends on the type of roughness, which ranges from uniform geometries like rivets to random structures like sandblasted metal. The following model is based on *equivalent sand grain roughness* [2],

$$\frac{1}{f^{1/2}} \approx 2.0 \log_{10}\left[\frac{Re_D f^{1/2}}{1 + 0.1(k/D)Re_D f^{1/2}}\right] - 0.8, \quad (9.30)$$

where it is common to use the Darcy friction factor,

$$f = 4C_f. \quad (9.31)$$

Two points can be made from this expression: First, if the relative roughness k/D is low enough, it doesn't have much of an effect on the equation. Scaling shows that roughness is not important if $(k/D)Re_D < 10$. On the other hand, if $(k/D)Re_D > 1000$, the roughness term dominates in the denominator, and the Reynolds number cancels; in other words, the friction factor is no longer dependent on the Re_D.

C.F. Colebrook [13] and C.M. White developed the following formula for commercial pipes,

$$\frac{1}{f^{1/2}} = -2.0 \log_{10}\left(\frac{k/D}{3.7} + \frac{2.51}{Re_D f^{1/2}}\right), \qquad (9.32)$$

with representative roughness values presented in Table 9.2. This function is what appears in the classic Moody chart [14] represented in Fig. 9.3.

Table 9.2

Type of pipe	Equivalent roughness, k (mm)
Glass	smooth
Drawn plastic tubing	0.0005 - 0.0025
Brass, aluminum, stainless steel, new	0.001 - 0.003
Concrete, smooth	0.02 - 0.08
Sheet metal	0.02 - 0.08
Cast iron	0.1 - 0.4
Concrete, rough	1 - 3

9.6 Momentum-Heat Transfer Analogies

We will now apply the analogy method to pipe flow. The development is applied to the case of a constant heat flux boundary condition. Strictly speaking, an analogy cannot be made for the case of a constant surface temperature. In spite of this fact, we find that resulting models approximately hold for this case as well.

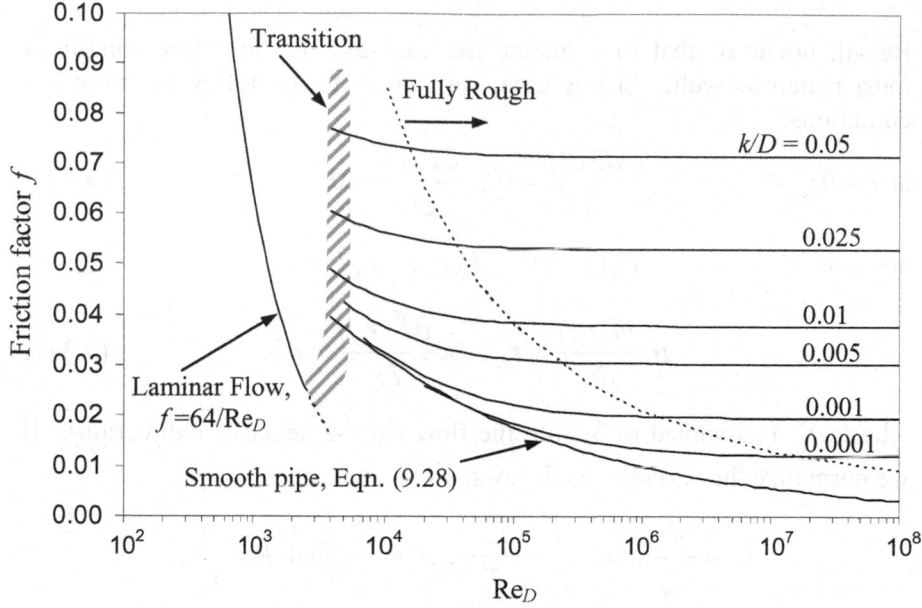

Fig. 9.3

The x-momentum equation (9.4) becomes, for hydrodynamically fully developed flow,

$$\frac{1}{\rho}\frac{dp}{dx} = \frac{1}{r}\frac{\partial}{\partial r}\left[r(\nu + \varepsilon_M)\frac{\partial \overline{u}}{\partial r}\right],\tag{9.33a}$$

while the energy equation (8.124) reduces to

$$\overline{u}\frac{\partial \overline{T}}{\partial x} = \frac{1}{r}\frac{\partial}{\partial r}\left[r(\alpha + \varepsilon_H)\frac{\partial \overline{T}}{\partial r}\right].\tag{9.33b}$$

We recall that an analogy is possible if the momentum and energy equations are identical. The right sides of the above equations look similar, but what about the left-hand sides? Note that in pipe flow the pressure gradient is non-zero, although constant with respect to x. To ensure an analogy, then, the left side of (9.33b) must then be constant. Fortunately, we found in Chapter 6 that, for thermally fully developed flow and a constant heat flux at the wall, the shape of the temperature profile is constant with respect to x, leading to

$$\frac{\partial \overline{T}}{\partial x} = \text{constant}.$$

So, it appears that a momentum-heat transfer analogy is possible.

Recall, however, that to complete the analogy, the boundary conditions must match as well. In this case, we can write the following boundary conditions:

at $r = 0$:
$$\frac{d\bar{u}(0)}{dr} = 0 , \quad \frac{\partial \bar{T}(0)}{\partial r} = 0 , \qquad (9.34a)$$

at $r = r_o$:
$$\bar{u}(r_o) = 0, \quad \bar{T}(r_o) = T_s(x) , \qquad (9.34b)$$

$$\mu \frac{d\bar{u}(r_o)}{dr} = \tau_o, \quad k \frac{\partial \bar{T}(r_o)}{\partial r} = q_o'', \qquad (9.34c)$$

where q_o'' is assumed to be into the flow (in the negative r-direction). If we normalize the variables as follows,

$$U = \frac{\bar{u}}{u_m} , \quad \theta = \frac{\bar{T} - T_s}{T_m - T_s} , \quad X = \frac{x}{L}, \text{ and } R = \frac{r}{r_o} ,$$

we can show that both the governing equations and the boundary conditions are identical in form.

9.6.1 Reynolds Analogy for Pipe Flow

If we assume that $\nu = \alpha$ $(Pr = 1)$ and $\varepsilon_H = \varepsilon_M$ $(Pr_t = 1)$, the same assumptions used to develop Reynold's analogy for a flat plate, then the governing equations (9.33a) and (9.33b) are identical. If we follow exactly the same process that we followed for the original derivation, we find that the Reynolds analogy is essentially identical for pipe flow,

$$St_D \equiv \frac{q_o''}{\rho u_m c_p (T_s - T_m)} = \frac{C_f}{2} \quad (Pr = 1),$$

or
$$St_D = \frac{Nu_D}{Re_D Pr} = \frac{C_f}{2} . \qquad (9.35)$$

Note that in this case the Stanton number is defined in terms of the mean velocity and bulk temperature, as is the wall shear stress: $\tau_o = \frac{1}{2} C_f \rho u_m^2$.

9.6.2 Adapting Flat-Plate Analogies to Pipe Flow

We have seen many similarities in the behavior of flat-plate flow and pipe flow, and in the last section we saw that Reynold's analogy is identical for both flows. It seems possible, then, that we could adapt other flat-plate analogies to pipe flow. This can be done, with some modifications to the original relations.

As an example, let's revisit the von Kármán analogy. The original development references conditions at the edge of the boundary layer, like V_∞ and T_∞. For pipe flow, we'll approximate these conditions with centerline properties,

$$V_\infty \approx u_{CL} \text{ and } T_\infty \approx T_{CL}.$$

These substitutions also affect the friction factor, which translates to

$$C_f \approx \frac{\tau_o}{\frac{1}{2}\rho u_{CL}^2}.$$

Following the development exactly as before, the result is almost identical:

$$\frac{q_o''}{\rho \bar{u}_{CL} c_p (T_s - T_{CL})} = \frac{C_f/2}{1+5\sqrt{\frac{C_f}{2}}\left\{(Pr-1)+\ln\left[\frac{5Pr+1}{6}\right]\right\}}. \tag{9.36}$$

The problem with this relation is that the left side of (9.36) and the friction factor are expressed in terms of centerline variables instead of the more common and convenient mean quantities u_m and T_m. We can correct this as follows:

$$\frac{q_o''}{\rho u_m c_p (T_s - T_m)}\left(\frac{u_m (T_s - T_m)}{u_{CL}(T_s - T_{CL})}\right) = \frac{(C_f/2)(u_m/u_{CL})^2}{1+5\frac{u_m}{u_{CL}}\sqrt{\frac{C_f}{2}}\left\{(Pr-1)+\ln\left[\frac{5Pr+1}{6}\right]\right\}},$$

where C_f is again defined in terms of the mean velocity, $C_f = \tau_o / \frac{1}{2}\rho u_m^2$, and the terms $q_o'' / \rho u_m c_p (T_s - T_m)$ are collectively the Stanton number for pipe flow. Simplifying,

$$St_D\left(\frac{T_s - T_m}{T_s - T_{CL}}\right) = \frac{(C_f/2)(u_m/u_{CL})}{1+5\left(\frac{u_m}{u_{CL}}\right)\sqrt{\frac{C_f}{2}}\left\{(Pr-1)+\ln\left[\frac{5Pr+1}{6}\right]\right\}}. \tag{9.37}$$

This is the von Kármán Analogy for pipe flow [developed in 15]. We can develop estimates for the ratios (u_m/u_{CL}) and $(T_s - T_m)/(T_s - T_{CL})$

using the definition of mean temperature, equation (9.9). We could estimate u_m and T_m using the 1/7th Law profiles, which for a circular pipe are:

$$\frac{\bar{u}}{u_{CL}} = \left(\frac{y}{r_o}\right)^{1/7} , \qquad (9.20)$$

and, similar to (8.111) for a flat plate,

$$\frac{\bar{T} - T_s}{T_{CL} - T_s} = \left(\frac{y}{r_o}\right)^{1/7} . \qquad (9.38)$$

Substituting these models into (9.8) and (9.9), we can show that

$$\frac{u_m}{\bar{u}_{CL}} = 0.817 , \qquad (9.39)$$

and

$$\frac{T_m - T_s}{T_{CL} - T_s} = 0.833 . \qquad (9.40)$$

9.6.3 Other Analogy-Based Correlations

A simple correlation for turbulent flow in a duct is based on the Colburn analogy. Beginning with the analogy, equation (8.96), and using equation (9.27) for the friction factor, we obtain

$$St_D = 0.023 Re_D^{-1/5} Pr^{-2/3} ,$$

or

$$Nu_D = 0.023 Re_D^{4/5} Pr^{1/3} . \qquad (9.41)$$

One of the most popular correlations is the Dittus-Boelter correlation [3], which is an empirical correlation based on the Colburn analogy:

$$Nu_D = 0.023 Re_D^{4/5} Pr^n , \qquad (9.42)$$

where $n = 0.4$ for heating ($T_s > T_m$) and $n = 0.3$ for cooling. The popularity of this correlation is undoubtedly due to its simplicity, as well as the fact that it compared well to a range of experimental data available at the time. However, the accuracy of this relation and the related Colburn correlation has been challenged in recent years (see discussion in Section 8.5.2), and models such as those by Petukhov and Gnielinski correlation (see Section 9.8) are preferred for their improved accuracy and range of applicability.

Analogies remain a common way to model the heat transfer in pipes, and models have been developed specifically for pipe flows. Two examples are by Reichardt [16] and Boelter, Martinelli, and Jonassen [17]. A recent analogy-based model was developed by Churchill and Zajic [18] in 2002, which the authors claim is to date the most accurate model for the internal flow. In an age of numerical simulation, there remains considerable usefulness and interest in these traditional approaches.

9.7 Algebraic Method Using Universal Temperature Profile

As we did for flow over a flat plate, we can use the universal temperature and velocity profiles to estimate the heat transfer in a circular duct. We begin again with the definition of the Nusselt number, which for flow in a duct can be expressed as

$$Nu_D \equiv \frac{hD}{k} = \frac{q_o'' D}{(T_s - T_m)k} , \qquad (9.43)$$

To invoke the universal temperature profile, we use the definition of T^+, equation 8.102, to define the mean temperature as

$$T_m^+ = (T_s - T_m)\frac{\rho c_p u^*}{q_o''} = (T_s - T_m)\frac{\rho c_p u_m \sqrt{C_f/2}}{q_o''} . \qquad (9.44)$$

Recall that for duct flow, the friction velocity u^* is defined in terms of the mean velocity. Substituting this expression into (9.43) for q_o'', and invoking the definitions of the Reynolds and Prandtl numbers,

$$Nu_D = \frac{Re_D Pr \sqrt{C_f/2}}{T_m^+} . \qquad (9.45)$$

As it turns out, there are several ways to proceed with the analysis. One approach is to evaluate T_m^+ using a dimensionless version of (9.33):

$$T_m^+ = \frac{2}{u_m^+ r_o^{+2}} \int_0^{r_o^+} T^+ u^+ (r_o^+ - y^+)dy^+ . \qquad (9.46)$$

Theoretically, it is a simple matter to substitute appropriate universal temperature and velocity profiles into (9.46) and integrate. Practically, this requires numerical integration. However, if we want a simpler, closed-form solution, a second approach can be taken. First, we rewrite the original Nusselt number relation (9.43) as follows,

$$Nu_D = \frac{q_o'' D}{(T_s - T_m)k} \frac{(T_s - T_{CL})}{(T_s - T_{CL})} ,$$

where T_{CL} is the centerline temperature. Then, substituting the definition of T^+ for the centerline temperature in the denominator, we obtain

$$Nu_D = \frac{Re_D \, Pr \sqrt{C_f / 2}}{T_{CL}^+} \frac{(T_s - T_{CL})}{(T_s - T_m)} . \qquad (9.47)$$

We can now use the universal temperature profile, equation (8.118), to evaluate T_{CL}^+ :

$$T_{CL}^+ = \frac{Pr_t}{\kappa} \ln y_{CL}^+ + 13 Pr^{2/3} - 7 . \qquad (9.48)$$

Now, just like in our analysis for flat plate flow, we can substitute the Law of the Wall velocity profile (8.59) for $\ln y_{CL}^+$:

$$u_{CL}^+ = \frac{1}{\kappa} \ln y_{CL}^+ + B . \qquad (9.49)$$

Substituting these into the Nusselt number relation,

$$Nu_D = \frac{Re_D Pr \sqrt{C_f / 2}}{\left[Pr_t (u_{CL}^+ - B) + 13 Pr^{2/3} - 7 \right]} \frac{(T_s - T_{CL})}{(T_s - T_m)} . \qquad (9.50)$$

We need expressions for u_{CL}^+ and $(T_s - T_{CL})/(T_s - T_m)$. For the centerline velocity, we can use the definition of u^+ for pipe flow:

$$u_{CL}^+ = \frac{u_{CL}}{u^*} = \frac{u_{CL}}{u_m} \sqrt{\frac{2}{C_f}} . \qquad (9.51)$$

It appears that, if we are to complete the analysis, we will need to evaluate the mean velocity and temperature after all. To avoid the complexity of the logarithmic velocity and temperature profiles, we could estimate these quantities using the much simpler $1/7^{th}$ Law profiles, which we saw in the last section yields

$$\frac{T_m - T_s}{T_{CL} - T_s} = 0.833 \quad \text{and} \quad \frac{u_m}{u_{CL}} = 0.817 .$$

Finally, using the definition of Stanton number, $St_D = Nu_D /(Re_D Pr)$, selecting $Pr_t = 0.9$ and $B = 5.0$, we can rearrange (9.50) obtain

$$St_D = \frac{C_f / 2}{0.92 + 10.8\left(Pr^{2/3} - 0.89\right)\sqrt{C_f / 2}} . \qquad (9.52)$$

Under what conditions is this equation applicable? The ultimate test would be to compare the expression to experimental data. However, since we invoked the $1/7^{th}$ power law, which is valid around 1×10^5, it might be reasonable as a first approximation to limit this model to $Re_D < 1 \times 10^5$.

9.8 Other Correlations for Smooth Pipe

Petukhov [19] followed a more rigorous theoretical development, invoking Reichardt's model for eddy diffusivity and velocity profile (9.15, 9.16). He obtained

$$St_D = \frac{C_f / 2}{1.07 + 12.7\left(Pr^{2/3} - 1\right)\sqrt{C_f / 2}} , \quad \left(\begin{array}{c} 0.5 \leq Pr \leq 2000 \\ 10^4 < Re_D < 5 \times 10^6 \end{array} \right) . (9.53)$$

which compares well to experimental data over a wide range of Prandtl and Reynolds numbers. He used the following model for friction factor, which he also developed:

$$\frac{C_f}{2} = (2.236 \ln Re_D - 4.639)^{-2} . \qquad (9.54)$$

Note the similarity between Petukhov's relation (9.53) and the algebraic result, equation (9.52). It seems as if we have captured the essential functionality even in our modest approach.

In 1976, Gnielinski [20] modified Petukhov's model slightly, extending the model to include lower Reynolds numbers:

$$Nu_D = \frac{(Re_D - 1000)PrC_f/2}{1 + 12.7(Pr^{2/3} - 1)\sqrt{C_f/2}}, \quad \left(\begin{array}{c} 0.5 \le Pr \le 2000 \\ 3 \times 10^3 < Re_D < 5 \times 10^6 \end{array}\right). \quad (9.55)$$

Again, Petukhov's friction model can be used in (9.55) for the friction factor. For all the above models, properties should be evaluated at the film temperature.

As was the case with the analogy-based correlations, these correlations are reasonable for channels with constant surface temperature as well as constant heat flux; the flows are relatively insensitive to boundary conditions.

9.9 Heat Transfer in Rough Pipes

We've discussed the effects of roughness on the heat transfer from flat plates in Section 8.5.6, and much of the same physical intuition applies to flow in channels. Norris [21,3] presents the following empirical correlation for flow through circular tubes:

$$\frac{Nu}{Nu_{smooth}} = \left(\frac{C_f}{C_{f,smooth}}\right)^n, \quad \left(\frac{C_f}{C_{f,smooth}} < 4\right), \quad (9.56)$$

where $n = 0.68Pr^{0.215}$. A correlation like Colebrook's (9.30) could be used to determine the rough-pipe friction factor.

The behavior of this relation reflects what we expect physically. Equation (9.56) shows that the Prandtl number influences the effect of roughness, and for very low-Pr fluids the roughness plays little role in the heat transfer. Regardless of Prandtl number, the influence of roughness size is limited: Norris reports that the effect of increasing roughness vanishes beyond $(C_f/C_{f,smooth}) \approx 4$, and so the equation reaches a maximum.

Although roughness enhances heat transfer, it also increases the friction, which can be expensive in terms of pumping costs. Neither the friction nor the heat transfer increase indefinitely with roughness size – both reach a limiting value. The application of roughness to increase the heat transfer (say, in heat exchangers) requires the designer to weigh the benefits against the costs.

REFERENCES

[1] Bejan, A., *Convection Heat Transfer*, 3rd Ed., John Wiley and Sons, Inc., 2004.

[2] White, F.M., *Viscous Fluid Flow*, 3rd Ed., McGraw-Hill, Boston, 2006.

[3] Kays, W.M., Crawford, M.E., and Weigand, B., *Convective Heat and Mass Transfer*, 4th Ed., McGraw-Hill, Boston, 2005.

[4] Burmeister, L.C., *Convective Heat Transfer*, 2nd Ed., John Wiley and Sons, New York, 1993.

[5] Reichardt, H., "Die Grundlagen des Turbulenten Wärmeübergange (Fundamentals of Turbulent Heat Transfer)," *Arch. Gesamte Waermetech.* Vol. 2, 1951, pp. 129-142.

[6] Blasius, H., *Forschungsarbeiten auf dem Gebiete des Ingenieurwesens*, No. 131, 1913.

[7] Prandtl, L., "Über die Ausgebildete Turbulenz (Investigations on Turbulent Flow)" *Z. Angew. Math. Mech.*, Vol. 5, 1925, pp. 136-139.

[8] von Kármán, T., "Über laminare und turbulente Reibung (On Laminar and Turbulent Friction)," *Z. Angew. Math. Mech.*, Vol. 1, 1921, pp. 233-252.

[9] Prandtl, L. and Tietjens, O.G., *Applied Hydro- and Aeromechanics*, Dover Publications, Inc., New York, 1934.

[10] Nikuradse, J., "Gesetzmässigkeit der turbulenten strömung in glatten Rohren (Law of Turbulent Flow in Smooth Pipes)," *Forschg. Arb. Ing.-Wes.* No. 356, 1932.

[11] Schlichting, H., *Boundary-Layer Theory*, 7th Ed., McGraw-Hill, New York, 1987.

[12] Prandtl, L., "On the Frictional Resistance of Air," *Göttinger Ergebinisse*, Vol. 3, p. 1, 1927.

[13] Colebrook, C.F., "Turbulent Flow in Pipes, with Particular Reference to the Transition Region Between Smooth and Rough Pipe Laws," *J. Inst. Civil Engineers*, 1939.

[14] Moody, L.F., "Friction Factors for Pipe Flow," *Trans. ASME*, Vol. 66, 1944, pp. 671-684.

[15] Arpaci, V.S. and Larsen, P.S., *Convection Heat Transfer*, Prentice-Hall, Inc., Englewood Cliffs, New Jersey, 1984.

[16] Reichardt, H., "Die Grundlagendes turbulenten Wärmeübertraganges (The Principles of Turbulent Heat Transfer)," *Archiv. Ges. Wärmetech.*, Vol. 2, 1951, pp. 129-142.

[17] Boelter, L.M.K., Martinelli, R.C., and Jonassen, F., "Remarks on the Analogy between Heat Transfer and Momentum Transfer," Trans. ASME, Vol. 63, 1941, pp. 447-456.

[18] Churchill, S.W. and Zajic, S.C., "Prediction of Fully Developed Turbulent Convection with Minimal Explicit Empiricism," *AIChE J. Fluid Mech. And Transport Phenomena*, Vol. 48, No. 5, 2002, pp. 927-940.

[19] Petukhov, B.S., "Heat Transfer and Friction in Turbulent Pipe Flow with Variable Physical Properties," *Advances in Heat Transfer*, Academic Press, New York, Vol. 6, 1970, pp. 503-564.

[20] Gnielinski, V., "New Equations for Heat and Mass Transfer in Turbulent Pipe and Channel Flow," *Int. Chem. Eng.*, Vol. 16, 1976, pp. 359-368.

[21] Norris, R.H., Augmentation of Convection Heat and Mass Transfer, ASME, New York, 1971.

PROBLEMS

9.1   Put a check mark in the appropriate column for each of the following statements.

	Statement	true	false	maybe
(a)	Sufficiently close to the wall, the u^+ profile for external flow over a flat plate is almost identical to that for pipe flow.			
(b)	The $1/7^{th}$ power law velocity profile is accurate for all velocities and pipe diameters.			
(c)	The $1/7^{th}$ power law velocity profile is accurate throughout the cross-sectional area of the pipe.			

(d)	The universal velocity profile for pipe flow more closely matches that of flat plate flow when there is an adverse pressure gradient.			
(e)	A momentum-heat transfer analogy is theoretically valid for pipe flow for the constant wall temperature boundary condition.			
(f)	The Reynolds analogy for pipe flow is the same form as that for flat plate flow.			
(g)	The same correlation for heat transfer can be used to model the heat transfer for both constant surface temperature and constant heat flux boundary conditions.			

9.2   Consider equation (9.1) or (9.2). All else being equal, how does a fluid's viscosity affect the hydrodynamic entrance length in a duct?

9.3   All else being equal, which fluid would result in a longer hydrodynamic entrance length in a duct, water or air? Explain your results in light of the answer to Problem 9.2.

9.4   For pipe flow with a 1/n power-law velocity profile, show that the ratio of mean velocity to centerline velocity is

$$\frac{u_m}{\overline{u}_{CL}} = \frac{2n^2}{(n+1)(2n+1)}.$$

Compare the ratio for $n = 6, 7, 8, 9$, and 10, and discuss.

9.5   For pipe flow with 1/n power-law temperature and velocity profiles, show that the normalized mean velocity is

$$\frac{T_m - T_s}{T_{CL} - T_s} = \frac{2n+1}{2(n+2)}.$$

Compare the values obtained for $n = 6, 7, 8, 9$, and 10, and discuss.

9.6   Put a check mark in the appropriate column for each of the following statements.

Statement		true	false	maybe
(a)	The friction factor in turbulent flow is higher than that for laminar flow.			
(b)	For turbulent flow in a pipe, the friction factor is higher for rough pipe than for smooth pipe.			
(c)	For laminar flow in a pipe, the friction factor is higher for rough pipe than for smooth pipe.			
(d)	For turbulent flow in a pipe, roughness has less of an influence on heat transfer for low-Pr fluids			
(e)	As the roughness in a pipe increases, the friction factor increases.			
(f)	As the flow rate increases in a rough pipe, the friction factor decreases.			
(g)	The effect of adding roughness to a pipe is to increase heat transfer as much as it increases the friction.			

9.7   Consider turbulent flow in a 6-cm-diameter smooth pipe with mean velocity $u_m = 10$ m/s and $Re_D = 4 \times 10^4$. What is the centerline velocity? What is the thickness of the viscous sublayer, in mm? What is the velocity there?

9.8   Recall from your undergraduate fluid mechanics textbook, the head loss h_l through a pipe of constant cross-sectional area is

$$\frac{p_1 - p_2}{\rho g} + z_1 - z_2 = h_l.$$

[a]   Apply the momentum equation to flow in a circular pipe of diameter D and length L oriented at some angle α to the horizontal, and show that the head loss can be expressed as

$$h_l = \frac{4\tau_o}{\rho g}\frac{L}{D} ,$$

noting that the head loss is a function only of wall friction and not elevation change.

[b]   Express the result in part [a] and [b] in terms of the Fanning friction factor, C_f , and the Darcy friction factor, f.

9.9   Water flows through a smooth, horizontal, 5-cm-diameter pipe of length 10 m at a rate of 1 kg/s. Assume the flow is fully developed.

[a]   Estimate the pressure drop through the pipe using Nikuradze's model, Prandtl's universal law and simplified version, and Petukhov's model.

[b]   Is it valid to assume that the flow is fully developed? How would the assumption of fully developed flow affect your prediction of friction factor?

9.10 Air flows will fully developed velocity through a tube of inside diameter 2.0 cm. The flow is fully developed with a mean velocity of 10 m/s. The surface is maintained at a uniform temperature of 90 °C. The inlet temperature is 30 °C. Determine the length of tube needed to increase the mean temperature to 70 °C.

9.11 Consider the conditions of Problem 9.9. The surface of the pipe is exposed to a uniform heat flux of 50 kW/m^2, and the water enters the pipe at 25 °C. Assuming the flow to be thermally fully developed,

[a]   Determine the mean temperature at the outlet, and the maximum temperature of the pipe. Use Petukhov's correlation.

[b]   Is it valid to assume that the flow is hydrodynamically and thermally fully developed? How would the assumption of fully developed flow affect your prediction of mean outlet temperature?

9.12 Water flows through a horizontal 4-cm-diameter, 1.5-m-long, smooth pipe with a mean velocity of 20 cm/s.

[a]   Assuming the flow to be fully developed, estimate the pressure drop through the pipe.

[b]   The water enters the pipe at 20 °C, and the wall of the pipe is held constant at 90 °C.   Assuming the flow is thermally fully developed, estimate the heat transfer coefficient, and determine the mean temperature at the exit of the pipe.

[c] Is it appropriate in this problem to neglect entrance effects?

9.13 Repeat Problem 9.12, assuming the pipe is made of a metal with roughness 0.05 mm. Compare the results of the two problems.

9.14 Water at an inlet temperature of 20 °C flows through a 5-cm-diameter pipe at a flow rate of 15 liters per minute. The 10-m-long pipe is made of cast iron.

[a] Estimate the pressure drop through the pipe.

[b] Determine the mean temperature of the outlet and the overall heat transfer to the water if the pipe surface is maintained at 60 °C.

9.15 A fluid with $Pr = 0.9$ flows through a smooth circular duct with constant surface heat flux. The Reynolds number is 2500, so the flow could be either laminar or turbulent. Assuming the flow is fully developed (hydrodynamically and thermally), calculate:

[a] The ratio of the friction factor for turbulent flow to that of laminar flow.

[b] The ratio of heat transfer coefficients.

[c] Repeat parts a and b under the condition of the pipe being rough with $k/D = 0.01$.

9.16 Air is heated in a 4cm×4cm sheet metal duct with a uniform surface heat flux of 590 W/m². The mean velocity is 3.2 m/s. The sheet metal has an average roughness 0.05 mm. If the air enters at 20 °C and leaves at a mean temperature of 120 °C, determine the following (ignore entrance effects):

[a]   The required length of the duct.

[b]   The maximum surface temperature.

CORRELATION EQUATIONS:
FORCED AND FREE CONVECTION

10.1 Introduction

There are many situations where analytic determination of the heat transfer coefficient h is difficult to obtain. As was shown in previous chapters, even after making many simplifying assumptions, analytical determination of h is generally not a simple mathematical problem. When complicating factors such as geometry, variable properties, turbulent flow, boiling, condensation, etc. are involved, the heat transfer coefficient is usually determined experimentally. This does not mean that each time there is a need for h for which there is no analytic solution we must conduct an experiment. Instead, we utilize the experimental results of other researchers. Experimental results are usually correlated and presented as dimensionless equations which are convenient to use. Such equations are known as *correlation equations*. They are extensively used in the solution of heat transfer problems and therefore deserve special attention.

In this chapter we will explain how correlation equations are obtained, discuss their selection and use and present common cases. Four topics will be considered: (1) external forced convection over plates, cylinders, and spheres, (2) internal forced convection through channels, (3) external free convection over plates, cylinders and spheres, and (4) free convection in enclosures.

10.2 Experimental Determination of Heat Transfer Coefficient h

To determine the heat transfer coefficient h it is common to work with Newton's law of cooling, which defines h as

$$h = \frac{q_s''}{T_s - T_\infty}. \qquad (10.1)$$

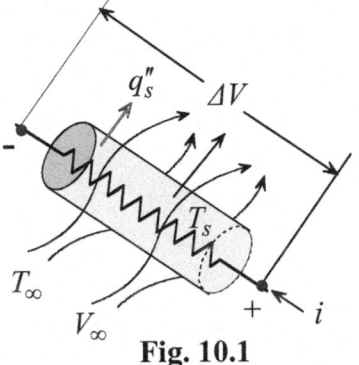

Fig. 10.1

By measuring surface temperature T_s, surface heat flux q_s'', and free stream temperature T_∞, equation (10.1) can be used to determine the heat transfer coefficient. A common method for heating a surface and calculating the flux is shown in Fig. 10.1. Heating is provided by an electric resistor. Measurement of current i and voltage drop ΔV provides data for calculating dissipated power and heat flux. Thermocouples are commonly used to measure surface temperature. A refinement of this experimental setup involves using multi-resistors and circuits to provide a prescribed surface flux or surface temperature.

Correlation equations are presented in dimensionless form. This is an effective and efficient way to organize and present experimental data. Instead of presenting equations for h, it is common to correlate data in terms of a dimensionless heat transfer coefficient called the *Nusselt number*. Since the parameters governing convection heat transfer are known from dimensional analysis, both the design of experiments for determining the Nusselt number and the form of correlation equations are based on this knowledge. For example, for constant properties forced convection heat transfer with no dissipation, we have shown that

$$Nu_x = f(x^*; Re, Pr). \qquad (2.52)$$

Thus, experiments are designed such that both the Reynolds and Prandtl numbers can be varied and measurements are made for calculating the Nusselt number at various locations x^*. The collected data is then correlated according to equation (2.52).

10.3 Limitations and Accuracy of Correlation Equations

All correlation equations have limitations which must be carefully noted before they are applied. First, geometry is an obvious factor. External flow over a tube is not the same as flow through a tube. Thus, each equation is valid for a specific configuration. Second, limitations on the range of parameters, such as the Reynolds, Prandtl and Grashof numbers, for which a correlation equation is valid, are determined by the availability of data and/or the extent to which an equation correlates the data.

Since correlation equations are based on experimentally determined data, they do not always provide very accurate predictions of h. Errors as high as 25% are not uncommon.

10.4 Procedure for Selecting and Applying Correlation Equations

To identify the appropriate correlation equation for a specific application, the following steps should be considered:

(1) Identify the geometry under consideration. Is it flow over a flat plate, over a cylinder, through a tube, or through a channel?

(2) Identify the classification of the heat transfer process. Is it forced convection, free convection, external flow, internal flow, entrance region, fully developed region, boiling, condensation, micro-gravity?

(3) Determine if the objective is finding the local heat transfer coefficient (local Nusselt number) or average heat transfer coefficient (average Nusselt number).

(4) Check the Reynolds number in forced convection. Is the flow laminar, turbulent or mixed?

(5) Identify surface boundary condition. Is it uniform temperature or uniform flux?

(6) Examine the limitations on the correlation equation to be used. Does your problem satisfy the stated conditions?

(7) Establish the temperature at which properties are to be determined. For external flow properties are usually determined at the film temperature T_f

$$T_f = (T_s + T_\infty)/2. \tag{10.2}$$

and for internal flow at the mean temperature $\overline{T}_m$. However, there are exceptions that should be noted.

(8) Use a consistent set of units in carrying out computations.

(9) Compare calculated values of h with those listed in Table 1.1. Large deviations from the range of h in Table 1.1 may mean that an error has been made.

10.5 External Forced Convection Correlations

10.5.1 Uniform Flow over a Flat Plate: Transition to Turbulent Flow

We consider boundary layer flow over a semi-infinite flat plate shown in Fig. 10.2. In the region close to the leading edge the flow is laminar. As the distance from the leading edge increases so does the Reynolds number. At some location downstream, $x = x_t$,

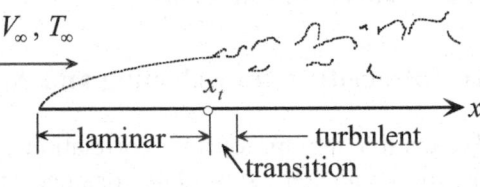

Fig. 10.2

turbulence begins to appear and transition from laminar to turbulent flow develops. The Reynolds number corresponding to this location is called the *transition* or *critical Reynolds number* Re_{x_t}. Its value, which is determined experimentally, depends on several factors including surface finish, pressure gradient, free stream turbulence, etc. For uniform flow over a flat plate the transition Reynolds number is approximately given by

$$Re_{x_t} = \frac{V_\infty x_t}{\nu} \approx 5 \times 10^5 . \tag{10.3}$$

It should be kept in mind that this value is not an exact criterion for this flow configuration. It may be lower or higher, with extreme values that differ by as much as orders of magnitude.

Correlation equations will be presented for the basic geometry of a semi-infinite flat plate with uniform upstream velocity and temperature. Laminar, turbulent and mixed flow conditions will be considered. We will examine various boundary conditions for this flat plate geometry.

(1) Plate at Uniform Surface Temperature. The *local heat transfer coefficient* is determined from the *local Nusselt number*. To proceed, establish if the flow is laminar or turbulent. For $x < x_t$ the flow is laminar and thus equation (4.72) is applicable. In the turbulent region, $x > x_t$, equations (8.71) and (8.98) give [1]

$$Nu_x = \frac{hx}{k} = 0.0296(Re_x)^{4/5}(Pr)^{1/3}.$$ (10.4a)

Equation (10.4a) is valid for:

flat plate, constant T_s

$5 \times 10^5 < Re_x < 10^7$ (10.4b)

$0.6 < Pr < 60$

properties at T_f

With the local heat transfer coefficient determined in the laminar and turbulent regions, we can construct the average heat transfer coefficient $\bar{h}$ for a plate of length L. For $L < x_t$ the flow is laminar. For $L > x_t$ the flow is mixed, being laminar for $0 < x < x_t$ and turbulent for $x_t < x < L$. Determining $\bar{h}$ for this case requires integration of the local value over both the laminar and turbulent regions. Starting with the definition of $\bar{h}$ in (2.50), we have

$$\bar{h} = \frac{1}{L}\int_0^L h(x)dx = \frac{1}{L}\left[\int_0^{x_t} h_L(x)dx + \int_{x_t}^L h_t(x)dx\right],$$ (10.5)

where $h_L(x)$ and $h_t(x)$ are the local heat transfer coefficients in the laminar and turbulent regions, respectively. The local laminar Nusselt number in (4.72) gives $h_L(x)$. Equation (10.4a) gives $h_t(x)$. Substituting (4.72b) and (10.4a) into (10.5) we obtain

$$\bar{h} = \frac{k}{L}\left[0.332\left(\frac{V_\infty}{v}\right)^{1/2}\int_0^{x_t}\frac{dx}{x^{1/2}} + 0.0296\left(\frac{V_\infty}{v}\right)^{4/5}\int_{x_t}^L\frac{dx}{x^{1/5}}\right](Pr)^{1/3}.$$

(10.6)

When the integration is carried out, the result is

$$\bar{h} = \frac{k}{L}\left\{0.664(Re_{x_t})^{1/2} + 0.037\left[(Re_L)^{4/5} - (Re_{x_t})^{4/5}\right]\right\}Pr^{1/3}.$$ (10.7a)

Or, expressed in terms of the average Nusselt number $\overline{Nu}_L$, equation (10.7a) gives

$$\overline{Nu}_L = \frac{\overline{h}L}{k} = \left\{0.664(Re_{x_t})^{1/2} + 0.037\left[(Re_L)^{4/5} - (Re_{x_t})^{4/5}\right]\right\}Pr^{1/3}.$$

$$(10.7b)$$

This result is limited to the assumptions leading to Pohlhausen's solution and the range of Pr and Re_x given in (10.4b).

(2) Plate at Uniform Surface Temperature with an Insulated Leading Section. This case is shown in Fig. 10.3. A leading section of length x_0 is insulated. Transition from laminar to turbulent flow can take place within or beyond this section. The laminar flow case was presented in Chapter 5 where the local Nusselt number is given by equation (5.21). For turbulent flow the local Nusselt number is given by [2]

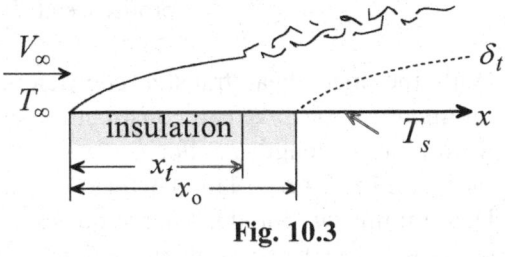

Fig. 10.3

$$Nu_x = \frac{hx}{k} = \frac{0.0296(Re_x)^{4/5}(Pr)^{1/3}}{\left[1-(x_0/x)^{9/10}\right]^{1/9}}.$$

$$(10.8)$$

(3) Plate with Uniform Surface Flux. Fig. 10.4 shows a plate which is heated uniformly along its surface. As with the case of uniform surface temperature, the flow is laminar for $0 < x < x_t$ and turbulent for $x > x_t$. For the laminar region the local Nusselt number is determined analytically using (5.36) or (5.37). In the turbulent region the local Nusselt number is [2]

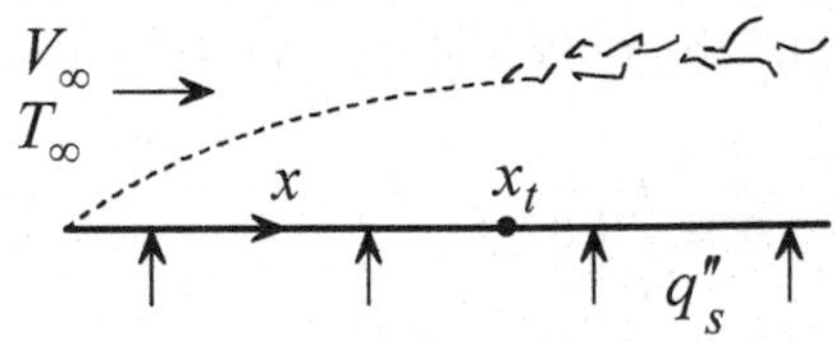

Fig. 10.4

$$Nu_x = \frac{hx}{k} = 0.030(Re_x)^{4/5} Pr^{1/3}.$$

$$(10.9)$$

Note that surface temperature $T_s(x)$ varies along the plate. The variation is determined by Newton's law of cooling using the local heat transfer coefficient (10.9). Properties are determined at the film temperature $T_f = (\overline{T}_s + T_\infty)/2$, where $\overline{T}_s$ is the average surface temperature.

Example 10.1: Power Dissipated by Chips

An array of 30×90 chips measuring 0.4 cm × 0.4 cm each are mounted flush on a plate. Surface temperature of the chips is $T_s = 76°C$. The array is cooled by forced convection of air at $T_\infty = 24°C$ flowing parallel to the plate with a free stream velocity $V_\infty = 35$ m/s. Determine the dissipated power in the array.

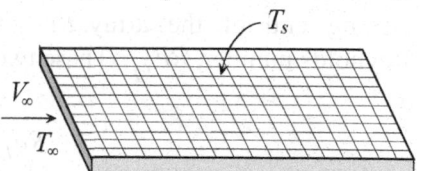

(1) Observations. (i) This is a forced convection problem over a flat plate. (ii) Surface temperature is uniform. (iii) The average heat transfer coefficient and Newton's law of cooling give the heat transfer rate from the surface to the air. (iv) The Reynolds number at the trailing end should be calculated to determine if the flow is laminar, turbulent or mixed.

(2) Problem Definition. Find the average heat transfer coefficient for flow over a semi-infinite flat plate.

(3) Solution Plan. Apply Newton's law of cooling to determine the heat transfer from the surface to the air. Calculate the Reynolds number to establish if the flow is laminar, turbulent or mixed. Use an analytic solution or a correlation equation to determine the average heat transfer coefficient.

(4) Plan Execution.

(i) **Assumptions.** (1) Continuum, (2) Newtonian, (3) steady state, (4) constant properties, (5) uniform upstream velocity and temperature, (6) uniform surface temperature, (7) negligible plate thickness, (8) negligible edge effects, (9) all dissipated power in chips is transferred to the air by convection, (10) no radiation, and (11) the array is oriented with its short side facing the flow.

(ii) **Analysis.** Applying Newton's law of cooling to the surface of the array gives

$$P = q_T = \overline{h} A (T_s - T_\infty), \tag{a}$$

where

A = surface area
$\overline{h}$ = average heat transfer coefficient, W/m²-°C

P = power dissipated by the chips, W

q_T = total heat transfer from surface = power dissipated in array, W

T_s = surface temperature = 76°C

T_∞ = free stream temperature = 24°C

To determine $\overline{h}$ it is necessary to establish if the flow is laminar, turbulent or mixed. This is determined by calculating the Reynolds number at the trailing end of the array, Re_L, and comparing it with the transition Reynolds number, Re_{x_t}. These two numbers are defined as

$$Re_L = \frac{V_\infty L}{\nu}, \tag{b}$$

and

$$Re_{x_t} = \frac{V_\infty x_t}{\nu} = 5 \times 10^5, \tag{c}$$

where

L = length of array = 90(chips)×0.4(cm/chip) = 36 cm = 0.36 m

V_∞ = free stream velocity = 35 m/s

ν = kinematic viscosity of air, m²/s

Properties are evaluated at the film temperature T_f given by

$T_f = (T_s + T_\infty)/2 = (76 + 24)(°C)/2 = 50°C$

Air properties at this temperature are

k = 0.02781 W/m-°C

Pr = 0.709

$\nu = 17.92 \times 10^{-6}$ m²/s

Substituting into (b)

$$Re_L = \frac{35(\text{m/s})0.36(\text{m})}{17.92 \times 10^{-6}(\text{m}^2/\text{s})} = 7.031 \times 10^5$$

Comparing this with the transition Reynolds number shows that the flow is turbulent at the trailing end. Therefore, the flow is mixed over the array and the average heat transfer coefficient is given by equation (10.7b)

$$\overline{h} = \frac{k}{L}\left\{0.664(Re_{x_t})^{1/2} + 0.037\left[(Re_L)^{4/5} - (Re_{x_t})^{4/5}\right]\right\}(Pr)^{1/3}. \tag{d}$$

This result is limited to the assumptions leading to Pohlhausen's solution and the range of Pr and Re_x given in (10.4b).

(iii) Computations. The area of the rectangular array is

$$A = 30(\text{chips}) \times 0.4(\text{cm/chip}) \times 90(\text{chips}) 0.4(\text{cm/chip}) = 432 \text{ cm}^2$$
$$= 0.0432 \text{ m}^2$$

Equations (c) and (d) give $\overline{h}$

$$\overline{h} = \frac{0.0278(\text{W/m}-^\circ\text{C})}{0.36(\text{m})} \left\{ 0.664(5 \times 10^5)^{1/2} + 0.037\left[(7.031 \times 10^5)^{4/5} - (5 \times 10^5)^{4/5}\right](0.703)^{1/3} \right\}$$

$$\overline{h} = 61.3 \text{ W/m}^2-^\circ\text{C}$$

Substituting into (a)

$$P = q_T = 61.3(\text{W/m}^2-^\circ\text{C}) \, 0.0432(\text{m}^2) \, (76 - 24)(^\circ\text{C}) = 137.7 \text{ W}$$

(iv) Checking. *Dimensional check*: Computations showed that equations (a), (b) and (d) are dimensionally consistent.

Quantitative check: The calculated value of $\overline{h}$ is within the range given in Table 1.1 for forced convection of gases.

(5) Comments. (i) Pohlhausen's solution (4.72b) for laminar flow and correlation equation (10.4a) for turbulent flow were used to solve this problem. The solution is limited to all the assumptions and restrictions leading to these two equations.

(ii) More power can be dissipated in the array if the boundary layer is tripped at the leading edge to provide turbulent flow over the entire array. The corresponding heat transfer coefficient can be obtained by setting $Re_{x_t} = 0$ in equation (d)

$$\overline{Nu}_L = \frac{\overline{h}L}{k} = 0.037 Re_L^{4/5} Pr^{1/3}.$$

Solving for $\overline{h}$

$$\overline{h} = \frac{0.02781(\text{W/m}-^\circ\text{C})}{0.36(\text{m})} (0.037)(7.031 \times 10^5)^{4/5} 0.709^{1/3} = 121.3 \text{ W/m}^2-^\circ\text{C}$$

Substituting into (a)

$$P = q_T = 121.3(\text{W/m}^2 - {}^\circ\text{C})0.0432(\text{m}^2)(76-24)({}^\circ\text{C}) = 272.5 \text{ W}$$

Thus, turbulent flow over the entire array almost doubles the maximum dissipated power.

10.5.2 External Flow Normal to a Cylinder

Fig.10.5 shows forced convection normal to a cylinder. Since the flow field varies in the angular direction θ, the heat transfer coefficient h also varies with θ. An equation which correlates the average heat transfer coefficient $\overline{h}$ over the circumference is given by [3]

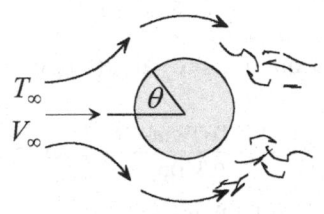

Fig. 10.5

$$\overline{Nu}_D = \frac{\overline{h}D}{k} = 0.3 + \frac{0.62 Re_D^{1/2} Pr^{1/3}}{\left[1 + \left(\dfrac{0.4}{Pr}\right)^{2/3}\right]^{1/4}} \left[1 + \left(\frac{Re_D}{282,000}\right)^{5/8}\right]^{4/5}.$$

(10.10a)

Valid for:

> flow normal to cylinder
> $Pe = Re_D Pr > 0.2$
> properties at T_f

(10.10b)

where Re_D is the Reynolds number based on diameter and Pe is the *Peclet number* defined as the product of the Reynolds and Prandtl numbers. For $Pe < 0.2$, the following is used [4]

$$\overline{Nu}_D = \frac{\overline{h}D}{k} = \frac{1}{0.8237 - 0.5 \ln Pe}.$$

(10.11a)

Valid for:

> flow normal to cylinder
> $Pe = Re_D Pr < 0.2$
> properties at T_f

(10.11b)

Equations (10.10) and (10.11) may also be applied to cylinders with uniform flux.

10.5.3 External Flow over a Sphere

The average Nusselt number for the flow over a sphere is given by [5]

$$\overline{Nu}_D = \frac{\overline{h}D}{k} = 2 + \left[0.4Re_D^{1/2} + 0.06Re_D^{2/3}\right]Pr^{0.4}\left(\frac{\mu}{\mu_s}\right)^{1/4}. \quad (10.12a)$$

Valid for:

$$\boxed{\begin{array}{l} \text{flow over sphere} \\ 3.5 > Re_D > 7.6\times10^4 \\ 0.71 < Pr < 380 \\ 1 < \mu/\mu_s < 3.2 \\ \text{properties at } T_\infty, \mu_s \text{ at } T_s \end{array}} \qquad (10.12b)$$

10.6 Internal Forced Convection Correlations

In Chapter 6, analytic determination of the heat transfer coefficient is presented for a few laminar flow cases. In Chapter 9 we presented analysis of turbulent flow through channels. We will now present correlation equations for the entrance and fully developed regions under both laminar and turbulent flow conditions. The criterion for transition from laminar to turbulent flow is expressed in terms of the Reynolds number Re_D, based on the mean velocity $\overline{u}$ and diameter D. The flow is considered laminar for $Re_D < Re_{D_t}$, where Re_{D_t} is the transition Reynolds number given by

$$Re_{D_t} = \frac{\overline{u}D}{\nu} \approx 2300. \qquad (10.13)$$

Properties for internal flow are generally evaluated at the mean temperature $\overline{T}_m$.

10.6.1 Entrance Region: Laminar Flow through Tubes at Uniform Surface Temperature

In considering heat transfer in the entrance region of tubes and channels, we must first determine if both velocity and temperature are developing simultaneously or if the velocity is already fully developed but the temperature is developing. This latter case is encountered where the heat transfer section of a tube is far away from the flow inlet section. Correlations for both cases will be presented for laminar flow in tubes at uniform surface temperature.

(1) Fully Developed Velocity, Developing Temperature: Laminar Flow. This case is encountered where the velocity profile develops prior to entering the thermal section as shown in Fig. 10.6. This problem was solved analytically using boundary layer theory. However, the form of the solution is not convenient to use. Results are correlated for the average Nusselt number for a tube of length L in the following form [6]:

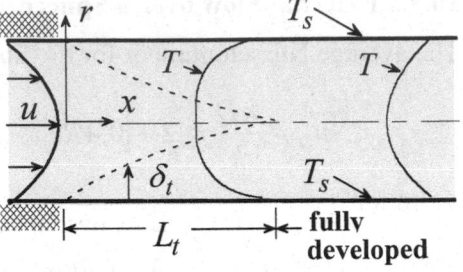

Fig. 10.6

$$\overline{Nu}_D = \frac{\overline{h}D}{k} = 3.66 + \frac{0.0668(D/L)\,Re_D\,Pr}{1+0.04\big[(D/L)\,Re_D\,Pr\big]^{2/3}}. \qquad (10.14a)$$

Valid for:

> entrance region of tube
> uniform surface temperature T_s
> fully developed laminar flow ($Re_D < 2300$)   (10.14b)
> developing temperature
> properties at $\overline{T}_m = (T_{mi} + T_{mo})/2$

where T_{mi} and T_{mo} are the mean temperatures at the inlet and outlet, respectively.

(2) Developing Velocity and Temperature: Laminar Flow. A correlation equation for this case is given by [5, 7]

$$\overline{Nu}_D = \frac{\overline{h}D}{k} = 1.86\big[(D/L)\,Re_D\,Pr\big]^{1/3}\left(\frac{\mu}{\mu_s}\right)^{0.14}. \qquad (10.15a)$$

Valid for:

> entrance region of tube
> uniform surface temperature T_s
> laminar flow ($Re_D < 2300$)
> developing velocity and temperature
> $0.48 < Pr < 16700$
> $0.0044 < \mu/\mu_s < 9.75$ (10.15b)
> $\left[\dfrac{D}{L}\,Re_D\,Pr\right]^{1/3}\left(\dfrac{\mu}{\mu_s}\right)^{0.14} > 2$
> properties at $\overline{T}_m = (T_{mi} + T_{mo})/2$, μ_s at T_s

Example 10.2: Force Convection Heating in a Tube

Water enters a tube with a uniform velocity $\bar{u}$ = 0.12 m/s and uniform temperature $T_{mi} = 18°C$.

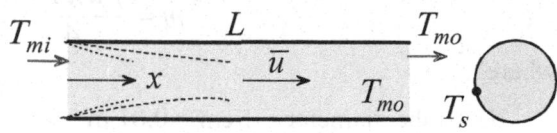

The surface of the tube is maintained at $T_s = 72°C$. The tube diameter is D = 1 cm and its length is L = 1.5 m. Determine the heat transfer rate to the water.

(1) Observations. (i) This is an internal flow problem through a tube at uniform surface temperature. (ii) Both velocity and temperature are developing. (iii) Entrance effects can be neglected if the tube is much longer than the developing lengths L_h and L_t.(iv) The Reynolds number establishes if the flow is laminar or turbulent. (v) Heat transfer to the water can be calculated if the outlet temperature is known.

(2) Problem Definition. Determine the outlet water temperature.

(3) Solution Plan. (i) Apply conservation of energy to the water to determine the rate of heat transfer q_s. (ii) Calculate the Reynolds number. (iii) Determine the hydrodynamic and thermal entrance lengths to establish if this is an entrance or fully developed flow problem.

(4) Plan Execution

(i) **Assumptions.** Anticipating the need to apply conservation of energy and to determine the heat transfer coefficient, the following assumptions are made: (1) Continuum, (2) negligible changes in kinetic and potential energy, (3) constant properties, (4) steady state, (5) no energy generation ($q''' = 0$), (6) negligible axial conduction ($Pe > 100$, to be verified), (7) axisymmetric flow, and (8) uniform surface temperature.

(ii) **Analysis.** Application of conservation of energy to the water between the inlet and outlet gives

$$q_s = mc_p(T_{mo} - T_{mi}), \tag{a}$$

where

c_p = specific heat, J/kg-°C
m = mass flow rate, kg/s
q_s = rate of heat transfer to water, W
T_{mi} = inlet temperature = 18°C
$T_{mo} = T_m(L)$ = outlet temperature, °C

The mass flow rate is given by

$$m = \frac{\rho \bar{u} \pi D^2}{4},$$ (b)

where

D = tube diameter = 1 cm = 0.01 m
$\bar{u}$ = mean velocity = 0.12 m/s
ρ = density, kg/m^3

Properties of water are evaluated at $\bar{T}_m$, defined as

$$\bar{T}_m = (T_{mi} + T_{mo})/2 .$$ (c)

The mean fluid temperature $T_m(x)$ at distance x from the inlet is given by equation (6.13). Setting $x = L$ in (6.13) gives the outlet temperature $T_m(L) = T_{mo}$

$$T_{mo} = T_s + (T_{mi} - T_s)\exp[-\frac{P\bar{h}}{\dot{m}c_p}L],$$ (d)

where

$\bar{h}$ = average heat transfer coefficient, W/m^2-°C
L = tube length = 1.5 m
P = tube perimeter = πD
T_s = surface temperature = 72°C

The problem now becomes one of finding the heat transfer coefficient $\bar{h}$. The Reynolds number is determined next to establish if the flow is laminar or turbulent. The Reynolds number is defined as

$$Re_D = \frac{\bar{u}D}{\nu},$$ (e)

where ν is the kinematic viscosity evaluated at the mean temperature $\bar{T}_m$. Since the outlet temperature T_{mo} is unknown, an iterative procedure is required to determine $\bar{T}_m$. An assumed value for T_{mo} is used to obtain approximate values for water properties needed to calculate T_{mo}. If the calculated T_{mo} is not close to the assumed value, the procedure is repeated until a satisfactory agreement is obtained. Assume T_{mo} = 42°C. Equation (c) gives

$$\bar{T}_m = (18 + 42)\ °C/2 = 30\ °C$$

Properties of water at this temperature are:

$c_p = 4180$ J/kg-°C

$k = 0.6150$ W/m-°C

$Pr = 5.42$

$\mu = 0.7978 \times 10^{-3}$ kg/s-m

$v = 0.8012 \times 10^{-6}$ m²/s

$\rho = 995.7$ kg/m³

Substituting into equation (e) gives the Reynolds number

$$Re_D = \frac{\bar{u}D}{v} = \frac{0.12(m/s)0.01(m)}{0.8012 \times 10^{-6}(m^2/s)} = 1497.8$$

Since this is smaller than the transition Reynolds number ($Re_{D_t} = 2300$), the flow is laminar. The Peclet number is calculated to verify assumption (6)

$$Pe = Re_D\, Pr = 1497.8 \times 5.42 = 8118$$

Thus neglecting axial conduction is justified. To determine if the flow is developing or fully developed, the hydrodynamic and thermal entrance lengths, L_h and L_t, are calculated using equations (6.5) and (6.6) and Table 6.1

$$L_h = 0.056 D Re_D = (0.056)(0.01 \text{ m})(1497.8) = 0.839 \text{ m}$$

$$L_t = 0.033 D Re_D Pr = (0.033)(0.01 \text{ m})(1497.8)(5.42) = 2.679 \text{ m}$$

Comparing these with the tube length, $L = 1.5$ m, shows that both velocity and temperature are developing. Therefore, entrance effects must be taken into consideration in determining $\bar{h}$. The applicable correlation equation for this case is (10.15a)

$$\overline{Nu}_D = \frac{\bar{h}D}{k} = 1.86\left[(D/L)\, Re_D Pr\right]^{1/3}\left(\frac{\mu}{\mu_s}\right)^{0.14}, \qquad \text{(f)}$$

where μ_s is the viscosity at surface temperature T_s. Before using equation (f), the conditions on its applicability, equation (10.15b), must be satisfied. Consideration is given to the 6th and 7th conditions in (10.15b).

$$\mu/\mu_s = 0.7978 \times 10^{-3}(kg/s\text{-}m)/0.394 \times 10^{-3}(kg/s\text{-}m) = 2.02$$

and

$$\left[(D/L)\,Re_D\,Pr\right]^{1/3}\left(\frac{\mu}{\mu_s}\right)^{0.14}=\left[\frac{0.01(m)}{1.5(m)}(1497.8)(5.42)\right]^{1/3}(2.02)^{0.14}=4.17$$

Therefore, all conditions listed in (10.15b) are satisfied.

(iii) **Computations.** Equation (b) gives m

$$m=\pi\frac{995.7(kg/m^3)0.12(m/s)(0.01)^2\,(m^2)}{4}=0.009384\ kg/s$$

Equation (f) gives $\overline{h}$

$$\overline{Nu}_D=\frac{\overline{h}D}{k}=1.86\left[\frac{0.01(m)}{1.5(m)}(1497.8)(5.42)\right]^{1/3}\left[\frac{0.7978\times10^{-3}\,(Kg/s-m)}{0.394\times10^{-3}\,(Kg/s-m}\right]^{0.14}$$

$$=7.766$$

$$\overline{h}=\frac{k}{D}\overline{Nu}_D=\frac{0.615\,(W/m\text{-}^\circ C)}{0.01(m)}7.766=477.6\ (W/m^2\text{-}^\circ C)$$

Substituting into (d) gives T_{mo}

$$T_{mo}=72(^\circ C)-(72-18)(^\circ C)\exp\left[-\frac{\pi\,477.6(W/m^2-^\circ C)0.01(m)1.5(m)}{0.009384\,(kg/s)4180\,(J/kg-^\circ C)}\right]$$

$$=41.6\,^\circ C$$

This is close to the assumed value of 42°C. Substituting into (a) gives q_s

$$q_s=0.009384(kg/s)\ 4180(J/kg\text{-}^\circ C)\ (41.6-18)(^\circ C)=925.7\ W$$

(iv) **Checking.** *Dimensional check*: Computations showed that equations (a), (b) and (d)-(f) are dimensionally consistent.

Quantitative check: (i) The calculated value of the heat transfer coefficient is within the range suggested in Table 1.1 for forced convection of liquids.

(ii) To check the calculated heat transfer rate q_s, assume that the water inside the tube is at a uniform temperature $\overline{T}_m$

$$\overline{T}_m=(T_{mi}+T_{mo})/2=(18+41.6)^\circ C/2=29.8\,^\circ C$$

Application of Newton's law of cooling gives

$$q_s=\overline{h}A(T_s-\overline{T}_m)=477.6(W/m^2\text{-}^\circ C)\ \pi\ 0.01\ (m)\ 1.5(m)\ (72-29.8)\ (^\circ C)$$
$$=949.8\ W$$

This is close to the exact answer of 925.7 W.

(5) Comments. (i) The determination of the Reynolds number is critical in solving this problem.

(ii) If we incorrectly assume fully developed flow, the Nusselt number will be 3.66, $\overline{h} = 225.09$ W/m^2–°C, $T_{mo} = 30.8°$C and $q_s = 502.1$ W. This is significantly lower than the value obtained for developing flow.

10.6.2 Fully Developed Velocity and Temperature in Tubes: Turbulent Flow

Unlike laminar flow through tubes, turbulent flow becomes fully developed within a short distance (10 to 20 diameters) from the inlet. Thus entrance effects in turbulent flow are sometimes neglected and the assumption that the flow is fully developed throughout is made. This is common in many applications such as heat exchangers. Another feature of turbulent flow is the minor effect that surface boundary conditions have on the heat transfer coefficient for fluids with Prandtl numbers greater than unity. Therefore, results for uniform surface temperature are close to those for uniform surface heat flux.

Because heat transfer in fully developed turbulent flow has many applications, it has been extensively investigated. As a result, there are many correlation equations covering different ranges of Reynolds and Prandtl numbers. Two correlation equations will be presented here.

(1) The Colburn Equation [8]: This is one of the earliest and simplest equations correlating the Nusselt number with the Reynolds and Prandtl numbers as

$$\overline{Nu}_D = \frac{\overline{h}D}{k} = 0.023(Re_D)^{4/5}(Pr)^{1/3} . \qquad (10.16a)$$

Valid for:

> fully developed turbulent flow
> smooth tubes
> $Re_D > 10^4$
> $0.7 < Pr < 160$
> $L/D > 60$
> properties at $\overline{T}_m = (T_{mi} + T_{mo})/2$

$\qquad\qquad (10.16b)$

This equation is not recommended since errors associated with it can be as high as 25%. Its accuracy diminishes as the difference in temperature between surface and fluid increases.

(2) The Gnielinski Equation [9, 10]: Based on a comprehensive review of many correlation equations for turbulent flow through tubes, the following equation is recommended:

$$\overline{Nu}_D = \frac{(f/8)(Re_D - 1000)Pr}{1 + 12.7(f/8)^{1/2}(Pr^{2/3} - 1)} \left[1 + (D/L)^{2/3}\right]. \quad (10.17a)$$

Valid for:

$$\begin{array}{|l|}\hline \text{developing or fully developed turbulent flow} \\ 2300 < Re_D < 5 \times 10^6 \\ 0.5 < Pr < 2000 \\ 0 < D/L < 1 \\ \text{properties at } \overline{T}_m = (T_{mi} + T_{mo})/2 \\ \hline \end{array} \qquad (10.17b)$$

The D/L factor in equation (10.17a) accounts for entrance effects. For fully developed flow set $D/L = 0$. The *friction factor* f is defined as

$$f = \frac{D}{L} \frac{\Delta p}{\rho \overline{u}^2 / 2}, \qquad (10.18)$$

where $\overline{u}$ is the mean fluid velocity and Δp is the pressure drop in a tube of length L. This factor depends on the Reynolds number and surface finish. It is obtained from the Moody chart [11]. For smooth tubes f may be approximated by [12]

$$f = (0.79 \ln Re_D - 1.64)^{-2} \cdot \qquad (10.19)$$

10.6.3 Non-circular Channels: Turbulent Flow

Correlation equations for turbulent flow through tubes can be applied to non-circular channels to provide a reasonable approximation for the average heat transfer coefficient. In such applications the diameter D appearing in the correlation equations is replaced by the hydraulic or equivalent diameter D_e defined as

$$D_e = \frac{4A_f}{P}, \qquad (10.20)$$

where A_f is the flow area and P is the wet perimeter.

10.7 Free Convection Correlations

10.7.1 External Free Convection Correlations

Correlation equations for several configurations are available. Some are based on analytical or numerical results while others are based on experimental data.

(1) Vertical Plate: Laminar Flow, Uniform Surface Temperature.

This is an important geometry since it can be used to model many applications. Fig.10.7 shows a vertical plate which is submerged in an infinite fluid at T_∞. Surface temperature T_s is uniform. If $T_s > T_\infty$, the fluid will rise along the surface forming viscous and thermal boundary layers. The viscous boundary layer can be laminar or turbulent. A solution to this problem was presented in Chapter 7 for laminar boundary layer flow [13]. The following equation correlates the results for the local Nusselt number to within 0.5% [14]

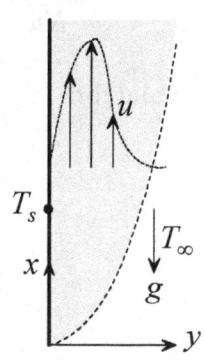

Fig. 10.7

$$Nu_x = \frac{hx}{k} = \frac{3}{4}\left[\frac{Pr}{2.435 + 4.884Pr^{1/2} + 4.953Pr}\right]^{1/4} (Ra_x)^{1/4}, \quad (10.21a)$$

where Ra_x is the local *Rayleigh number* defined in (7.2) with L replaced by the variable x. To determine the average Nusselt number for a plate of length L, equation (10.21a) is substituted into (2.50) to give the average heat transfer coefficient $\overline{h}$ and $\overline{Nu}_L$

$$\overline{Nu}_L = \frac{\overline{h}L}{k} = \left[\frac{Pr}{2.435 + 4.884Pr^{1/2} + 4.953Pr}\right]^{1/4} (Ra_L)^{1/4}. \quad (10.21b)$$

Equations (10.21a) and (10.21b) are valid for:

vertical plate

uniform surface temperature T_s

laminar, $10^4 < Ra_L < 10^9$ (10.21c)

$0 < Pr < \infty$

properties at T_f

(2) Vertical Plates: Laminar and Turbulent, Uniform Surface Temperature. A single equation which correlates experimental data for the average Nusselt number for laminar, transition and turbulent flow was developed by Churchill and Chu [15]

$$\overline{Nu}_L = \frac{\overline{h}L}{k} = \left\{ 0.825 + \frac{0.387(Ra_L)^{1/6}}{\left[1 + (0.492 / Pr)^{9/16}\right]^{8/27}} \right\}^2 .\quad (10.22a)$$

Valid for:

vertical plate

uniform surface temperature T_s

laminar, transition, and turbulent

$10^{-1} < Ra_L < 10^{12}$ (10.22b)

$0 < Pr < \infty$

properties at T_f

Although (10.22a) can be applied in the laminar range, $Ra_L < 10^9$, better accuracy is obtained using (10.21b).

(3) Vertical Plates: Laminar Flow, Uniform Surface Heat Flux. Of interest in this case is the determination of surface temperature $T_s(x)$ which varies along the plate. The local Nusselt number for laminar flow is given by [16]

$$Nu_x = \frac{hx}{k} = \left[\frac{Pr^2}{4 + 9Pr^{1/2} + 10Pr} Gr_x^* \right]^{1/5} ,\quad (10.23)$$

where the local heat transfer coefficient $h(x)$ is expressed in terms of surface heat flux q_s'' as

$$h(x) = \frac{q_s''}{T_s(x) - T_\infty} .\quad (10.24)$$

The modified Grashof number Gr_x^* in (10.23) is defined as

$$Gr_x^* = \frac{\beta g q_s''}{k v^2} x^4 . \qquad (10.25)$$

Substituting (10.24) and (10.25) into (10.23) and solving for $(T_s - T_\infty)$, we obtain

$$T_s(x) - T_\infty = \left[\frac{4 + 9 Pr^{1/2} + 10 Pr}{Pr^2} \left(\frac{v^2}{\beta g} \right) \left(\frac{q_s''}{k} \right)^4 x \right]^{1/5} . \qquad (10.26a)$$

Equations (10.23) and (10.26a) are valid for:

> vertical plate
> uniform surface flux q_s''
> laminar, $10^4 < Gr_x^* Pr < 10^9$
> $0 < Pr < \infty$

$(10.26b)$

However, properties in (10.26a) depend on surface temperature $T_s(x)$, which is not known a priori. A solution can be obtained using an iterative procedure. An assumed value for the surface temperature at the mid-point, $T_s(L/2)$, is used to calculate the film temperature at which properties are determined. Equation (10.26a) is then used to calculate $T_s(L/2)$. If the calculated value does not agree with the assumed temperature, the procedure is repeated until a satisfactory agreement is obtained.

(4) Inclined Plates: Laminar Flow, Uniform Surface Temperature. We consider a plate of length L which is tilted at an angle θ from the vertical. Fig. 10.8a shows an inclined plate with its heated surface facing downward while Fig. 10.8b shows a plate with its cooled surface facing upward. Note that the flow field is identical for both cases and consequently the same solution holds for both. Note further that gravity component for the inclined plate is $g\cos\theta$ while for the vertical

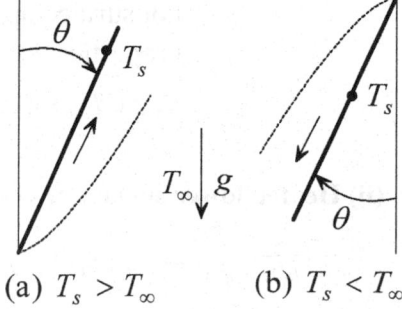

(a) $T_s > T_\infty$ (b) $T_s < T_\infty$

Fig. 10.8

plate it is g. It is reasonable to use the correlation equation for the vertical plate, with g replaced by $g\cos\theta$. However, this approximation deteriorates at large values of θ. Thus equations (10.21a), (10.21b) and (10.22a) can be used for inclined plates with Rayleigh number modified to

$$Ra_x = \frac{\beta g \cos\theta \, (T_s - T_\infty)\, x^3}{\alpha\, v}.$$ (10.27)

This approximation is valid for:

> inclined plate
> uniform surface temperature T_s
> laminar, $Ra_L < 10^9$
> $0 \le \theta \le 60$ (10.28)

For an inclined plate with its heated surface facing up or cooled surface facing down, the flow is complicated by transition and three-dimensional effects. Correlation equations for this case are given in [17, 18].

(5) Horizontal Plates: Uniform Surface Temperature. The recommended correlations for the following two arrangements are [19-21]:

(i) Heated upper surface or cooled lower surface

$$\overline{Nu}_L = 0.54\left(Ra_L\right)^{1/4}, \quad \text{for } 10^5 < Ra_L < 2 \times 10^7,$$ (10.29a)

$$\overline{Nu}_L = 0.14\left(Ra_L\right)^{1/3}, \quad \text{for } 2 \times 10^7 < Ra_L < 3 \times 10^{10}.$$ (10.29b)

Valid for:

> horizontal plate
> hot surface up or cold surface down
> properties, except β, at T_f
> β at T_f for liquids, T_∞ for gases   (10.29c)

(ii) Heated lower surface or cooled upper surface

$$\overline{Nu}_L = 0.27\left(Ra_L\right)^{1/4}, \quad \text{for } 3 \times 10^5 < Ra_L < 3 \times 10^{10},$$ (10.30a)

Valid for:

> horizontal plate
> hot surface down or cold surface up
> properties, except β, at T_f
> β at T_f for liquids, T_∞ for gases

$$(10.30b)$$

Properties in equations (10.29) and (10.30) are determined at the film temperature T_f. The characteristic length L is defined as [18]

$$L = \frac{\text{surface area}}{\text{perimeter}} . \qquad (10.31)$$

Although equations (10.29) and (10.30) are for uniform surface temperature, they are applicable to uniform surface flux. In this case the flux is specified while surface temperature is unknown. Surface temperature is determined following the procedure used in vertical plates at uniform flux, described in case (3) above.

(6) Vertical Cylinders. Correlation equations for vertical plates can be applied to vertical cylinders if the effect of boundary layer curvature is negligible. This approximation is valid if the thermal boundary layer thickness δ_t is small compared to the diameter of the cylinder D. The condition for $\delta_t/D \ll 1$ is

$$\frac{D}{L} > \frac{35}{(Gr_L)^{1/4}} , \quad \text{for } Pr \geq 1. \qquad (10.32)$$

(7) Horizontal Cylinders. This case has many engineering applications such as heat loss from steam pipes, refrigeration lines and fins. Fig. 10.9 shows free convection over a horizontal cylinder. Due to flow asymmetry, the local heat transfer coefficient varies along the circumference. The following equation correlates the average Nusselt number for a wide range of Rayleigh numbers [22]

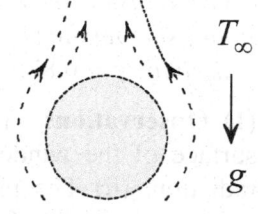

Fig. 10.9

$$\overline{Nu}_D = \frac{\overline{h}D}{k} = \left\{ 0.60 + \frac{0.387(Ra_D)^{1/6}}{\left[1 + (0.559/Pr)^{9/16}\right]^{8/27}} \right\}^2 . \qquad (10.33a)$$

Valid for:

> horizontal cylinder
> uniform surface temperature or flux
> $10^{-5} < Ra_D < 10^{12}$
> properties at T_f

(10.33b)

Note that the characteristic length in the Rayleigh number is the diameter D of the cylinder.

(8) Spheres. The average Nusselt number for a sphere is given by [23]

$$\overline{Nu}_D = \frac{\overline{h}D}{k} = 2 + \frac{0.589(Ra_D)^{1/4}}{\left[1 + \left(\dfrac{0.469}{Pr}\right)^{9/16}\right]^{4/9}}.$$ (10.34a)

Valid for:

> sphere
> uniform surface temperature or flux
> $Ra_D < 10^{11}$
> $Pr > 0.7$
> properties at T_f

(10.34b)

Example 10.3: Free Convection Heat Loss from a Window

Estimate the heat loss to a 2.5 m high and 1.25 m wide glass window. The average inside surface temperature is 7°C. Room air temperature is 23°C. Room surroundings is at 28°C and window surface emissivity $\varepsilon = 0.87$.

(1) Observations. (i) Heat transfer to the inside surface of the window is by free convection and radiation. (ii) The problem can be modeled as a vertical plate at uniform surface temperature. (iii) Newton's law of cooling gives the heat transfer rate. (iv) The Rayleigh number should be checked to establish if the flow is laminar or turbulent. (iiv) Stefan-Boltzmann relation (1.12) can be used to estimate radiation heat loss.

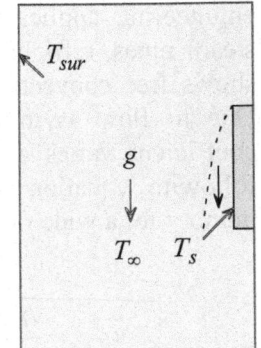

(2) Problem Definition. Determine the average free convection heat transfer coefficient $\overline{h}$ for a vertical plate at uniform surface temperature.

(3) Solution Plan. (i) Apply Newton's law of cooling. (ii) Use a free convection correlation equation for an isothermal vertical plate to determine the heat transfer coefficient. Apply Stefan-Boltzmann relation (1.12) to estimate radiation heat rate.

(4) Plan Execution.

(i) Assumptions. (1) Continuum, (2) vertical plate, (3) uniform surface temperature, (4) quiescent ambient air, (5) negligible edge effects, (6) window is small compared to room, and (7) ideal gas.

(ii) Analysis. The total heat transfer rate q is given by

$$q = q_c + q_r, \tag{a}$$

where q_c and q_r are heat transfer rates by convection and radiation, respectively. Newton's law of cooling q_c

$$q_c = \overline{h}A(T_\infty - T_s), \tag{b}$$

where

A = surface area of glass = 2.5(m)×1.25(m) = 3.125 m^2
$\overline{h}$ = average heat transfer coefficient, W/m^2-°C
T_s = surface temperature = 7°C = 280.15 K
T_∞ = room air temperature = 23°C = 296.15 K

Radiation heat rate is determined using Stefan-Boltzmann relation (1.12)

$$q_r = \varepsilon\sigma A(T_{sur}^4 - T_s^4), \tag{c}$$

where

T_{sur} = surroundings temperature = 28°C = 301.15 K
ε = emissivity = 0.87
σ = 5.67 x 10^{-8} W/m^2-K^4

The Rayleigh number is calculated to determine the appropriate correlation equation for the average heat transfer coefficient $\overline{h}$. The Rayleigh number is defined as

$$Ra_L = \frac{\beta g(T_\infty - T_s)L^3}{\nu^2}Pr, \tag{d}$$

where

g = gravitational acceleration = 9.81 m/s^2
L = length scale in the direction of gravity = 2.5 m
Pr = Prandtl number
β = coefficient of thermal expansion, 1/K
v = kinematic viscosity, m^2/s

Air properties are evaluated at the film temperature T_f defined as

$$T_f = (T_s + T_\infty)/2 = (7°C + 23°C)/2 = 15°C$$

Air properties at this temperature are

k = 0.02526 W/m-°C
Pr = 0.7145
v = 14.64 × 10^{-6} m^2/s

For an ideal gas β is given by

$$\beta = \frac{1}{T_f}, \tag{e}$$

where T_f in this equation is in degrees kelvin. Thus

$$\beta = 1/(15 + 273.15)K = 0.003471/K$$

Substituting into (d)

$$Ra_L = \frac{0.00347(1/K)9.81(m/s^2)(23-7)(°C)(m^3)}{(14.64\times10^{-6})^2(m^4/s^2)}0.7145 = 28.373\times10^9$$

Thus the flow is turbulent and the appropriate correlation equation is (10.24a)

$$\overline{Nu}_L = \frac{\overline{h}L}{k} = \left\{0.825 + \frac{0.387(Ra_L)^{1/6}}{\left[1+(0.492/Pr)^{9/16}\right]^{8/27}}\right\}^2, \tag{f}$$

The conditions for the applicability of (f), listed in (10.24b) must be satisfied.

(iii) Computations. Equation (f) gives

$$\overline{Nu}_L = \frac{\overline{h}L}{k} = \left\{0.825 + \frac{0.387(28.373\times10^9)^{1/6}}{\left[1+(0.492/0.7145)^{9/16}\right]^{8/27}}\right\}^2 = 351.57$$

$$\overline{h} = \frac{351.57 \times 0.02526(\text{W/m-}^\circ\text{C})}{2.5(\text{m})} = 3.55 \text{ W/m}^2\text{-}^\circ\text{C}$$

Substituting into (b) gives

$$q_c = 3.55(\text{W/m}^2\text{-}^\circ\text{C})\, 3.125(\text{m}^2)\,(23\text{-}7)(^\circ\text{C}) = 177.5\text{W}$$

Equation (c) gives radiation heat loss

$$q_r = 0.87(5.67 \times 10^{-8})(\text{W/m}^2\text{-K}^4)3.125(\text{m}^2) \times$$

$$\left[(301.15)^4(\text{K}^4) - (280.15)^4(\text{K}^4)\right] = 318.4\text{W}$$

Total heat loss to the window is

$$q = 177.5 \text{ W} + 318.4 \text{ W} = 495.9 \text{ W}$$

(iv) Checking. *Dimensional check*: Computations show that equations (b), (c), (d) and (f) are dimensionally consistent.

Quantitative check: The magnitude of $\overline{h}$ is in line with typical free convection values for air given in Table 1.1.

Validity of correlation equation (10.26a): Conditions listed in equation (10.24b) are satisfied.

(5) Comments. (i) Radiation heat loss is very significant. It accounts for 64% of the total heat loss.

(ii) The use of the simplified radiation model of equation (1.12) is justified since the window has a small area compared to the walls, floor and ceiling of the room.

(iii) No information on the glass thickness and outside heat transfer coefficient is needed to solve this problem because the inside surface temperature is given.

10.7.2 Free Convection in Enclosures

Examples of free convection in enclosures are found in double-glazed windows, solar collectors, building walls, concentric cryogenic tubes and electronic packages. A fluid in an enclosed space experiences free convection if the walls of the enclosure are not at a uniform temperature. A buoyancy force causes the fluid to circulate in the enclosure transferring heat from the hot side to the cold side. If buoyancy forces are not large enough to overcome viscous forces, circulation will not occur and heat transfer across the enclosure will essentially be by conduction. Heat flux due to circulation is determined from Newton's law

$$q'' = h(T_h - T_c),$$ (10.35)

where h is the heat transfer coefficient, T_c and T_h are the cold and hot surface temperatures. The heat transfer coefficient is obtained from Nusselt number correlation equations. Such equations depend on configuration, orientation, geometric aspect ratio, Rayleigh number Ra_L, and Prandtl number Pr. We will consider selected common examples.

(1) Vertical Rectangular Enclosures. Consider a rectangular cavity with one side at T_h and the opposite side at T_c, shown in Fig. 10.10. The top and bottom surfaces are insulated. The fluid adjacent to the hot surface rises while that near the cold wall falls. This sets up circulation in the cavity resulting in the transfer of heat from the hot to the cold side. Boundary layers form on the side walls while the core remains stagnant. The aspect ratio L/δ is one of the key parameters governing the Nusselt number. Another parameter is the Rayleigh number based on the spacing δ and defined as

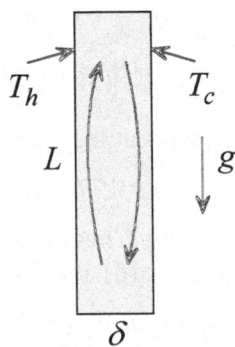

Fig. 10.10

$$Ra_\delta = \frac{\beta g (T_h - T_c)\delta^3}{v^2} Pr.$$

(10.36)

For $1 < L/\delta < 40$, the following correlation equations are recommended [24-26].

$$\overline{Nu}_\delta = \frac{\overline{h}\delta}{k} = 0.18 \left[\frac{Pr}{0.2 + Pr} Ra_\delta \right]^{0.29}.$$ (10.37a)

Valid for

vertical rectangular enclosure

$1 < \dfrac{L}{\delta} < 2$

$10^{-3} < Pr < 10^5$

$\dfrac{Pr}{0.2 + Pr} Ra_\delta > 10^3$

properties at $\overline{T} = (T_c + T_h)/2$

(10.37b)

$$\overline{Nu}_\delta = \frac{\overline{h}\delta}{k} = 0.22 \left[\frac{Pr}{0.2 + Pr} Ra_\delta \right]^{0.28} \left[\frac{L}{\delta} \right]^{-0.25} . \quad (10.38a)$$

Valid for

> vertical rectangular enclosure
>
> $2 < \dfrac{L}{\delta} < 10$
>
> $Pr < 10^5$
>
> $10^3 < Ra_\delta < 10^{10}$
>
> properties at $\overline{T} = (T_c + T_h)/2$

(10.38b)

$$\overline{Nu}_\delta = \frac{\overline{h}\delta}{k} = 0.046 \left[Ra_\delta \right]^{1/3} . \quad (10.39a)$$

Valid for

> vertical rectangular enclosure
>
> $1 < \dfrac{L}{\delta} < 40$
>
> $1 < Pr < 20$
>
> $10^6 < Ra_\delta < 10^9$
>
> properties at $\overline{T} = (T_c + T_h)/2$

(10.39b)

$$\overline{Nu}_\delta = \frac{\overline{h}\delta}{k} = 0.42 \left[Pr \right]^{0.012} \left[Ra_\delta \right]^{0.25} \left[\frac{L}{\delta} \right]^{-0.3} . \quad (10.40a)$$

Valid for

> vertical rectangular enclosure
>
> $10 < \dfrac{L}{\delta} < 40$
>
> $1 < Pr < 2 \times 10^4$
>
> $10^4 < Ra_\delta < 10^7$
>
> properties at $\overline{T} = (T_c + T_h)/2$

(10.40b)

(2) Horizontal Rectangular Enclosures. Fig. 10.11 shows a horizontal enclosure heated from below. At low Rayleigh numbers the fluid remains stagnant and heat transfer through the cavity is by conduction. At a critical value of the Rayleigh number, $Ra_{\delta c}$, a cellular flow pattern develops. This Rayleigh number is given by

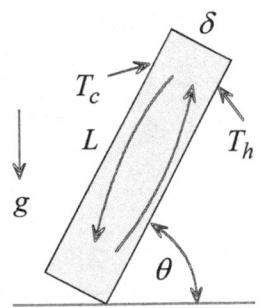

$$Ra_{\delta c} = 1708.$$

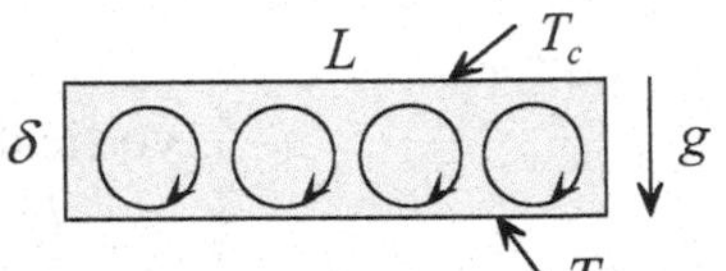

Fig. 10.11

The Nusselt number for cellular flow is given by [27]

$$\overline{Nu}_\delta = \frac{\overline{h}\delta}{k} = 0.069[Ra_\delta]^{1/3}[Pr]^{0.074}. \qquad (10.41a)$$

Valid for

> horizontal rectangular enclosure heated from below
>
> $3 \times 10^5 < Ra_\delta < 7 \times 10^9$
>
> properties at $\overline{T} = (T_c + T_h)/2$
>
> (10.41b)

(3) Inclined Rectangular Enclosures.

An important application of this geometry is solar collectors. To maximize solar energy absorption the collector is tilted an angle θ from the horizontal, as shown in Fig. 10.12. However, energy is lost from the collector to the ambient air due to convection. To estimate this loss it is necessary to determine the heat transfer coefficient in the collector's enclosure. Correlation equations for the Nusselt number depend on the aspect ratio L/δ and inclination angle θ. For $0^o < \theta < 90^o$ the lower surface is heated and the upper surface is cooled. This relationship is reversed for $90^o < \theta < 180^o$. Within $0^o < \theta < 90^o$ the average Nusselt number passes through a minimum value at a critical angle θ_c which varies with aspect ratio according to Table 10.1. Due to the changing flow pattern

Fig. 10.12

Table 10.1 Critical tilt angle					
L/δ	1	3	6	12	>12
θ_c	25°	53°	60°	67°	70°

with inclination and aspect ratio, a single correlation equation is not available. The following equations are recommended (28-31):

$$\overline{Nu}_\delta = \frac{\overline{h}\delta}{k} = 1 + 1.44\left[1 - \frac{1708}{Ra_\delta\cos\theta}\right]\left[1 - \frac{1708(1.8\sin\theta)^{1.6}}{Ra_\delta\cos\theta}\right] + \left[\frac{(Ra_\delta\cos\theta)^{1/3}}{18} - 1\right] .$$

$$(10.42a)$$

Valid for

> inclined rectangular enclosure
> $L/\delta \geq 12$
> $0 < \theta < \theta_c$
> set $[\ ]^* = 0$ when negative
> properties at $\overline{T} = (T_c + T_h)/2$ (10.42b)

$$\overline{Nu}_\delta = \frac{\overline{h}\delta}{k} = \overline{Nu}_\delta(0^\circ)\left[\frac{\overline{Nu}_\delta(90^\circ)}{\overline{Nu}_\delta(0^\circ)}(\sin\theta_c)^{0.25}\right]^{\theta/\theta_c} . \quad (10.43a)$$

Valid for

> inclined rectangular enclosure
> $L/\delta \leq 12$
> $0 < \theta \leq \theta_c$
> properties at $\overline{T} = (T_c + T_h)/2$ (10.43b)

$$\overline{Nu}_\delta = \frac{\overline{h}\delta}{k} = \overline{Nu}_\delta(90^\circ)[\sin\theta]^{0.25}. \quad (10.44a)$$

Valid for

> inclined rectangular enclosure
> all L/δ
> $\theta_c < \theta < 90^\circ$
> properties at $\overline{T} = (T_c + T_h)/2$ (10.44b)

$$\overline{Nu}_\delta = \frac{\overline{h}\delta}{k} = 1 + \left[\overline{Nu}_\delta(90^\circ) - 1\right]\sin\theta.\qquad (10.45a)$$

Valid for

> inclined rectangular enclosure
> all L/δ
> $90^\circ < \theta < 180^\circ$
> properties at $\overline{T} = (T_c + T_h)/2$

$(10.45b)$

Example 10.4: Advertising Display

A proposed device for an advertising display is based on observing fluid circulation in a rectangular enclosure. The idea is to fill the enclosure with colored water and many small reflective particles of the same density as water. The particles move with the fluid providing visual observation of the flow patterns. The enclosure is 70 cm long, 5 cm wide and 70 cm deep. The heated side is to be maintained at $27^\circ C$ and the cold side at $23^\circ C$. The design allows the inclination angle θ of the cavity to be varied from 0° to 180°. Estimate the power requirement for the device when the inclination angle is 30°.

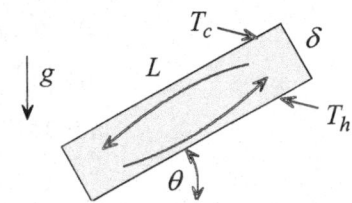

(1) Observations. (i) Power requirement is equal to the heat transfer rate through the enclosure. (ii) The problem can be modeled as an inclined rectangular cavity at specified hot and cold surface temperatures. (iii) Newton's law of cooling gives the heat transfer rate. (iv) The aspect ratio and critical inclination angle should be computed to determine the applicable correlation equation for the Nusselt number.

(2) Problem Definition. Determine the average free convection heat transfer coefficient $\overline{h}$ for an inclined rectangular enclosure.

(3) Solution Plan. (i) Apply Newton's law of cooling. (ii) Compute the aspect ratio and critical inclination angle. Select an appropriate Nusselt number correlation equation for convection in an inclined rectangular cavity.

(4) Plan Execution.

(i) Assumptions. (1) Continuum, (2) uniform hot and cold surface temperatures, (3) insulated top and bottom surfaces, (4) negligible radiation and (5) properties of the water-particles mixture are the same as those of water.

(ii) Analysis. Newton's law of cooling gives

$$P = q = \bar{h}A(T_h - T_c),$$

(a)

where

A = surface area of rectangle = 0.7(m)×0.7(m) = 0.49 m²
$\bar{h}$ = average heat transfer coefficient, W/m²-°C
P = power requirement, W
q = heat transfer rate through cavity, W
T_h = hot surface temperature = 27°C
T_c = cold surface temperature = 23°C

The aspect ratio is defined as

$$\text{aspect ratio} = \frac{L}{\delta},$$

(b)

where

L = length of rectangle = 0.7 m
δ = width of rectangle = 0.05 m

Equation (b) gives

$$\frac{L}{\delta} = \frac{0.7(\text{m})}{0.05(\text{m})} = 14$$

According to Table 10.1, the critical angle is $\theta_c = 70°$. Since $L/\delta > 12$ and $0 < \theta < \theta_c$, it follows that the applicable correlation equation for the Nusselt number is

$$\overline{Nu}_\delta = \frac{\bar{h}\delta}{k} = 1 + 1.44\left[1 - \frac{1708}{Ra_\delta\cos\theta}\right]^*\left[1 - \frac{1708(1.8\sin\theta)^{1.6}}{Ra_\delta\cos\theta}\right] + \left[\frac{(Ra_\delta\cos\theta)^{1/3}}{18} - 1\right]^*.$$

(10.42a)

The Rayleigh number is defined as

$$Ra_\delta = \frac{\beta g(T_h - T_c)\delta^3}{v^2}Pr,$$

(c)

where

g = gravitational acceleration = 9.81 m/s^2
Pr = Prandtl number
β = coefficient of thermal expansion, 1/K
v = kinematic viscosity, m^2/s

Water properties are evaluated at the film temperature $\overline{T}$ defined as

$$\overline{T} = (T_h + T_c)/2. \tag{d}$$

(iii) Computations. Equation (d) gives

$$\overline{T} = \frac{(27 + 23)(^\circ C)}{2} = 25^\circ C$$

Properties of water at this temperature are:

k = thermal conductivity = 0.6076 W/m$-^\circ$C
Pr = 6.13
$\beta = 0.259 \times 10^{-3}$ 1/K
$v = 0.8933 \times 10^{-6}$ m^2/s

Substituting into (c)

$$Ra_\delta = \frac{0.259 \times 10^{-3}(1/K)9.81(m/s^2)(27-23)(^\circ C)(0.05)^3(m^3)}{(0.8933 \times 10^{-6})^2(m^4/s^2)}6.13$$

$$= 9.75898 \times 10^6$$

Substituting into (10.42a)

$$\overline{Nu}_\delta = \frac{\overline{h}\delta}{k} = 1 + 1.44\left[1 - \frac{1708}{9.75898 \times 10^6 \cos 30^\circ}\right]^* \times$$

$$\left[1 - \frac{1708(1.8 \sin 30^\circ)^{1.6}}{9.75898 \times 10^6 \cos 30^\circ}\right] + \left[\frac{(9.75898 \times 10^6 \cos 30)^{1/3}}{18} - 1\right]^* = 12.755$$

$$\overline{h} = 12.755\frac{k}{\delta} = 12.755\frac{0.6076(W/m-^\circ C)}{0.05(m)} = 155 \ W/m^2-^\circ C$$

Equation (a) gives the required power

$$P = 155(W/m^2-^\circ C)(0.49)(m^2)(27-23)(^\circ C) = 303.8 \ W$$

(iv) Checking. *Dimensional check*: Computations showed that the Nusselt number and Rayleigh number are dimensionless.

Quantitative check: The magnitude of $\bar{h}$ is in line with typical free convection values for liquids given in Table 1.1.

Validity of correlation equation (10.42a): Conditions listed in equation (10.42b) are satisfied.

(5) Comments. (i) If the device is to be used continuously, the estimate power requirement is relatively high. Decreasing the temperature difference between the hot and cold surfaces will reduce the power requirement.

(ii) The ambient temperature plays a role in the operation of the proposed device. The design must take into consideration changing ambient temperature.

(iii) Changing the inclination angle will change the power requirement.

(4) Horizontal Concentric Cylinders. Fig. 10.13 shows two long concentric cylinders. The inner cylinder of diameter D_i is maintained at uniform temperature T_i. The outer cylinder of diameter D_o is maintained at uniform temperature T_o. If $T_i > T_o$, buoyancy force sets up two flow circulation cells in the annular space as shown in Fig. 2.13 . Flow direction is reversed for $T_i < T_o$. In both cases flow circulation results in an enhancement of the thermal conductivity. The one-dimensional heat transfer rate per unit cylinder length, q', is given by

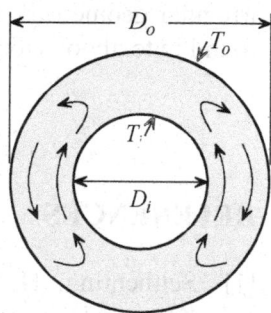

Fig. 10.13

$$q' = \frac{2\pi k_{eff}}{\ln(D_o / D_i)}(T_i - T_o) .$$

(10.46)

Correlation equation for the effective conductivity k_{eff} is given by [32]

$$\frac{k_{eff}}{k} = 0.386\left[\frac{Pr}{0.861 + Pr} Ra^*\right]^{1/4} ,$$

(10.47a)

where

$$Ra^* = \frac{[\ln(D_o / D_i)]^4}{\delta^3\left[(D_i)^{-3/5} + (D_o)^{-3/5}\right]^5} Ra_\delta .$$

(10.47b)

$$\delta = \frac{D_o - D_i}{2}.$$

(10.47c)

Valid for

concentric cylinders
$$10^2 < Ra^* < 10^7$$
properties at $\overline{T} = (T_i + T_o)/2$

(10.47d)

10.8 Other Correlations

In the previous sections, correlation equations have been presented for limited processes and configurations. It should be emphasized that the above treatment is highly abridged. There are many other correlation equations for topics such as condensation, boiling, high speed flow, jet impingement, dissipation, liquid metals, enhancements, finned geometries, irregular geometries, micro-gravity, non-Newtonian fluids, etc.. Some are found in textbooks, others in handbooks and journals.

REFERENCES

[1] Schlichting, H., *Boundary Layer Theory*, 6th ed., translated by J. Kestin, McGraw-Hill, New York, 1968, p. 600.

[2] Kays, W.M., and M.E. Crawford, *Convection Heat and Mass Transfer*, 3rd ed., McGraw Hill, New York, 1993, pp. 281-283.

[3] Churchill, S.W., and M. Bernstein, "A Correlation Equation for Forced Convection from Gases and Liquids to a Circular Cylinder in Cross Flow," J. Heat Transfer, Trans. ASME, Vol. 94, 1977, pp. 300-306.

[4] Nakai, S., and T. Okazaki, "Heat Transfer from a Horizontal Circular Wire at Small Reynolds and Grashof Numbers-I Pure Convection," Int. J. Heat and Mass Transfer, Vol. 18, 1975, pp. 387-396.

[5] Whitaker, S., "Forced Convection Heat Transfer Correlations for Flow in Pipes, Past Flat Plates, Single Cylinders, Single Spheres, and Flow in Packed Beds and Tube Bundles," AIChE J. Vol. 18, 1972, pp. 361-371.

[6] Hausen, H., "Darstellung des Wärmeüberganges in Rohren Durch Vergallgemeinerte Potenzbeziehungen," Z. VDI Beihefte Verfahrenstech., No. 4, 1943, pp. 91-98.

[7] Sieder, E.N., and G.E. Tate, "Heat Transfer and Pressure Drop of Liquids in Tubes," Ind. Eng. Chem., Vol. 28, 1936, pp. 1429-1435.

[8] Colburn, A. P., "A Method of Correlating Forced Convection Heat Transfer Data and a Comparison with Liquid Friction," Trans. AIChE, Vol. 29, 1933, pp. 174-210; also in Int. J. Heat Mass Transfer, Vol. 7, 1964, pp. 1359-1384.

[9] Gnielinski, V., "Forced Convection in Ducts," in *Handbook of Heat Exchanger Design*, ed. G.F. Hewitt, Begell House, New York, 1992.

[10] Gnielinski, V., "New Equations for Heat and Mass Transfer in Turbulent Pipe and Channel Flow," Int. Chem. Eng. Vol. 16, 1976, pp. 359-368.

[11] Moody, L.F., "Friction Factors for Pipe Flow," Trans. ASME J., Vol. 66, 1944, pp. 671-684.

[12] Petukhov, B.S., "Heat Transfer and Friction in Turbulent Pipe Flow with Variable Physical Properties," in *Advances in Heat Transfer*, Vol. 6, eds. T. F. Irvine, and J.P. Hartnett, Academic Press, New York, 1970.

[13] Ostrach, S., "An Analysis of Laminar Free Convection Flow and Heat Transfer about a Flat Plate Parallel to the Direction of the Generating Body Force," NACA Report 1111, 1953.

[14] LeFevre, E.J., "Laminar Free Convection from a Vertical Plane Surface," Proc. 9th Int. Congress Applied Mechanics, Vol. 4, 1956, pp. 168-174.

[15] Churchill, S.W., and H.H.S. Chu, "Correlation Equations for Laminar and Turbulent Free Convection from a Vertical Plate," Int. J. Heat Mass Transfer, Vol. 18, 1975, pp. 1323-1329.

[16] Fujii, T., and M. Fujii, "The Dependence of Local Nusselt Number on Prandtl Number in the Case of Free Convection Along a Vertical Surface with Uniform Heat Flux," Int. J. Heat Mass Transfer, Vol.19, 1976, pp.121-122.

[17] Fujii, T., and H. Imura, "Natural Convection Heat Transfer from a Plate with Arbitrary Inclination," Int. J. Heat Mass Transfer, Vol. 15,1972,pp.755-765.

[18] Vliet, G.C., "Natural Convection Local Heat Transfer on Constant Heat Flux Inclined Surfaces," J. Heat Transfer, Vol. 91, 1969, pp. 511-516.

[19] McAdams, W.H. *Heat Transmission*, 3rd ed., McGraw-Hill, New York, 1954.

[20] Goldstein, R.J., E.M. Sparrow, and D.C. Jones, "Natural Convection Mass Transfer Adjacent to Horizontal Plates," Int. J. Heat Mass Transfer, Vol. 16, 1973, pp.1025-1035.

[21] Lloyd, J.R., and W.R. Moran, "Natural Convection Adjacent to Horizontal Surfaces of Various Plan Forms," ASME Paper 74-WA/HT-66, 1974.

[22] Churchill, S.W., and H.H.S. Chu, "Correlation Equations for Laminar and Turbulent Free Convection from a Horizontal Cylinder," Int. J. Heat Mass Transfer, Vol. 18, 1975, pp. 1049-1053.

[23] Churchill, S.W., "Free Convection around Immersed Bodies," *Heat Exchanger Design Handbook*, ed. E.U. Schlunder, Hemisphere Publishing Corp. New York, 1983.

[24] Catton, I., "Natural Convection in Enclosures," Proceedings of Sixth International Heat Transfer Conference, Toronto, Canada, 1978, Vol. 6, pp.13-31.

[25] Berkovsky, B.M. and V.K. Polevikov, "Numerical Study of Problems on High-Intensive Free convection." In Heat Transfer and Turbulent buoyancy Convection, ed. D.B. Spalding and N. Afgan, 1977, pp. 443-445, Hemisphere, Washington, D.C.

[26] MacGregor, R.K. and A. P. Emery, "Free Convection through Vertical Plane Layers: Moderate and High Prandtl Number Fluids," J. Heat Transfer, Vol. 91, 1969, p.391.

[27] Globe, S. and D. Dropkin, "Natural Convection Heat Transfer in Liquids confined by Horizontal Plates and Heated from Below," J. Heat Transfer, Vol. 18, 1959, pp.24-28.

[28] Catton, I., "Natural Convection in Enclosures," 6th Int. Heat Transfer Conference, Toronto, Hemisphere Publishing, Washington, D.C., 1978, Vol. 6, pp. 13-31.

[29] Hollands, K.G.T., T.E. Emery, G.D. Raithby and L.J. Konicek, "Free Convection Heat Transfer across Inclined Air Layers," J. Heat Transfer, Vol. 98, 1976, pp. 189-193.

[30] Arnold, J.N., P.N. Bonaparte, I.Catton and D.K. Edwards, "Experimental Investigation of Natural Convection in a Finite Rectangular

Regions Inclined at Various Angles from $^{\circ}0$ to $^{\circ}180$, " Proc. 1974 Heat Transfer and Fluid Mech. Inst., Corvallis, Oregon, Stanford University Press, Stanford, CA, 1974, pp. 321-329.

[31] Ayyaswamy, P.S. and I. Catton, "The Boundary Layer Regime for Natural Convection in Differentially Heated Tilted Rectangular Cavity," J. Heat Transfer, Vol. 95, 1973, pp. 543-545.

[32] Raithby, C.D. and K.G.T. Hollands, "A General Method of Obtaining Approximate Solutions to Laminar and Turbulent Free Convection Problems," in *Advances in Heat Transfer*, Vol. 11, 1975, eds. J.P. Hartnett and T.F. Irvine, Jr., Academic Press, New York.

PROBLEMS

10.1 Water at 120°C boils inside a channel with a flat surface measuring 45cm×45 cm. Air at 62 m/s and 20°C flows over the channel parallel to the surface. Determine the heat transfer rate to the air. Neglect wall resistance.

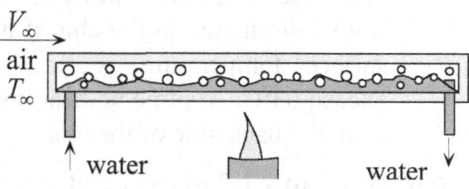

10.2 Steam at 105°C flows inside a specially designed narrow channel. Water at 25°C flows over the channel with a velocity of 0.52 m/s. Assume uniform outside surface temperature $T_s = 105\,°C$.

[a] Determine surface heat flux at 20 cm and 70 cm down-stream from the leading edge of the channel.

[b] Determine the total heat removed by the water if the length is $L = 80$ cm and the width is $W = 100$ cm.

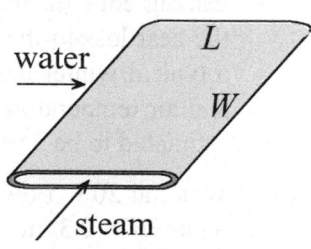

10.3 Electronic components are mounted on one side of a circuit board. The board is cooled on the other side by air at 23°C flowing with a velocity of 10 m/s. The length of the board is $L = 20$ cm and its width is $W = 25$ cm. Assume uniform board temperature.

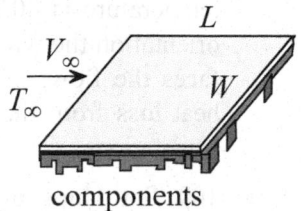

[a] Determine the maximum power that can be dissipated in the package if surface temperature is not to exceed 77°C. Assume that all dissipated power is conducted through the plate to the air.

[b] To increase the maximum power without increasing surface temperature, it is recommended that the boundary layer be tripped to turbulent flow very close to the leading edge. Is this a valid recommendation? Substantiate your view.

10.4   Water at 15°C flows with a velocity of 0.18 m/s over a plate of length $L = 20$ cm and width $W = 25$ cm. Surface temperature is 95°C. Determine the heat transfer rate from the leading and trailing halves of the plate.

10.5   A chip measuring 5 mm × 5 mm is placed flush on a flat plate 18 cm from the leading edge. The chip is cooled by air at 17°C flowing with a velocity of 56 m/s. Determine the maximum power that can be dissipated in the chip if its surface temperature is not to exceed 63°C. Assume no heat loss from the back side of the chip.

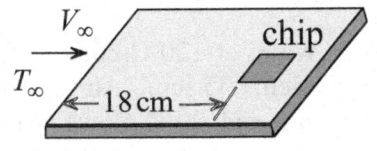

10.6   A 1.2 m × 1.2 m solar collector is mounted flush on the roof of a house. The leading edge of the collector is located 5 m from the leading edge of the roof. Estimate the heat loss to the ambient air on a typical winter day when wind speed parallel to the roof is 12 m/s and air temperature is 5°C. Outside collector surface temperature is estimated to be 35°C.

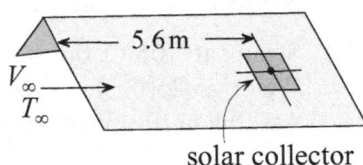

solar collector

10.7   Water at 20°C flows over a rectangular plate of length $L = 1.8$ m and width $W = 0.3$ m. The upstream velocity is 0.8 m/s and surface temperature is 80°C. Two orientations are considered. In the first orientation the width W faces the flow and in the second the length L faces the flow. Which orientation should be selected to minimize heat loss from the plate? Determine the heat loss ratio of the two orientations.

10.8   100 flat chips are placed on a 10 cm × 10 cm circuit board and cooled by forced convection of air at 27°C. Each chip measures

1 cm × 1 cm and dissipates 0.13 W. The maximum allowable chip temperature is 83°C. Free stream air velocity is 5 m/s. Tests showed that several chips near the trailing end of the board exceeded the allowable temperature. Would you recommend tripping the boundary layer to turbulent flow at the leading edge to solve the overheating problem? Substantiate your recommendation.

10.9 Water at 27°C flows normally over a tube with a velocity of 4.5 m/s. The outside diameter of the tube is 2 cm. Condensation of steam inside the tube results in a uniform outside surface temperature of 98°C. Determine the length of tube needed to transfer 250,000 W of energy to the water.

10.10 A proposed steam condenser design for marine applications is based on the concept of rejecting heat to the surrounding water while a boat is in motion. The idea is to submerge a steam-carrying tube in the water such that its axis is normal to boat velocity. Estimate the rate of steam condensation for a 75 cm long tube with an outside diameter of 2.5 cm. Assume a condensation temperature of 90°C and a uniform surface temperature of 88°C. Ambient water temperature is 15°C and boat speed is 8 m/s.

10.11 An inventive student wanted to verify the speed of a boat using heat transfer analysis. She used a 10 cm long electrically heated tube with inside and outside radii of 1.1 cm and 1.2 cm, respectively. She immersed the tube in the water such that its axis is normal to boat velocity. She recorded the following measurements:

Water temperature = 16.5°C
Outside surface temperature of tube = 23.5°C
Electric energy dissipated in tube = 480 W

Determine the speed of the boat.

10.12 A thin electric heater is wrapped around a rod of diameter 3 cm. The heater dissipates energy uniformly at a rate of 1300 W/m. Air at 20°C flows normal to the rod with a velocity of 15.6 m/s. Determine the steady state surface temperature of the heater.

10.13 A fluid velocity measuring instrument consists of a wire which is heated electrically. By positioning the axis of the wire normal to flow direction and measuring surface temperature and dissipated electric power, fluid velocity can be estimated. Determine the velocity of air at 25°C for a wire diameter of 0.5 mm, dissipated power 35 W/m and surface temperature 40°C.

10.14 Students were asked to devise unusual methods for determining the height of a building. One student designed and tested the following system. A thin walled copper balloon was heated to 133°C and parachuted from the roof of the building. Based on aerodynamic consideration, the student reasoned that the balloon dropped at approximately constant speed. The following measurements were made:

D = balloon diameter = 13 cm
M = mass of balloon = 150 grams
T_f = balloon temperature at landing = 47°C
T_∞ = ambient air temperature = 20°C
U = balloon velocity = 4.8 m/s

Determine the height of the building.

10.15 A 6 cm diameter sphere is used to study skin friction characteristics at elevated temperatures. The sphere is heated internally with an electric heater and placed in a wind tunnel. To obtain a nearly uniform surface temperature the sphere is made of copper. Specify the required heater capacity to maintain surface temperature at 140°C. Air velocity in the wind tunnel is 18 m/s and its temperature is 20°C.

10.16 A hollow aluminum sphere weighing 0.2 kg is initially at 200°C. The sphere is parachuted from a building window 100 m above street level. You are challenged to catch the sphere with your bare hands as it reaches the street. The sphere drops with an average velocity of 4.1 m/s. Its diameter is 40 cm and the ambient air temperature is 20°C. Will you accept the challenge? Support your decision.

10.17 Steam condenses on the outside surface of a 1.6 cm diameter tube. Water enters the tube at 12.5°C and leaves at 27.5°C. The mean water velocity is 0.405 m/s. Outside surface temperature is 34 °C. Neglecting wall thickness, determine tube length.

10.18 A 150 cm long tube with 8 mm inside diameter passes through a laboratory chamber. Air enters the tube at 12°C with fully developed velocity and a flow rate 0.0005 kg/s. Assume uniform surface temperature of 25°C, determine outlet air temperature.

10.19 Water enters a tube with a fully developed velocity and uniform temperature T_{mi} = 18°C. The inside diameter of the tube is 1.5 cm and its surface temperature is uniform at T_s = 125°C. Neglecting wall thickness, determine the length of the tube needed to heat the water to 82°C at a flow rate of 0.002 kg/s.

10.20 Cold air is supplied to a research apparatus at a rate of 0.14 g/s. The air enters a 20 cm long tube with uniform velocity and uniform temperature of −20°C. The inside diameter of the tube is 5 mm. The inside surface is maintained at 30°C. Determine the outlet air temperature.

10.21 Water flows through a tube of inside diameter 2.5 cm. The inside surface temperature is 230°C and the mean velocity is 3 cm/s. At a section far away from the inlet the mean temperature is 70°C.

[a] Calculate the heat flux at this section.

[b] What will the flux be if the mean velocity is increased by a factor of ten?

10.22 Air flows through a tube of inside diameter 5 cm. At a section far away from the inlet the mean temperature is 30°C. At another section further downstream the mean temperature is 70°C. Inside surface temperature is 90°C and the mean velocity is 4.2 m/s. Determine the length of this section.

10.23 Two identical tubes have inside diameters of 6 mm. Air flows through one tube at a rate of 0.03 kg/hr and through the other at a rate of 0.4 kg/hr. Far away from the inlets of the tubes the mean temperature is 120°C for both tubes. The air is heated at a uniform surface temperature which is identical for both tubes. Determine the ratio of the heat flux of the two tubes at this section.

10.24 Two concentric tubes of diameters 2.5 cm and 6.0 cm are used as a heat exchanger. Air flows through the inner tube with a mean velocity of 2 m/s and mean temperature of 190°C. Water flows in the annular space between the two tubes with a mean velocity of 0.5 m/s and a mean temperature of 30°C. Determine the inside and outside heat transfer coefficients.

10.25 A heat exchanger consists of a tube and square duct. The tube is placed co-axially inside the duct. Hot water flows through the tube while cold water passes through the duct. The inside and outside diameters are 5 cm and 5.2 cm, respectively. The side of the duct is 10

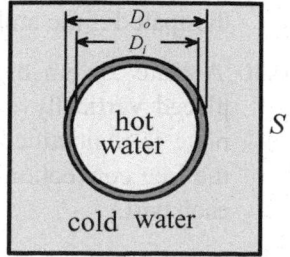

cm. At a section far away from the inlet the mean hot water temperature is 90°C and the mean cold water temperature is 30°C. The mean hot water velocity is 1.32 m/s and the mean cold water velocity is 0.077 m/s. Determine the inside and outside heat transfer coefficient.

10.26 In designing an air conditioning system for a pizza restaurant an estimate of the heat added to the kitchen from the door of the pizza oven is needed. The rectangular door is 50 cm × 120 cm with its short side along the vertical direction. Door surface temperature is 110°C. Ambient air and surroundings temperatures are 20°C and 24°C, respectively. Door surface emissivity is 0.08. Estimate the heat loss from the door.

10.27 To compare the rate of heat transfer by radiation with that by free convection, consider the following test case. A vertical plate measuring 12 cm × 12 cm 12 is maintained at a uniform surface temperature of 125°C. The ambient air and the surroundings are at 25°C. Compare the two modes of heat transfer for surface emissivities of 0.2 and 0.9.

10.28 A sealed electronic package is designed to be cooled by free convection. The package consists of components which are mounted on the inside surfaces of two cover plates measuring 10 cm × 10 cm each. Because the plates are made of high conductivity material, surface temperature may be assumed uniform. The maximum allowable surface temperature is 70°C. Determine the maximum power that can be dissipated in the package without violating design constraints. Ambient air temperature is 20°C. Neglect radiation heat exchange.

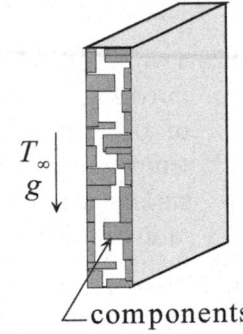

components

10.29 Assume that the electronic package of Problem 10.28 is to be used in an undersea application. Determine the maximum power that can be dissipated if the ambient water temperature is 10°C.

10.30 A plate 20 cm high and 25 cm wide is placed vertically in water at 29.4°C. The plate is maintained at 70.6°C. Determine the free convection heat transfer rate from each half.

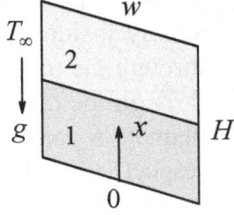

10.31 Consider laminar free convection from a vertical plate at uniform surface temperature. Two 45° triangles are drawn on the plate as shown.

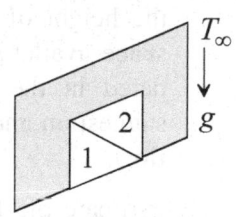

[a] Explain why free convection heat transfer from triangle 1 is greater than that from the triangle 2.

[b] Determine the ratio of the heat transfer from two triangles.

10.32 A vertical plate measuring 21 cm × 21 cm is at a uniform surface temperature of 80°C. The ambient air temperature is 25°C. Determine the free convection heat flux at 1 cm, 10 cm and 20 cm from the lower edge.

10.33 200 square chips measuring 1 cm × 1 cm each are mounted on both sides of a thin vertical board measuring 10 cm × 10 cm. The chips dissipate 0.035 W each. Assume uniform surface heat flux. Determine the maximum surface temperature in air at 22°C. Neglect heat exchange by radiation.

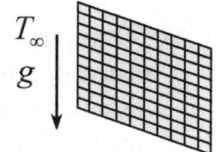

10.34 An apparatus is designed to determine surface emissivity of materials. The apparatus consists of an electrically heated cylindrical sample (disk) of diameter D and thickness δ. The disk is insulated along its heated side and rim. It is placed horizontally with its heated surface facing down in a large chamber whose surface is maintained at uniform temperature T_{sur}. The sample is cooled by free convection and radiation from its upper surface. To determine the emissivity of a sample, measurements are made of the diameter D, electric power input P, surface temperature T_s, surroundings temperature T_{sur} and ambient temperature T_{∞}. Determine the emissivity of a sample using the following data:

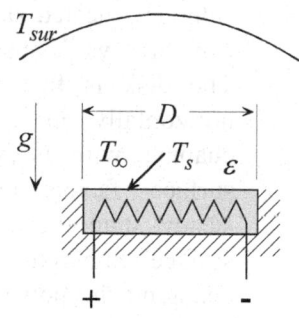

$D = 12$ cm, $\delta = 0.5$ cm, $P = 13.2$ W, $T_s = 98°C$, $T_{sur} = 27°C$, $T_{\infty} = 22°C$.

10.35 It is desired to increase heat loss by free convection from a wide vertical plate without increasing its surface temperature. Increasing the height of the plate is ruled out because of the limited vertical space available. It is suggested that a taller plate can be accommodated in the same vertical space by tilting it 45°. Explore this suggestion and make appropriate recommendations. Assume laminar flow.

10.36 Estimate the free convection heat transfer rate from five sides of a cubical ceramic kiln. Surface temperature of each side is assumed uniform at 70°C and the ambient air temperature is 20°C. Each side measures 48 cm.

10.37 Determine the surface temperature of a single burner electric stove when its power supply is 75 W. The diameter of the burner is 18 cm and its emissivity is 0.32. The ambient air temperature is $30\,^{\circ}C$ and the surroundings temperature is $25\,^{\circ}C$.

10.38 A test apparatus is designed to determine surface emissivity of material. Samples are machined into disks of diameter D. A sample disk is heated electrically on one side and allowed to cool off on the opposite side. The heated side and rim are well insulated. The disk is first placed horizontally in a large chamber with its exposed surface facing up. At steady state the exposed surface temperature is measured. The procedure is repeated, without changing the power supplied to the disk, with the exposed surface facing down. Ambient air temperature in the chamber is recorded.

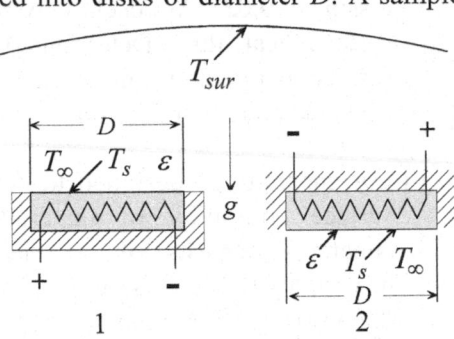

[a] Show that surface emissivity is given by

$$\varepsilon = \frac{\overline{h}_1(T_{s1}-T_\infty)-\overline{h}_2(T_{s2}-T_\infty)}{\sigma(T_{s2}^4-T_{s1}^4)},$$

where subscripts 1 and 2 refer to the exposed surface facing up and down, respectively, and

$\overline{h}$ = average heat transfer coefficient, $W/m^2-^{\circ}C$
T_s = surface temperature, K
T_∞ = ambient temperature, K

σ = Stefan-Boltzmann constant, W/m^2-K^4

[b] Calculate the emissivity for the following case:

$D = 14$ cm, $T_{s1} = 533$ K, $T_{s2} = 573$ K, $T_\infty = 293$ K.

10.39 A hot water tank of diameter 65 cm and height 160 cm loses heat by free convection. Estimate the free convection heat loss from its cylindrical and top surfaces. Assume a surface temperature of 50°C and an ambient air temperature of 20°C.

10.40 Hot gases from a furnace are discharged through a round horizontal duct 30 cm in diameter. The average surface temperature of a 3 m duct section is 180°C. Estimate the free convection heat loss from the duct to air at 25°C.

10.41 A 6 m long horizontal steam pipe has a surface temperature of 120°C. The diameter of the pipe is 8 cm. It is estimated that if the pipe is covered with a 2.5 cm thick insulation material its surface temperature will drop to 40°C. Determine the free convection heat loss from the pipe with and without insulation. The ambient air temperature is 20°C.

10.42 An electric wire dissipates 0.6 W/m while suspended horizontally in air at 20°C. Determine its surface temperature if the diameter is 0.1 mm. Neglect radiation.

10.43 The diameter of a 120 cm long horizontal section of a neon sign is 1.5 cm. Estimate the surface temperature in air at 25°C if 12 watts are dissipated in the section. Neglect radiation heat loss.

10.44 An air conditioning duct passes horizontally a distance of 2.5 m through the attic of a house. The diameter is 30 cm and the average surface temperature is 10°C. The average ambient air temperature in the attic during the summer is 42°C. Duct surface emissivity is 0.1. Estimate the rate of heat transfer to the cold air in the duct.

10.45 Estimate the surface temperature of a light bulb if its capacity is 150 W and the ambient air is at 23°C. Model the bulb as a sphere of diameter 9 cm. Neglect radiation.

10.46 A sphere of radius 2.0 cm is suspended in a very large water bath at 25°C. The sphere is heated internally using an electric coil. Determine the rate of electric power that must be supplied to the sphere so that its average

water

surface temperature is 85°C. Neglect radiation.

10.47 A fish tank at a zoo is designed to maintain water temperature at $4°C$. Fish are viewed from outdoors through a glass window L = 1.8 m high and w = 3 m wide. The average ambient temperature during summer months is $26°C$. To reduce water cooling load it is proposed to create an air enclosure over the entire window

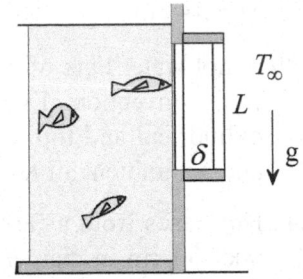

using a pexiglass plate. Estimate the reduction in the rate of heat transfer to the water if the air gap thickness is $\delta = 6$ cm. Neglect radiation. Assume that the cold side of the enclosure is at the same temperature as the water and the warm side is at ambient temperature.

10.48 It is proposed to replace a single pane observation window with double pane. On a typical winter day the inside and outside air temperatures are $T_i = 20°C$ and $T_o = -10°C$. The inside and outside heat transfer coefficients are $h_i = \text{W/m}^2-°C$ and $h_o = 37\,\text{W/m}^2-°C$. The height of the window is $L = 0.28\,\text{m}$ and its width is w = 3 m. The thickness of glass is t = 0.3 cm and its conductivity is $k_g = 0.7\,\text{W/m}-°C$. Estimate the savings in energy if the single pane window is replaced. Note that for the single pane window there are three resistances in series and the heat transfer rate q_1 is given by

$$q_1 = \frac{A(T_i - T_o)}{\dfrac{1}{h_i} + \dfrac{t}{k_g} + \dfrac{1}{h_o}}.$$

For the double pane window, two additional resistances are added. The width of the air space in the double pane is $\delta = 3\,\text{cm}$. In determining the heat transfer coefficient in the cavity, assume that enclosure surface temperatures are the same as the inside and outside air temperatures.

10.49 To reduce heat loss from an oven, a glass door with a rectangular air cavity is used. The cavity has a baffle at its center. Door height is $L = 65\,\text{cm}$ and its width is $w = 70\,\text{cm}$. The

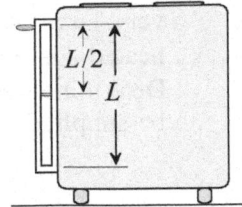

air space thickness is $\delta = 1.5$ cm. Estimate the heat transfer rate through the door if the inside and outside surface temperatures of the cavity are $198°$ C and $42°$ C.

10.50 The ceiling of an exhibit room is designed to provide natural light by using an array of horizontal skylights. Each unit is rectangular with an air gap $\delta = 6.5$ cm thick. The length

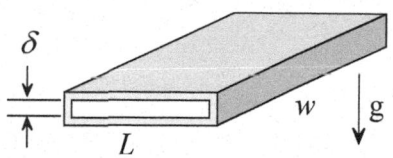

and width of each unit are $L = 54$ cm and $w = 120$ cm. On a typical day the inside and outside glass surface temperatures are $15°$ C and $-15°$ C. Estimate the rate of heat loss from each unit.

10.51 Repeat Example 10.4 using inclination angles of $0°$, $60°$, $90°$, $120°$, $150°$ and $175°$. Plot heat transfer rate q vs. inclination angle θ.

10.52 A rectangular solar collector has an absorber plate of length $L = 2.5$ m and width $w = 4.0$ m. A protection cover is used to form a rectangular air enclosure of thickness $\delta = 4$ cm to provide insulation. Estimate the heat loss by convection from the plate when the enclosure inclination angle is $45°$ and its surfaces are at $28°$ C and $72°$ C.

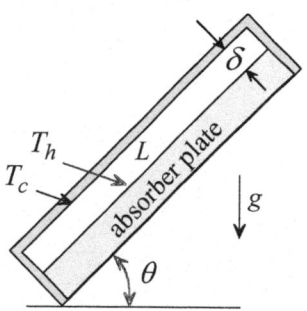

10.53 A liquid-vapor mixture at $T_i = -20°$ C flows inside a tube of diameter $D_i = 4$ cm and length $L = 3$ m. The tube is placed concentrically inside another tube of diameter $D_o = 6$ cm. Surface temperature of the outer tube is at $T_o = 10°$ C. Air fills the annular space. Determine the heat transfer rate from the mixture.

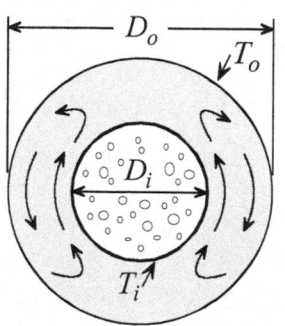

11

Convection in Microchannels

11.1 Introduction

Research on fluid flow and heat transfer in microchannels was partly driven by miniaturization of microelectronic devices. The need for efficient cooling methods for high heat flux components focused attention on the cooling features of microchannels. Microchannels are used in a variety of engineering and scientific applications. The inkjet printer is a classic example. Extensive use is found in medical applications and in mico-elecro-mechanical systems (MEMS) such as micro heat exchangers, mixers, pumps, turbines, sensors and actuators.

11.1.1 Continuum and Thermodynamic Hypothesis

The analysis and results of all previous chapters are based on two fundamental assumptions: (1) *continuum*, and (2) *thermodynamic equilibrium*. The continuity equation, Navier-Stokes equations, and the energy equation are applicable as long as the continuum assumption is valid. The *no-velocity slip* and *no-temperature jump* at a solid boundary, imposed in previous chapters, are valid as long as thermodynamic or quasi-thermodynamic equilibrium can be justified. In Chapter 1, the Knudsen number was used to establish a criterion for the validity of the continuum and thermodynamic assumptions. The Knudsen number is defined in terms of the molecular mean free path λ as

$$Kn = \frac{\lambda}{D_e}, \qquad (1.2)$$

where D_e is a characteristic length, such as channel equivalent diameter. The continuum model is valid for [1]

$$Kn < 0.1. \tag{1.3a}$$

Thus as channel size becomes smaller the Knudsen number increases and the continuum assumption begins to fail at approximately $Kn = 0.1$. On the other hand, departure from thermodynamic equilibrium leads to the failure of the no-velocity slip and no-temperature jump boundary conditions. This begins to take place at a much smaller Knudsen number of $Kn \approx 001$. Thus the no-velocity slip and no-temperature jump conditions are valid for

$$Kn \leq 001. \tag{1.3b}$$

It should be understood that departure from continuum behavior takes place progressively as the Knudsen number is increased. Microchannels are characterized by their relatively small size. A legitimate question is, how small must a channel be to be classified as micro? The answer to this question is not obvious since the mean free path depends on the fluid as well as on its temperature and pressure. Noting that the mean free path of liquids is much smaller than that of gases, liquid flow in a small channel may be within the continuum domain while gas flow may be outside it. Thus classification of microchannels by size is inherently arbitrary.

11.1.2 Surface Forces

As channel size becomes smaller the ratio of surface area to volume becomes larger. This can be illustrated for the case of a tube of diameter D and length L. The ratio of surface area A to volume V is

$$\frac{A}{V} = \frac{\pi D L}{\pi D^2 L / 4} = \frac{4}{D}. \tag{11.1}$$

Equation (11.1) shows that the smaller the diameter, the larger is A/V. Consequently, the role of surface forces becomes more dominant as the diameter decreases. As an example, for a tube with $D = 1$ m, equation (11.1) gives $A/V = 4$ m^{-1}. On the other hand, for $D = 1\,\mu m$, $A/V = 4 \times 10^6$ m^{-1}. This represents a 10^6 fold increase in A/V. Thus conditions at the boundaries may depart from the continuum behavior and take on different forms. This has important implications in the analysis of microchannel problems. Under certain conditions, continuum governing equations for flow and energy can still be applied while boundary conditions must be modified. Another size effect on gas flow in microchannels is the increase in pressure drop in long channels. This results in significant density changes along channels. Consequently, unlike flow in macrochannels, compressibility becomes an important factor and must be taken into consideration.

11.1.3 Chapter Scope

This chapter presents an introduction to convection heat transfer in microchannels. To lay the foundation for the treatment of microchannel convection, topic classification and definitions are presented. This includes: distinction between gases and liquids, microchannel classification, rarefaction and compressibility, velocity slip and temperature jump phenomena. The effect of compressibility and axial conduction will be examined. Analytic solutions to Couette and Poiseuille flows and heat transfer will be detailed. Attention will be focused on convection of gases in microchannels. The treatment will be limited to single phase shear driven laminar flow between parallel plates (Couette flow) and pressure driven flow (Poiseuille flow) through rectangular channels and tubes.

Although extensive research on fluid flow and heat transfer in microchannels has been carried out during the past two decades, much remains unresolved. Due to the complex nature of the phenomena, the role of various factors such as channel size, Reynolds number, Knudsen number, surface roughness, dissipation, axial conduction, and thermophysical properties, is not fully understood. As with all new research areas, discrepancies in findings and conclusions are not uncommon. Conflicting findings are attributed to the difficulty in making accurate measurements of channel size, surface roughness, pressure distribution, as well as uncertainties in entrance effects and the determination of thermophysical properties.

11.2 Basic Considerations

11.2.1 Mean Free Path

The mean free path λ of a fluid is needed to establish if the continuum assumption is valid or not. For gases, λ is given by [2]

$$\lambda = \frac{\mu}{p}\sqrt{\frac{\pi}{2}RT} \,, \qquad\qquad (11.2)$$

where p is pressure, R is gas constant, T is absolute temperature, and μ is viscosity. Since λ is very small, it is expressed in terms of *micrometers*, $\mu = 10^{-6}$ m.. This unit is also known as *micron*. For liquids, λ is much smaller than for gases. It is clear from (11.2) that as the pressure decreases the mean free path increases. Application of (11.2) to air at $300\,\mathrm{K}$ and atmospheric pressure ($p = 101,330\,\mathrm{N/m}^2$) gives $\lambda = 0.067\,\mu\mathrm{m}$. Properties and the mean free path of various gases are listed in Table 11.1. Pressure drop in channel flow results in an axial increase in λ. There-fore the Knudsen number increases in the flow direction. Upper atmospheric air cannot be treated as continuum and is referred to as *rarefied gas* due to low pressure and large λ.

Table 11.1

gas	R J/kg–K	ρ kg/m³	$\mu \times 10^7$ kg/s–m	λ μm
Air	287.0	1.1614	184.6	0.067
Helium	2077.1	0.1625	199.0	0.1943
Hydrogen	4124.3	0.08078	89.6	0.1233
Nitrogen	296.8	1.1233	178.2	0.06577
Oxygen	259.8	1.2840	207.2	0.07155

11.2.2 Why Microchannels?

We return to the continuum, no-slip solution for laminar fully developed convection in tubes. For constant surface temperature, we learned in Chapter 6 that the Nusselt number is constant in the fully developed region. This is true for tubes as well as channels of other cross section geometries. Equation (6.57) gives

$$Nu_D = \frac{hD}{k} = 3.657 \,, \qquad (6.57)$$

where D is diameter, h is heat transfer coefficient, and k is fluid thermal conductivity. Solving (6.57) for h, gives

$$h = 3.657\frac{k}{D}. \qquad (11.3)$$

Examination of (11.3) shows that the smaller the diameter, the larger the

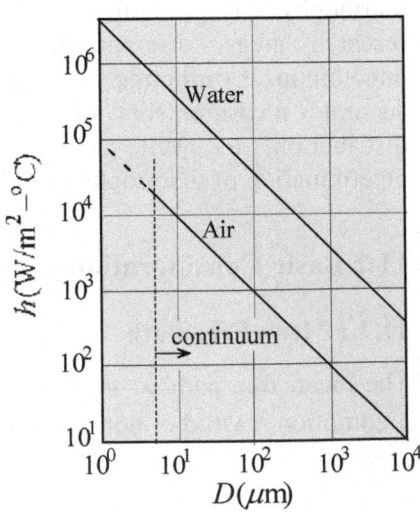

Fig. 11.1

heat transfer coefficient. Fig. 11.1 shows the variation of h with D for air
and water. The conductivity k is determined at a mean fluid temperature of
40°C. The dramatic increase in h as the diameter is decreased has
motivated numerous studies aimed at the development of efficient cooling
methods to maintain pace with the rapid miniaturization of microelectronic
devices during the past three decades. Early studies have analytically and
experimentally demonstrated the potential of microchannels for cooling
high power density devices using water [3]. Typically, grooves are
machined in a sink to form fins to
enhance heat transfer. The heat sink
is attached to a substrate and forms
flow channels as shown in Fig. 11.2.
Due to fabrication constraints,
microchannels usually have rectangular or trapezoidal cross-sections. It
should be noted that although
microchannels have high heat
transfer coefficients, pressure drop

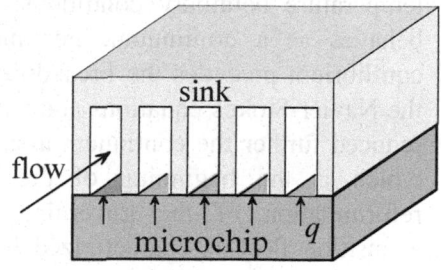

Fig. 11.2

through them increases as channel size becomes smaller.

11.2.3 Classification

A common classification of flow in microchannels is based on the Knudsen
number. The continuum and thermodynamic equilibrium assumptions hold
as long as channel size is large compared to the mean free path λ. As
channel size approaches λ, flow and temperature phenomena begin to
change. The following flow classification is recommended for gases [1]:

$$
\begin{aligned}
Kn &< 0.001 & &continuum,\ no-slip\ flow \\
0.001 &< Kn < 0.1 & &continuum,\ slip\ flow \\
0.1 &< Kn < 10 & &transition\ flow \\
10 &< Kn & &free\ molecular\ flow
\end{aligned}
\tag{11.4}
$$

To appreciate the classification of (11.4) attention is focused on four
factors: (i) continuum, (ii) thermodynamic equilibrium, (iii) velocity slip,
and (iv) temperature jump. If there is no relative velocity between the fluid
and a surface, the condition is referred to as *no-slip*. Similarly, if there is
no-temperature discontinuity at a surface (fluid and surface are at the same
temperature), the condition is described as *no-temperature jump*. Macro-

scale analysis (Chapters 1-10) is based on the assumptions of continuum, thermodynamic equilibrium, no-velocity slip and no-temperature jump. These conditions are valid in the first regime of (11.4) where $Kn < 0.001$. Recall that under these conditions solutions are based on the continuity, Navier-Stokes equations, and energy equation. As device or channel size is reduced the Knudsen number increases. At the onset of the second regime, $Kn > 0.001$, thermodynamic equilibrium begins to fail, leading to velocity slip and temperature jump. This requires reformulation of the velocity and temperature boundary conditions. Away from the boundaries the fluid behaves as a continuum. For most gases, failure of thermodynamic equilibrium precedes the breakdown of the continuum assumption. Thus, the Navier-Stokes equations and energy equation are still valid. As size is reduced further the continuum assumption fails. This occurs at $Kn \geq 0.1$, which is the beginning of the transition flow range. This requires reformulation of the governing equations and boundary conditions. Transition flow is characterized by total departure from thermodynamic equilibrium and the continuum model. It is commonly analyzed using statistical methods to examine the behavior of a group of molecules. As device size becomes an order of magnitude smaller than the mean free path, $Kn \geq 10$, the free molecular flow mode begins. This flow is analyzed using kinetic theory where the laws of mechanics and thermodynamics are applied to individual molecules.

It should be noted that regime limits in (11.4) are arbitrary. Furthermore, transition from one flow regime to another takes place gradually. In this chapter we will limit ourselves to the slip flow regime.

11.2.4 Macro and Microchannels

Since channel size has significant effect on flow and heat transfer, channels can also be classified according to size. However, size alone does not establish if the continuum assumption is valid or not. Nevertheless, channels that function in the continuum domain, with no velocity slip and temperature jump, and whose flow and heat transfer behavior can be predicted from continuum theory or correlation equations, are referred to as macrochannels. On the other hand, channels for which this approach fails to predict their flow and heat transfer characteristics are known as microchannels. It should be emphasized that for microchannels the continuum assumption may or may not hold. Various factors contribute to distinguishing microchannel flow phenomena from macrochannels. These factors include two and three dimensional effects, axial conduction, dissipation, temperature dependent properties, velocity slip and temperature jump at the boundaries and the increasingly dominant role of surface forces as channel size is reduced.

11.2.5 Gases vs. Liquids

In the analysis of macro flow and heat transfer no distinction is made between gases and liquids. Solutions to gas and liquid flows for similar geometries are identical as long as the governing parameters (Reynolds number, Prandtl number, Grashof number, etc.) and boundary conditions are the same for both. This is not the case under micro scale conditions. The following observations are made regarding gas and liquid characteristics in microscale applications [4].

(1) Because the mean free paths of liquids are much smaller than those of gases, the continuum assumption may hold for liquids but fail for gases. Thus, despite the small size of typical MEMS applications, the continuum assumption is valid for liquid flows.

(2) While the Knudsen number provides a criterion for the validity of thermodynamic equilibrium and the continuum model for gases, it does not for liquids.

(3) The onset of failure of thermodynamic equilibrium and continuum is not well defined for liquids. Thus the range of validity of the no-slip, no-temperature jump, linearity of stress-rate of strain relation, (2.7), and linearity of Fourier's heat flux-temperature relation, (1.8), are unknown.

(4) As device size becomes smaller, surface forces become more important. In addition, the nature of surface forces in liquids differs from that of gases. Consequently, boundary conditions for liquids differ from those for gases.

(5) Liquid molecules are much closer to each other than gas molecules. Thus liquids are almost incompressible while gases are compressible.

In general, the physics of liquid flow in microdevices is not well known. Analysis of liquid flow and heat transfer is more complex for liquids than for gases and will not be considered here.

11.3 General Features

As channel size is reduced, flow and heat transfer behavior change depending on the domain of the Knudsen number in condition (11.4). Knudsen number effect is referred to as *rarefaction*. Density change due to pressure drop along microchannels gives rise to *compressibility* effects. Another size effect is *viscous dissipation* which affects temperature distribution. Of particular interest is the effect of channel size on the velocity profile, flow rate, friction factor, transition Reynolds number, and Nusselt number. Consideration will be limited to the variation of these factors for fully developed microchannel gas flow as the Knudsen number increases from the continuum through the slip flow domain.

11.3.1 Flow Rate

Fig. 11.3 shows the velocity profiles for fully developed laminar flow. The no-slip and slip profiles are shown in Fig. 11.3a and 11.3b, respectively. Velocity slip at the surface results in an increase in the flow rate Q as conditions depart from thermodynamic equilibrium. Thus

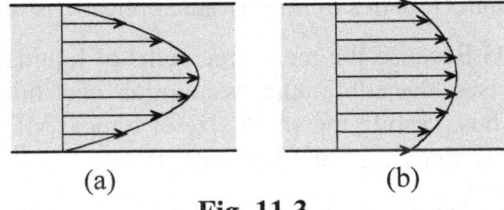

(a) (b)

Fig. 11.3

$$\frac{Q_e}{Q_t} > 1, \qquad (11.5)$$

where subscript e refers to the experimentally determined slip flow rate and subscript t represents the flow rate determined from macrochannel theory or correlation equations. This notation will be used throughout to refer to other performance characteristics such as friction factor and Nusselt number.

11.3.2 Friction Factor

The friction coefficient C_f for channel flow is defined as

$$C_f = \frac{\tau_w}{(1/2)\rho u_m^2}, \qquad (4.37a)$$

where τ_w is wall shear stress and u_m is the mean velocity. For fully developed flow through tubes, C_f can be expressed in terms of pressure drop and is referred to as the *friction factor f*

$$f = \frac{1}{2} \frac{D}{L} \frac{\Delta p}{\rho u_m^2}, \qquad (11.6)$$

where D is diameter, L is length and Δp is pressure drop. Thus, measurements of Δp can be used to determine f. For fully developed laminar flow in macrochannels f is independent of surface roughness. Furthermore, the product of f and Reynolds number is constant. That is

$$f\ Re = Po,\qquad(11.7)$$

where Po is known as the *Poiseuille number*. For example, for continuum flow through tubes, $Po = 64$. For rectangular channels the Poiseuille number depends on channel aspect ratio. For fully developed laminar flow in macrochannels Po is independent of Reynolds number. To examine the accuracy of theoretical models, the Poiseuille number has been computed using extensive experimental data on microchannels. Applying (11.7) to experimentally determined Po and normalizing it with respect to the theoretical value, gives

$$\frac{(Po)_e}{(Po)_t} = C*.\qquad(11.8)$$

When $(Po)_t$ is determined from macroscopic (continuum) theory or correlation equations, the departure of $C*$ from unity represents the degree to which macroscopic theory fails to predict microscopic conditions.

The behavior of f depends on the Knudson number as well as on the nature of the fluid. Extensive experimental data on gases and liquids by several investigators have been compiled and evaluated [5-7]. The compiled data covers a wide range of Reynolds numbers, hydraulic diameters, and aspect ratios. Because pressure drop in the reported experiments is usually measured between channel inlet and outlet, the computed friction factor does not always correspond to fully developed flow. Reported values for C^* range from much smaller than unity to much larger than unity. Nevertheless, reviewing the experimental data on friction factors in microchannels suggests the following preliminary conclusions:

(1) The Poiseuille number Po appears to depend on the Reynolds number. This is in contrast to macrochannels where Po is independent of Reynolds number for fully developed flow.

(2) Both increase and a decrease in the friction factor are reported.

(3) The conflicting findings are attributed to the difficulty in making accurate measurements of channel size, surface roughness, pressure distribution, as well as uncertainties in entrance effects, transition to turbulent flow, and the determination of thermophysical properties.

11.3.3 Transition to Turbulent flow

The Reynolds number is used as the criterion for transition from laminar to turbulent flow. In macrochannels, transition Reynolds number depends on cross-section geometry and surface roughness. For flow through smooth tubes it is given by

$$Re_t = \frac{\bar{u}D}{v} \approx 2300. \tag{6.1}$$

However, for microchannels, reported transition Reynolds numbers ranged from 300 to 16,000 [7]. One of the factors affecting the determination of transition Reynolds number in microchannels is fluid property variation. Outlet Reynolds number can be significantly different from inlet. The effect of size and surface roughness on the transition Reynolds number is presently not well established.

11.3.4 Nusselt number

As shown in equation (6.57), the Nusselt number for fully developed laminar flow in macrochannels is constant, independent of Reynolds number. However, the constant depends on channel geometry and thermal boundary conditions. As with the friction factor, the behavior of the Nusselt number for microchannels is not well understood, resulting in conflicting published conclusions. Nevertheless, there is agreement that microchannel Nusselt number depends on surface roughness and Reynolds number. However, the following demonstrates the widely different reported results for the Nusselt number [8, 11]

$$0.21 < \frac{(Nu)_e}{(Nu)_t} < 100. \tag{11.9}$$

Difficulties in accurate measurements of temperature and channel size, as well as inconsistencies in the determination of thermophysical properties, partly account for the discrepancies in the reported values of the Nusselt number.

11.4 Governing Equations

It is generally accepted that in the slip-flow domain, $0.001 < Kn < 0.1$, the continuity, Navier Stokes equations, and energy equation are valid throughout the flow field [1, 10]. However, common assumptions made in the analysis of macrochannels require reconsideration. Macrochannel

solutions of Chapter 6 are based on negligible compressibility, axial
conduction, and dissipation.

11.4.1 Compressibility

The level of compressibility is expressed in terms of Mach number M
which is defined as the ratio of fluid velocity and the speed of sound.
Incompressible flow is associated with Mach numbers that are small
compared to unity. Compressibility in microchannel flow results in non-
linear pressure drop [8, 11]. Its effect depends on Mach number as well as
the Reynolds number [12, 13]. The friction factor increases as the Mach
number is increased. For example, at $M = 0.35$ Poiseuille number ratio for
tube flow is $C* = 1.13$. On the other hand, the Nusselt number decreases
from $Nu = 3.5$ at $M = 0.01$ to $Nu = 1.1$ at $M = 0.1$ [8].

11.4.2 Axial Conduction

In examining axial conduction in channel flow a distinction must be made
between conduction in the channel wall and conduction in the fluid. In
Section 6.6.2 fluid axial conduction in macrochannels was neglected for
Peclet numbers greater than 100. However, microchannels are typically
operated at low Peclet numbers where axial conduction in the fluid may be
important. A study on laminar fully developed gas flow through microchannels and tubes showed that the effect of axial conduction is to increase
the Nusselt number in the velocity-slip domain [14]. However, the increase
in Nusselt number diminishes as the Knudsen number is increased. The
maximum increase is of order 10%, corresponding to $Kn = 0$.

11.4.3. Dissipation

To examine the role of dissipation we return to the dimensionless form of
the energy equation (2.41a)

$$\frac{DT^*}{Dt^*} = \frac{1}{RePr}\nabla^{*2}T^* + \frac{Ec}{Re}\Phi^*, \qquad (2.41a)$$

where $\Phi*$ is the dissipation function and Ec is the Eckert number defined
as

$$Ec = \frac{V_\infty^2}{c_p(T_s - T_\infty)}. \qquad (2.43)$$

Since Ec is proportional to V_∞^2, it can be shown that it is proportional to the square of Mach number, M^2. Thus as long as M is small compared to unity, the effect of dissipation can be neglected in microchannels.

11.5 Velocity Slip and Temperature Jump Boundary Conditions

To obtain solutions in the slip-flow domain, fluid velocity and thermal conditions must be specified at the boundaries. Unlike the no-slip case, the velocity does not vanish at stationary surfaces and fluid temperature departs from surface temperature. An approximate equation for the velocity slip for gases is referred to as the Maxwell slip model and is given by [1]

$$u(x,0) - u_s = \frac{2-\sigma_u}{\sigma_u} \lambda \frac{\partial u(x,0)}{\partial n}, \tag{11.10}$$

where

$u(x,0)$ = fluid axial velocity at the surface
u_s = surface axial velocity
x = axial coordinate
n = normal coordinate measured from the surface
σ_u = tangential momentum accommodating coefficient

Gas temperature at a surface is approximated by [1]

$$T(x,0) - T_s = \frac{2-\sigma_T}{\sigma_T} \frac{2\gamma}{1+\gamma} \frac{\lambda}{Pr} \frac{\partial T(x,0)}{\partial n}, \tag{11.11}$$

where

$T(x,0)$ = fluid temperature at the surface
T_s = surface temperature
$\gamma = c_p / c_v$, specific heat ratio
σ_T = energy accommodating coefficient

The accommodating coefficients, σ_u and σ_T, are empirical factors that reflect the interaction between gas molecules and a surface. They depend on the gas as well as the geometry and nature of the impingement surface. Their values range from zero (perfectly smooth) to unity. Experimentally determined values of σ_u and σ_T are very difficult to obtain. Nevertheless, there is general agreement that their values for various gases flowing over several surfaces are close to unity [15]. Two observations are made regarding (11.10) and (11.11):

(1) They are valid for gases. Liquid flow through microchannels gives rise to different surface phenomena and boundary conditions.

(2) They represent first order approximation of the velocity slip and temperature jump. Additional terms in (11.10) and (11.11) provide second order correction which extend the limits of their applicability to *Kn* = 0.4 [10, 16].

11.6 Analytic Solutions: Slip Flows

In previous chapters we considered both Couette and Poiseuille flows. Analysis and solutions to these two basic flows, subject to slip conditions, will be presented in this section. In Couette flow the fluid is set in motion inside a channel by moving an adjacent surface. This type of flow is also referred to as *shear driven* flow. On the other hand, fluid motion in Poiseuille flow is generated by an axial pressure gradient. This class of flow problems is referred to as *pressure driven* flow. Both flows find extensive applications in MEMS. An example of shear driven flow is found in the electrostatic comb-drive used in microactuators and microsensors. Fig. 11.4 shows a schematic diagram of such a device. The lateral motion of the comb drives the fluid in the channel formed between the stationary and moving parts. Typical channel length is 100 μm and width is 2 μm. A model for this application is Couette flow between two infinite plates.

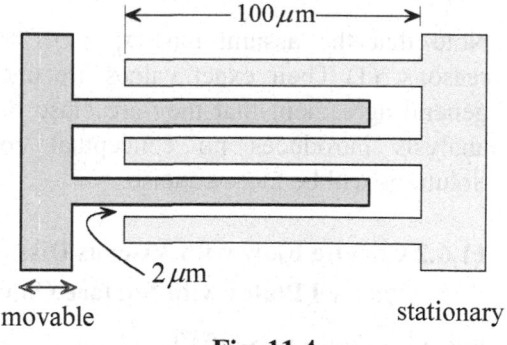

Fig. 11.4

Other examples of shear driven flows are found in lubrication of micromotors, rotating shafts and microturbines. A simplified model for this class of problems is shown in Fig. 11.5. Angular motion of the fluid in the gap between the inner cylinder (rotor) and the housing (stator) is shear driven by the rotor. Poiseuille flow is encountered in many MEMS devices such as micro heat exchangers and mixers. Fig. 11.2 shows a typical example of fluid cooled micro heat sink. A major concern in Poiseuille flow is the large pressure drop associated with microchannels.

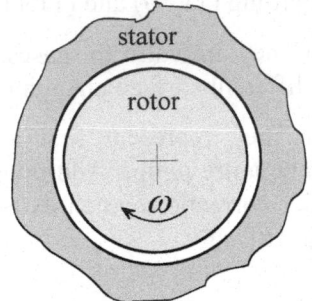

Uniform surface temperature and uniform surface heat flux are two basic boundary conditions that will be considered in heat transfer analysis of Couette and Poiseuille flows. Consideration will also be given to surface convection, compressibility, and internal heat dissipation.

Fig. 11.5

11.6.1 Assumptions.

Analytical solution will be based on common simplifying assumptions. These assumptions are:

(1) Steady state, (2) laminar flow, (3) two-dimensional, (4) idea gas, (5) slip flow, regime ($0.001 < Kn < 0.1$), (6) constant viscosity, conductivity, and specific heats, (7) negligible lateral variation of density and pressure, (8) negligible dissipation (unless otherwise stated), (9) negligible gravity, and (10) the accommodation coefficients are assumed to be equal to unity, ($\sigma_u = \sigma_T = 1.0$).

Note that the assumption $\sigma_u = \sigma_T = 1.0$ is made for the following reasons: (1) Their exact values are uncertain, and furthermore, there is general agreement that they are close to unity. (1) Including them in the analysis introduces no conceptual complications or difficulties. (3) Solutions will be more concise.

11.6.2 Couette Flow with Viscous Dissipation: Parallel Plates with Surface Convection

Fig. 11.6 shows two infinitely large parallel plates separated by a distance H. The upper plate moves axially with uniform velocity u_s. The lower plate is insulated while the upper plate exchanges heat with

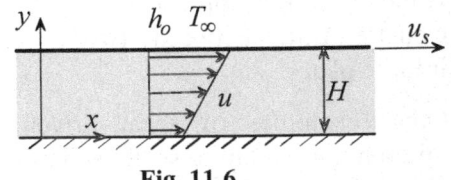

Fig. 11.6

the ambient by convection. The ambient temperature is T_∞ and the heat transfer coefficient along the exterior surface of the moving plate is h_o. This is an example of shear driven flow in which the fluid is set in motion by the plate. Taking into consideration dissipation and slip conditions, we wish to determine the following:

(1) The velocity distribution

(2) The mass flow rate

(3) The Nusselt number

Thus the problem is finding the flow field and temperature distribution in the moving fluid.

Flow Field.   The vector form of the Navier-Stokes equations for compressible, constant viscosity flow is given by (2.11)

$$\rho \frac{D\vec{V}}{Dt} = \rho \vec{g} - \nabla \vec{p} + \frac{1}{3} \mu \nabla (\nabla \cdot \vec{V}) + \mu \nabla^2 \vec{V} . \qquad (2.11)$$

The axial component is

$$\rho \left(\frac{\partial u}{\partial t} + u \frac{\partial u}{\partial x} + v \frac{\partial u}{\partial y} \right) = \rho g_x - \frac{\partial p}{\partial x} + \frac{4}{3} \mu \frac{\partial^2 u}{\partial x^2} + \mu \frac{\partial^2 u}{\partial y^2} + \frac{\mu}{3} \frac{\partial}{\partial x} \left(\frac{\partial v}{\partial y} \right) . \qquad (a)$$

For steady state and negligible gravity

$$\frac{\partial u}{\partial t} = g_x = 0 . \qquad (b)$$

In addition, since the plates are infinitely long and the boundary conditions are uniform, it follows the all derivatives with respect to x must vanish. That is

$$\frac{\partial}{\partial x} = 0 . \qquad (c)$$

Thus (a) simplifies to

$$\rho v \frac{\partial u}{\partial y} = \mu \frac{\partial^2 u}{\partial y^2} . \qquad (d)$$

The continuity equation (2.2a) is now used to provide information on the vertical velocity component v

$$\frac{\partial \rho}{\partial t} + \frac{\partial}{\partial x} (\rho u) + \frac{\partial}{\partial y} (\rho v) + \frac{\partial}{\partial z} (\rho w) = 0 . \qquad (2.2a)$$

Introducing the above assumptions into (2.2a), gives

$$\frac{\partial}{\partial y}(\rho v) = 0.$$ (e)

However, since density variation in the lateral y-direction is assumed negligible, (e) yields

$$\frac{\partial v}{\partial y} = 0.$$ (f)

Integration of (f) and using the condition that v vanishes at the surfaces shows that $v = 0$ everywhere and thus streamlines are parallel (see Example 3.1, Section 3.3.1). Substituting (f) into (d)

$$\frac{d^2 u}{dy^2} = 0.$$ (11.12)

Equation (11.10) provides two boundary conditions on (11.12). Setting $\sigma_u = 1$ and noting that for the lower plate $n = y = 0$, (11.10) gives

$$u(x,0) = \lambda \frac{du(x,0)}{dy}.$$ (g)

For the upper plate, $n = H - y$, (11.10) gives

$$u(x,H) = u_s - \lambda \frac{du(x,H)}{dy}.$$ (h)

The solution to (11.12) is
$$u = Ay + B.$$ (i)

Boundary conditions (g) and (h) give the two constants of integration A and B

$$A = \frac{u_s}{H + 2\lambda}, \quad B = \frac{u_s \lambda}{H + 2\lambda}.$$

Substituting into (i)

$$u = \frac{u_s}{1 + 2(\lambda/H)}\left(\frac{y}{H} + \frac{\lambda}{H}\right). \qquad (j)$$

Defining the Knudsen number as

$$Kn = \frac{\lambda}{H}. \qquad (11.13)$$

Solution (j) becomes

$$\frac{u}{u_s} = \frac{1}{1 + 2Kn}\left(\frac{y}{H} + Kn\right). \qquad (11.14)$$

The following observations are made regarding this result:

(1) Fluid velocity at the moving plate, $y = H$, is

$$\frac{u(H)}{u_s} = \frac{1 + Kn}{1 + 2Kn} < 1.$$

Thus the effect of slip is to decrease fluid velocity at the moving plate and increase it at the stationary plate.

(2) Setting $Kn = 0$ in (11.14) gives the limiting case of no-slip.

(3) For the no-slip case ($Kn = 0$), the velocity distribution is linear

$$\frac{u}{u_s} = \frac{y}{H}. \qquad (k)$$

This agrees with equation (3.8) of Example 3.1

Mass Flow Rate. The flow rate, m, for a channel of width W is given by

$$m = W \int_0^H \rho u \, dy. \qquad (11.15)$$

Substituting (11.14) into (11.15) and noting that ρ is assumed constant along y, gives

$$m = \rho W \int_0^H \frac{u_s}{1 + 2Kn}\left(\frac{y}{H} + Kn\right) dy.$$

Evaluating the integral

$$m = \rho W H \frac{u_s}{2}. \tag{11.16}$$

It is somewhat surprising that the flow rate is independent of the Knudsen number. To compare this result with the flow rate through macro-channels, $\dot{m}_o$, solution (k) is substituted into (11.15). This yields

$$m_o = \rho W H \frac{u_s}{2}. \tag{11.17}$$

This is identical to (11.16). Thus

$$\frac{m}{m_o} = 1. \tag{11.18}$$

This result indicates that the effect of an increase in fluid velocity at the lower plate is exactly balanced by the decrease at the moving plate.

Nusselt Number. The Nusselt number for a parallel plate channel, based on the equivalent diameter $D_e = 2H$, is defined as

$$Nu = \frac{2Hh}{k}. \tag{l}$$

The heat transfer coefficient h for channel flow is defined as

$$h = \frac{-k \dfrac{\partial T(H)}{\partial y}}{T_m - T_s}.$$

Substituting into (l)

$$Nu = -2H \frac{\dfrac{\partial T(H)}{\partial y}}{T_m - T_s}, \tag{11.19}$$

where

$k =$ thermal conductivity of fluid
$T =$ fluid temperature function (variable)
$T_m =$ fluid mean temperature

T_s = plate temperature

It is important to note the following: (1) The heat transfer coefficient in microchannels is defined in terms of surface temperature rather than fluid temperature at the surface. (2) Because a temperature jump develops at the surface of a microchannel, fluid temperature at the moving plate, $T(x, H)$, is not equal to surface temperature T_s. (3) Surface temperature is unknown in this example. It is determined using temperature jump equation (11.11). (4) Care must be taken in applying (11.11) to Fig. 11.6. For $n = H - y$ and $\sigma_T = 1$, (11.11) gives

$$T_s = T(x, H) + \frac{2\gamma}{1+\gamma} \frac{\lambda}{Pr} \frac{\partial T(x, H)}{\partial y}. \tag{11.20}$$

The mean temperature T_m, as defined in Section 6.6.2, is

$$mc_p T_m = W \int_0^H \rho c_p u T \, dy. \tag{11.21}$$

Noting that c_p and ρ are independent of y, and using (11.16) for the mass flow rate m, the above gives

$$T_m = \frac{2}{u_s H} \int_0^H u T \, dy, \tag{11.22}$$

where u is given in (11.14). Examination of equations (11.19)-(11.22) shows that the determination of the heat transfer coefficient requires the determination of the temperature distribution of the moving fluid. Temperature distribution is governed by the energy equation. For two-dimensional, constant conductivity flow, (2.15) gives

$$\rho c_p \left(\frac{\partial T}{\partial t} + u \frac{\partial T}{\partial x} + v \frac{\partial T}{\partial y} \right) = k \left(\frac{\partial^2 T}{\partial x^2} + \frac{\partial^2 T}{\partial y^2} \right) + \beta T \left(u \frac{\partial p}{\partial x} + v \frac{\partial p}{\partial y} \right) + \mu \Phi. \tag{2.15}$$

However, noting that all derivatives with respect to x must vanish, and in addition, $v = \partial / \partial t = 0$, the above simplifies to

$$k \frac{\partial^2 T}{\partial y^2} + \mu \Phi = 0. \tag{11.23}$$

The dissipation function $\varPhi$ is given by

$$\varPhi = 2\left[\left(\frac{\partial u}{\partial x}\right)^2 + \left(\frac{\partial v}{\partial y}\right)^2 + \left(\frac{\partial w}{\partial z}\right)^2\right]$$

$$+ \left[\left(\frac{\partial u}{\partial y} + \frac{\partial v}{\partial x}\right)^2 + \left(\frac{\partial v}{\partial z} + \frac{\partial w}{\partial y}\right)^2 + \left(\frac{\partial w}{\partial x} + \frac{\partial u}{\partial z}\right)^2\right] - \frac{2}{3}\left(\frac{\partial u}{\partial x} + \frac{\partial v}{\partial y} + \frac{\partial w}{\partial z}\right)^2 .$$

$$(2.17)$$

This simplifies to

$$\varPhi = \left(\frac{\partial u}{\partial y}\right)^2 . \qquad (11.24)$$

Substituting (11.24) into (11.23)

$$\frac{d^2T}{dy^2} = -\frac{\mu}{k}\left(\frac{du}{dy}\right)^2 . \qquad (11.25)$$

Note that T is independent of x. This energy equation requires two boundary conditions. They are:

$$\frac{dT(0)}{dy} = 0 , \qquad (m)$$

and

$$-k\frac{dT(H)}{dy} = h_o(T_s - T_\infty) .$$

Using (11.20) to eliminate T_s in the above, gives the second boundary condition

$$-k\frac{dT(H)}{dy} = h_o\left[T(x,H) + \frac{2\gamma}{1+\gamma}\frac{\lambda}{Pr}\frac{\partial T(x,H)}{\partial y} - T_\infty\right] . \qquad (n)$$

To solve (11.25) for the temperature distribution, the velocity solution (11.14) is substituted into (11.25)

$$\frac{d^2T}{dy^2} = -\frac{\mu}{k}\left[\frac{u_s}{H(1+2Kn)}\right]^2 . \qquad (o)$$

Defining the constant φ as

$$\varphi = \frac{\mu}{k} \left[\frac{u_s}{H(1+2Kn)} \right]^2 . \tag{p}$$

Substituting (p) into (o)

$$\frac{d^2 T}{dy^2} = -\varphi . \tag{q}$$

Integration of (q) gives

$$T = -\frac{\varphi}{2} y^2 + Cy + D , \tag{r}$$

where C and D are constants of integration. Application of boundary conditions (m) and (n) gives the two constants:

$$C = 0 .$$

and

$$D = \frac{Hk\varphi}{h_o} + \frac{H^2 \varphi}{2} + \frac{2\gamma}{\gamma+1} \frac{Kn}{Pr} H^2 \varphi + T_\infty . \tag{s}$$

Substituting into (r)

$$T = -\frac{\varphi}{2} y^2 + \frac{kH\varphi}{h_o} + \frac{H^2 \varphi}{2} + \frac{2\gamma}{\gamma+1} \frac{Kn}{Pr} H^2 \varphi + T_\infty . \tag{11.26}$$

To determine the Nusselt number using (11.19), equation (11.26) is used to formulate T_s, $dT(H)/dy$, and T_m. Differentiating (11.26)

$$\frac{dT(H)}{dy} = -H\varphi . \tag{t}$$

Equation (11.26) and (11.20) give T_s

$$T_s = \frac{kH\varphi}{h_o} + T_\infty . \tag{u}$$

Finally, T_m is determined by substituting (11.14) and (11.26) into (11.22)

$$T_m = \frac{2}{H(1+2Kn)} \int_0^H (\frac{y}{H} + Kn)(-\frac{\varphi}{2}y^2 + D)\, dy, \qquad \text{(v)}$$

where D is defined in (s). Evaluating the integral, gives

$$T_m = \frac{2}{1+2Kn}\left[-\frac{1}{8}H^2\varphi - \frac{1}{6}KnH^2\varphi + (\frac{1}{2} + Kn)D\right].$$

Substituting (s) into the above

$$T_m = \frac{1}{1+2Kn}\left[\frac{1}{4}H^2\varphi + \frac{2}{3}KnH^2\varphi\right] + \frac{kH\varphi}{h_o} + \frac{2\gamma}{\gamma+1}\frac{Kn}{Pr}H^2\varphi + T_\infty. \qquad \text{(w)}$$

Using (t), (u) and (w) into (11.191) gives the Nusselt number

$$Nu = \frac{2H^2\varphi}{\dfrac{1}{1+2Kn}\left[\dfrac{H^2}{4}\varphi + \dfrac{2}{3}KnH^2\varphi\right] + \dfrac{kH\varphi}{h_o} + \dfrac{2\gamma H^2}{\gamma+1}\dfrac{Kn}{Pr}\varphi + T_\infty - \dfrac{kH\varphi}{h_o} - T_\infty}.$$

This simplifies to

$$Nu = \frac{8(1+2Kn)}{1 + \dfrac{8}{3}Kn + \dfrac{8\gamma}{\gamma+1}\dfrac{(1+2Kn)Kn}{Pr}}. \qquad (11.27)$$

We make the following remarks:

(1) The Nusselt number is independent of Biot number. This means that changing the heat transfer coefficient h_o does not affect the Nusselt number.

(2) The Nusselt number is independent of the Reynolds number. This is also the case with macrochannel flows.

(3) Unlike macrochannels, the Nusselt number depends on the fluid.

(4) The second term in the denominator of (11.27) represents the effect of rarefaction (Knudsen number) while the third term represents the effect of temperature jump. Both act to reduce the Nusselt number.

(5) The corresponding Nusselt number for macrochannel flow, Nu_o, is determined by setting $Kn = 0$ in (11.27) to obtain

$$Nu_o = 8 \, . \tag{11.28}$$

Taking the ratio of (11.27) and (11.28)

$$\frac{Nu}{Nu_o} = \frac{1 + 2Kn}{1 + \dfrac{8}{3}Kn + \dfrac{8\gamma}{\gamma+1}\dfrac{(1+2Kn)Kn}{Pr}} \, . \tag{11.29}$$

This result shows that the ratio is less than unity.

(6) If dissipation is neglected ($\varphi = 0$), equation (11.26) gives the corresponding temperature solution as

$$T = T_\infty \, .$$

Thus, the temperature is uniform and no heat transfer takes place. Consequently, equation (11.27) for the Nusselt number is not applicable to this limiting case.

Example 11.1: Micro Shaft Temperature

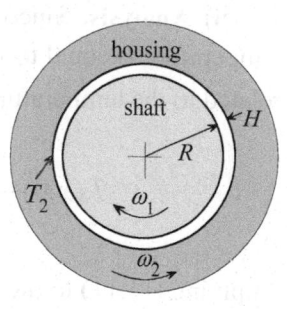

A micro shaft rotates clockwise with angular velocity ω_1 *inside a housing which rotates counterclockwise with an angular velocity* ω_2. *The radius of the shaft is* R *and the clearance between it and the housing is H. The fluid in the clearance is air and the inside surface temperature of the housing is* T_2. *Consider slip flow domain and assume that* $H/R \ll 1$, *set up the governing equations and boundary conditions for the determination of the maximum shaft temperature. List all assumptions.*

(1)   **Observations.**   (i) The effect of dissipation must be included; otherwise the entire system will be at uniform temperature T_2. (ii) The shaft is at uniform temperature. Thus maximum shaft temperature is equal to shaft surface temperature. (iii) Velocity slip and temperature jump take place at both boundaries of the flow channel.   (iv) For $H/R \ll 1$, the problem can be modeled as shear driven Couette flow between two parallel plates moving in the opposite direction. (v) To determine temperature distribution it is necessary to determine the velocity distribution. (vi) No heat is conducted through the shaft. Thus its surface is insulated.

(2) Problem Definition. Formulate the Navier-Stokes equations, energy equation, velocity slip, and temperature jump boundary conditions for shear driven Couette flow between parallel plates.

(3) Solution Plan. Model the flow as shear driven Couette flow between two parallel plates. To formulate the governing equations, follow the analysis of Section 11.6.2. Modify velocity slip at the shaft and thermal boundary condition at the housing surface.

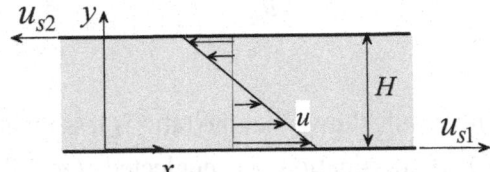

(4) Plan Execution.

(i) **Assumptions.** (1) Steady state, (2) laminar flow, (3) $H/R \ll 1$, (4) one-dimensional (no variation with axial distance x and normal distance z), (5) slip flow regime ($0.001 < Kn < 0.1$), (6) ideal gas, (7) constant viscosity, conductivity and specific heats, (8) negligible lateral variation of density and pressure, (9) the accommodation coefficients are assumed equal to unity, $\sigma_u = \sigma_T = 1.0$, and (10) negligible gravity.

(ii) **Analysis.** Since the shaft is at uniform temperature, its maximum temperature is equal to its surface temperature, T_{s1}. Surface temperature is related to the temperature jump given by (11.11)

$$T(x,0) - T_s = \frac{2-\sigma_T}{\sigma_T} \frac{2\gamma}{1+\gamma} \frac{\lambda}{Pr} \frac{\partial T(x,0)}{\partial n}. \qquad (11.11)$$

Applying (11.11) to the shaft surface, $n = y = 0$, and setting $\sigma_T = 1$, gives

$$T(0) - T_{s1} = \frac{2\gamma}{1+\gamma} \frac{\lambda}{Pr} \frac{dT(0)}{dy}, \qquad (a)$$

where $T(y)$ is the fluid temperature distribution. Thus, the problem becomes one of determining $T(y)$. To determine temperature distribution, it is necessary to determine the flow field.

Flow Field. Following the analysis of Section 11.6.2, the axial component of the Navier-Stokes equations is given by (11.12)

$$\frac{d^2u}{dy^2} = 0.$$

(11.12)

Boundary conditions for (11.12) are given by (11.10)

$$u(x,0) - u_s = \frac{2 - \sigma_u}{\sigma_u} \lambda \frac{\partial u(x,0)}{\partial n}.$$

(11.10)

Applying (11.10) to the lower surface, $n = y = 0$, and setting $\sigma_u = 1$

$$u(0) - u_{s1} = \lambda \frac{du(0)}{dy},$$

(b)

where

$$u_{s1} = \omega_1 R.$$

(c)

For the upper surface, $n = H - y$, (11.10) gives

$$u(H) - u_{s2} = -\lambda \frac{du(H)}{dy},$$

(d)

where u_{s2} is the velocity of the upper surface, given by

$$u_{s2} = -\omega_2 (R + H).$$

(e)

Temperature Distribution. The energy equation for this configuration is given by (11.25)

$$\frac{d^2T}{dy^2} = -\frac{\mu}{k} \left(\frac{du}{dy} \right)^2.$$

(11.25)

The boundary condition at $y = 0$ is

$$\frac{dT(0)}{dy} = 0.$$

(f)

At the upper surface, $n = H - y$, surface temperature is specified. Thus, (11.11) gives

$$T(H) - T_2 = -\frac{2\gamma}{1 + \gamma} \frac{\lambda}{Pr} \frac{dT(H)}{dy}.$$

(g)

(iii) Checking. *Dimensional check*: Equations (11.25), (b), (d) and (g) are dimensionally correct.

Limiting Check: For the limiting case of no-slip, fluid and surface must have the same velocity and temperature. Setting $\lambda = 0$ in (b), (d) and (g) gives $u(0) = u_{s1}$, $u(H) = u_{s2}$, and $T(H) = T_2$. These are the correct boundary conditions for the no-slip case.

(5) Comments. (i) The problem is significantly simplified because no angular variations take place in velocity, pressure, and temperature. (ii) The effect of slip is to decrease fluid velocity at the upper and lower surfaces.

11.6.3 Fully Developed Poiseuille Channel Flow: Uniform Surface Flux

We consider heat transfer in microchannels under pressure driven flow conditions. Fig. 11.7 shows two infinitely large parallel plates separated by a distance H. This configuration is often used to model flow and heat transfer in rectangular channels with large aspect

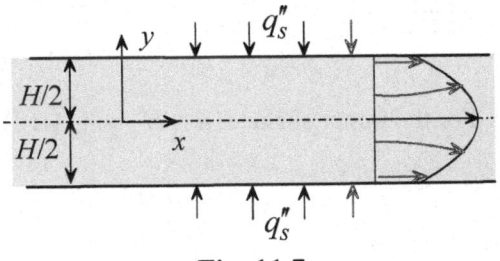

Fig. 11.7

ratios. Velocity and temperature are assumed to be fully developed. Inlet and outlet pressures are p_i and p_o, respectively. The two plates are heated with uniform and equal flux q_s''. We wish to determine the following:

(1) Velocity distribution
(2) Pressure distribution
(3) Mass flow rate
(4) Nusselt number

Poiseuille flow differs from Couette flow in that axial pressure gradient in Poiseuille flow does not vanish. It is instructive to examine how microchannel Poiseuille flow differs from fully developed, no-slip macrochannel flow. Recall that incompressible fully developed Poiseuille flow in macrochannels is characterized by the following: (1) parallel streamlines, (2) zero lateral velocity component ($v = 0$), (3) invariant axial velocity with axial distance ($\partial u / \partial x = 0$), and (4) linear axial pressure ($dp / dx = $ constant). However, in microchannels, compressibility and rarefaction change this flow pattern, and consequently none of these conditions hold. Because of the large pressure drop in microchannels, density change in gaseous flows becomes appreciable and the flow can no longer be assumed incompressible. Another effect is due to rarefaction. According to equation (11.2), a decrease in pressure in microchannels results in an increase in the mean free path λ. Thus the Knudsen number increases along a microchannel in Poiseuille flow. Consequently, axial velocity varies with axial distance, lateral velocity component does not vanish, streamlines are not parallel, and pressure gradient is not constant.

Poiseuille flow and heat transfer have been extensively studied experimentally and analytically. The following analysis presents a first order solution to this problem [16-19].

Assumptions. We invoke the 10 assumptions listed in Example 11.1. Additional assumptions will be made as needed.

Flow Field. Following the analysis of Section 11.6.2, the axial component of the Navier-Stokes equation for constant viscosity and compressible flow is given by

$$\rho\left(\frac{\partial u}{\partial t} + u\frac{\partial u}{\partial x} + v\frac{\partial u}{\partial y}\right) = \rho g_x - \frac{\partial p}{\partial x} + \frac{4}{3}\mu\frac{\partial^2 u}{\partial x^2} + \mu\frac{\partial^2 u}{\partial y^2} + \frac{\mu}{3}\frac{\partial}{\partial x}\left(\frac{\partial v}{\partial y}\right).\text{(a)}$$

For steady state and negligible gravity

$$\frac{\partial u}{\partial t} = g_x = 0. \tag{b}$$

To simplify (a) further. the following additional assumptions are made:

(11) Isothermal flow. This assumption eliminates temperature as a variable in the momentum equations. In addition, density can be expressed in terms of pressure using the ideal gas law.

(12) Negligible inertia forces. With this assumption the inertia terms $\rho(u\partial u/\partial x + v\partial u/\partial y)$ can be neglected. This approximation is justified for low Reynolds numbers. The Reynolds number in most microchannels is indeed small because of the small channel spacing or equivalent diameter.

(13) The dominant viscous force is $\mu(\partial^2 u/\partial y^2)$. Scale analysis shows that this term is of order H^{-2} while $\mu(\partial^2 u/\partial x^2)$ and $\mu\partial/\partial x(\partial v/\partial y)$ are of order L^{-2}. Thus these two terms can be neglected. Using (b) and the above assumptions, equation (a) simplifies to

$$-\frac{\partial p}{\partial x} + \mu \frac{\partial^2 u}{\partial y^2} = 0.$$ (c)

Since pressure is assumed independent of y, this equation can be integrated directly to give the axial velocity u. Thus

$$u = \frac{1}{2\mu}\frac{dp}{dx}y^2 + Ay + B,$$ (d)

where A and B are constants of integration obtained from boundary conditions on u. Symmetry at $y = 0$ gives using (11.10).

$$\frac{\partial u(x,0)}{\partial y} = 0.$$ (e)

Applying (11.10) to the upper plate, $n = H - y$, and setting $\sigma_u = 1$ gives the second boundary condition

$$u(x, H/2) = -\lambda \frac{\partial u(x, H/2)}{\partial y}.$$ (f)

Applying (e) and (f) to (d), and using the definition of the Knudsen number in (11.13), give A and B

$$A = 0, \quad B = -\frac{H^2}{8\mu}\frac{dp}{dx}\left[1 + 4Kn(p)\right].$$ (g)

Substituting (g) into (d)

$$u = -\frac{H^2}{8\mu}\frac{dp}{dx}\left[1 + 4Kn(p) - 4\frac{y^2}{H^2}\right].$$ (11.30)

Note the following: (1) The Knudsen number, which varies with pressure along the channel, represents rarefaction effect on axial velocity. (2) Pressure gradient is unknown and must be determined to complete the solution. (3) Setting $Kn = 0$ in (11.30) gives the no-slip solution to Poiseuille flow in macrochannels.

To complete the flow field solution, the lateral velocity component v and pressure distribution p must be determined. The continuity equation for compressible flow, (2.2a), is used to determine v

$$\frac{\partial \rho}{\partial t} + \frac{\partial}{\partial x}(\rho u) + \frac{\partial}{\partial y}(\rho v) + \frac{\partial}{\partial z}(\rho w) = 0.$$ (2.2a)

Introducing the above assumptions into (2.2a), we obtain

$$\frac{\partial}{\partial x}(\rho u) + \frac{\partial}{\partial y}(\rho v) = 0.$$ (h)

Integration of this equation gives v. The density ρ is eliminated using the ideal gas law

$$\rho = \frac{p}{RT},$$ (11.31)

where R is the gas constant and T is temperature. Substituting (11.31) into (h), assuming constant temperature, and rearranging

$$\frac{\partial}{\partial y}(pv) = -\frac{\partial}{\partial x}(pu).$$ (i)

Substituting (11.30) into (i)

$$\frac{\partial}{\partial y}(pv) = \frac{H^2}{8\mu}\frac{\partial}{\partial x}\left[p\frac{dp}{dx}\left(1 + 4Kn(p) - 4\frac{y^2}{H^2}\right)\right].$$ (j)

Flow symmetry with respect to y gives the following boundary condition on v

$$v(x,0) = 0.$$ (k)

A second condition is obtained by requiring that the lateral velocity vanishes at the wall. Thus

$$v(x, H/2) = 0.$$ (l)

Multiplying (j) by dy, integrating from $y = 0$ to $y = y$, and using (k)

$$\int_0^y d(pv) = \frac{H^2}{8\mu}\frac{\partial}{\partial x}\left[p\frac{dp}{dx}\int_0^y\left(1 + 4Kn(p) - 4\frac{y^2}{H^2}\right)dy\right].$$ (m)

Evaluating the integrals and solving for v

$$v = \frac{H^3}{8\mu} \frac{1}{p} \frac{\partial}{\partial x} \left\{ p \frac{dp}{dx} \left[[1 + 4Kn(p)] \frac{y}{H} - \frac{4}{3} \frac{y^3}{H^3} \right] \right\} . \qquad (11.32)$$

It remains to determine the pressure $p(x)$. Application of boundary condition (l) to (11.32) gives a differential equation for p

$$\frac{\partial}{\partial x} \left\{ p \frac{dp}{dx} \left[[1 + 4Kn(p)] \frac{y}{H} - \frac{4}{3} \frac{y^3}{H^3} \right] \right\}_{y=H/2} = 0 . \qquad (n)$$

To integrate (n), the Knudsen number must be expressed in terms of pressure. Equations (11.2) and (11.13) give

$$Kn = \frac{\lambda}{H} = \frac{\mu}{H} \sqrt{\frac{\pi}{2} RT} \frac{1}{p} . \qquad (11.33)$$

Evaluating (n) at $y = H/2$, substituting (11.33) into (n) and integrating

$$p \frac{dp}{dx} \left[\frac{1}{3} + \frac{\mu}{H} \sqrt{2\pi RT} \frac{1}{p} \right] = C .$$

Integrating again noting that T is assumed constant

$$\frac{1}{6} p^2 + \frac{\mu}{H} \sqrt{2\pi RT} \, p = Cx + D , \qquad (o)$$

where C and D are constants of integration. The solution to this quadratic equation is

$$p(x) = -3 \frac{\mu}{H} \sqrt{2\pi RT} + \sqrt{18\pi RT \frac{\mu^2}{H^2} + 6Cx + 6D} . \qquad (p)$$

The constants C and D are determined by specifying channel inlet and outlet pressures. Let

$$p(0) = p_i, \quad p(L) = p_o , \qquad (q)$$

where L is channel length. Application of (q) to (p) gives C and D

$$C = \frac{1}{6L}(p_o^2 - p_i^2) + \frac{\mu}{HL}\sqrt{2\pi RT}(p_o - p_i),$$

$$D = \frac{p_i^2}{6} + \frac{\mu}{H}\sqrt{2\pi RT}\, p_i.$$

Substituting the above into (p) and normalizing the pressure by p_o, give

$$\frac{p(x)}{p_o} = -\frac{3\mu}{Hp_o}\sqrt{2\pi RT} +$$

$$\sqrt{\frac{18\mu^2}{H^2}\frac{\pi RT}{p_o^2} + \left[1 - \frac{p_i^2}{p_o^2} + \frac{6\mu}{Hp_o}\sqrt{2\pi RT}(1 - \frac{p_i}{p_o})\right]\frac{x}{L} + \frac{p_i^2}{p_o^2} + \frac{6\mu}{Hp_o}\sqrt{2\pi RT}\frac{p_i}{p_o}}.$$

$$\text{(r)}$$

This result can be expressed in terms of the Knudsen number at the outlet using (11.2) and (11.13)

$$Kn_o = \frac{\lambda(p_o)}{H} = \frac{\mu}{Hp_o}\sqrt{\frac{\pi}{2}RT_o}, \qquad (11.34)$$

where T in (r) is approximated by the outlet temperature T_o. Equation (r) becomes

$$\frac{p(x)}{p_o} = -6Kn_o + \sqrt{\left[6Kn_o + \frac{p_i}{p_o}\right]^2 + \left[(1 - \frac{p_i^2}{p_o^2}) + 12Kn_o(1 - \frac{p_i}{p_o})\right]\frac{x}{L}}.$$

$$\text{(11.35)}$$

Note the following regarding (11.35): (1) unlike macrochannel Poiseuille flow, pressure variation along the channel is non-linear. (2) Knudsen number terms represent rarefaction effect on the pressure distribution. (3) The terms $(p_i/p_o)^2$ and $[1-(p_i/p_o)^2](x/L)$ represent the effect of compressibility. (4) Application of (11.35) to the limiting case of $Kn_o = 0$ gives

$$\frac{p(x)}{p_o} = \sqrt{\frac{p_i^2}{p_o^2} + (1 - \frac{p_i^2}{p_o^2})\frac{x}{L}}. \qquad (11.36)$$

This result represents the effect of compressibility alone. Axial pressure distribution for this case is also non-linear.

Mass Flow Rate. The flow rate m for a channel of width W is

$$m = 2W \int_0^{H/2} \rho u\, dy. \tag{s}$$

Using (11.30), the above yields

$$m = -2W \frac{H^2}{8\mu} \rho \frac{dp}{dx} \int_0^{H/2} \left[1 + 4Kn(p) - 4\frac{y^2}{H^2} \right] dy.$$

Since ρ and p are assumed uniform along y, they are treated as constants in the above integral. Evaluating the integral, gives

$$m = - \frac{WH^3}{12\mu} \left[1 + 6Kn(p) \right] \rho \frac{dp}{dx}. \tag{t}$$

The density ρ is expressed in terms of pressure using the ideal gas law

$$\rho = \frac{p}{RT}. \tag{11.37}$$

Substituting (11.33) and (11.37) into (t), gives

$$m = - \frac{WH^3}{12\mu RT} \left[p + 6\frac{\mu}{H}\sqrt{\frac{\pi}{2}RT} \right] \frac{dp}{dx}. \tag{11.38}$$

Using (11.35) to formulate the pressure gradient, substituting into (11.38), assuming constant temperature ($T \cong T_o$), and rearranging, gives

$$m = \frac{1}{24} \frac{WH^3 p_o^2}{\mu LRT_o} \left[\frac{p_i^2}{p_o^2} - 1 + 12\,Kn_o\left(\frac{p_i}{p_o} - 1\right) \right]. \tag{11.39}$$

It is instructive to compare this result with the corresponding no-slip macrochannel case where the flow is assumed incompressible. The mass flow rate for this case is given by

$$m_o = \frac{1}{12} \frac{W H^3 p_o^2}{\mu L R T_o} \left[\frac{p_i}{p_o} - 1 \right].$$ (11.40)

Taking the ratio of the two results

$$\frac{m}{m_o} = \frac{1}{2} \left[\frac{p_i}{p_o} + 1 + 12 \, Kn_o \right].$$ (11.41)

We make the following observations:

(1) The mass flow rate in microchannels, (11.39), is very sensitive to channel height H. This partly explains the difficulty in obtaining accurate data where channel height is typically measured in microns.

(2) Equation (11.39) shows the effect of rarefaction (slip) and compressibility on the mass flow rate. To examine the effect of compressibility alone (long channels with no-slip), set $Kn_o = 0$ in (11.39).

(3) Since $p_i / p_o > 1$, equation (11.41) shows that neglecting the effect of compressibility and rarefaction underestimates the mass flow rate.

Nusselt Number. Following the analysis of Section 11.6.2, the Nusselt number is defined as

$$Nu = \frac{2hH}{k}.$$ (u)

The heat transfer coefficient h for uniform surface flux q_s'' is

$$h = \frac{q_s''}{T_s - T_m}.$$

Substituting into (u)

$$Nu = \frac{2Hq_s''}{k(T_s - T_m)},$$ (v)

where

   T_m = fluid mean temperature
   T_s = plate temperature

As usual, the heat transfer coefficient in microchannels is defined in terms of surface temperature rather than fluid temperature at the surface. Plate surface temperature T_s is given by (11.11), which for the coordinate y shown, takes the form

$$T_s = T(x, H/2) + \frac{2\gamma}{1+\gamma} \frac{\lambda}{Pr} \frac{\partial T(x, H/2)}{\partial y}. \qquad (11.42)$$

The mean temperature T_m is defined in Section 6.6.2. Since density and specific heat are assumed invariant with respect to y, the mean temperature takes the form

$$T_m = \frac{\displaystyle\int_0^{H/2} uT\,dy}{\displaystyle\int_0^{H/2} u\,dy}. \qquad (11.43)$$

Thus, equations (v), (11.42) and (11.43) show that axial velocity, $u(x, y)$, and temperature distribution, $T(x, y)$, are required for the determination of the Nusselt number. We consider first velocity distribution $u(x, y)$. The solution obtained above, equation (11.30), is limited to isothermal flow, as indicated in assumption (11) listed above. To proceed with the solution, additional assumptions are made. However, in the heat transfer aspect of this problem the temperature is not uniform. We assume that the effect of temperature variation on the velocity distribution is negligible. Continuing with the list of assumptions, we add

(14) Axial velocity distribution is approximated by the solution to the isothermal case.

Temperature distribution is governed by energy equation (2.15)

$$\rho c_p \left(\frac{\partial T}{\partial t} + u \frac{\partial T}{\partial x} + v \frac{\partial T}{\partial y} \right) = k \left(\frac{\partial^2 T}{\partial x^2} + \frac{\partial^2 T}{\partial y^2} \right) + \beta T \left(u \frac{\partial p}{\partial x} + v \frac{\partial p}{\partial y} \right) + \mu \Phi.$$

$$(2.15)$$

To simplify this equation, additional assumptions are made:

(15) Negligible dissipation, $\Phi = 0$
(16) Negligible axial conduction, $\partial^2 T / \partial x^2 \ll \partial^2 T / \partial y^2$
(17) Negligible effect of compressibility on the energy equation
(18) Nearly parallel flow, $v = 0$

Equation (2.15) becomes

$$\rho c_p u \frac{\partial T}{\partial x} = k \frac{\partial^2 T}{\partial y^2}.$$

(11.44)

This equation requires two boundary conditions. They are:

$$\frac{\partial T(x,0)}{\partial y} = 0,$$

(w)

and

$$k \frac{\partial T(x, H/2)}{\partial y} = q_s''.$$

(x)

To proceed with the solution to (11.44), we follow the analysis of Chapter 6 and introduce the following important assumption:

(19) Fully developed temperature. Introducing the dimensionless temperature ϕ

$$\phi = \frac{T(x, H/2) - T(x, y)}{T(x, H/2) - T_m(x)}.$$

(11.45)

Fully developed temperature is defined as a profile in which ϕ is independent of x. That is

$$\phi = \phi(y).$$

(11.46)

Thus

$$\frac{\partial \phi}{\partial x} = 0.$$

(11.47)

Equations (11.45) and (11.46) give

$$\frac{\partial \phi}{\partial x} = \frac{\partial}{\partial x} \left[\frac{T(x, H/2) - T(x, y)}{T(x, H/2) - T_m(x)} \right] = 0.$$

Expanding and using the definition of ϕ in (11.45)

$$\frac{dT(x, H/2)}{dx} - \frac{\partial T}{\partial x} - \phi(y) \left[\frac{dT(x, H/2)}{dx} - \frac{dT_m(x)}{dx} \right] = 0. \quad (11.48)$$

The relationship between the three gradients, $\partial T(x,y)/\partial x$, $dT(x,H/2)/dx$, and $dT_m(x)dx$ will be determined. The heat transfer coefficient h, is given by

$$h = \frac{-k\dfrac{\partial T(x,H/2)}{\partial y}}{T_m(x)-T_s(x)},$$ (y)

where $T_s(x)$ is given in (11.42). Temperature gradient in (y) is obtained from (11.45)

$$T(x,y) = T(x,H/2) - [T(x,H/2) - T_m(x)]\phi.$$

Differentiating the above and evaluating the derivative at $y = H/2$

$$\frac{\partial T(x,H/2)}{\partial y} = -[T(x,H/2)-T_m(x)]\frac{d\phi(H/2)}{dy}.$$ (z)

Substituting (z) into (y) and using (11.42) for $T_s(x)$

$$h = -\frac{k[T(x,H/2)-T_m(x)]}{T_s(x)-T_m(x)}\frac{d\phi(H/2)}{dy}.$$ (11.49)

Newton's law of cooling gives another equation for h

$$h = \frac{q_s''}{T_s(x)-T_m(x)}.$$

Equating the above with (11.411) and rearranging

$$T(x,H/2)-T_m(x) = -\frac{q_s''}{k\dfrac{d\phi(H/2)}{dy}} = \text{constant}.$$ (11.50)

Differentiating

$$\frac{\partial T(x,H/2)}{\partial x} - \frac{\partial T_m(x)}{\partial x} = 0.$$

Combining this with (11.48), gives

$$\frac{dT(x, H/2)}{dx} = \frac{dT_m(x)}{dx} = \frac{\partial T}{\partial x}. \tag{11.51}$$

This is an important result since $\partial T / \partial x$ in differential equation (11.44) can be replaced with dT_m / dx. The next step is to formulate an equation for the mean temperature gradient dT_m / dx by applying conservation of energy. For the element shown in Fig. 11.8, conservation of energy gives

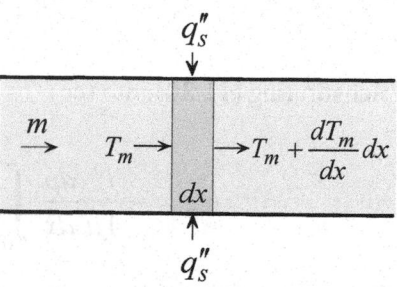

Fig. 11.8

$$2q_s'' W dx + mc_p T_m = mc_p \left[T_m + \frac{dT_m}{dx} dx \right].$$

Simplifying

$$\frac{dT_m}{dx} = \frac{2Wq_s''}{mc_p} = \text{constant.} \tag{aa}$$

However,

$$m = WH\rho u_m, \tag{bb}$$

where u_m is the mean axial velocity. Substituting (bb) into (aa)

$$\frac{dT_m}{dx} = \frac{2q_s''}{\rho c_p u_m H} = \text{constant.} \tag{11.52}$$

Substituting (11.52) into (11.51)

$$\frac{dT(x, H/2)}{dx} = \frac{dT_m(x)}{dx} = \frac{\partial T}{\partial x} = \frac{2q_s''}{\rho c_p u_m H}. \tag{11.53}$$

Equation (11.53) shows that $T(x, r)$, $T_m(x)$ and $T_s(x)$ vary linearly with axial distance x. Substituting (11.53) into (11.44)

$$\frac{\partial^2 T}{\partial y^2} = \frac{2q_s''}{kH} \frac{u}{u_m}. \tag{11.54}$$

The velocity u is given by (11.30) and the mean velocity is defined as

$$u_m = \frac{2}{H} \int_0^{H/2} u \, dy \, . \tag{cc}$$

Substituting (11.30) into (cc)

$$u_m = -\frac{H}{4\mu} \frac{dp}{dx} \int_0^{H/2} \left[1 + 4Kn - 4\frac{y^2}{H^2}\right] dy \, .$$

Integration gives

$$u_m = -\frac{H^2}{12\mu} \frac{dp}{dx} \left[1 + 6Kn\right] . \tag{11.55}$$

Combining (11.30) and (11.55)

$$\frac{u}{u_m} = \frac{6}{1 + 6Kn} \left[\frac{1}{4} + Kn - \frac{y^2}{H^2}\right] . \tag{11.56}$$

(11.56) into (11.54)

$$\frac{\partial^2 T}{\partial y^2} = \frac{12}{1 + 6Kn} \frac{q_s''}{kH} \left[\frac{1}{4} + Kn - \frac{y^2}{H^2}\right] . \tag{11.57}$$

Integrating twice

$$T(x, y) = \frac{12 q_s''}{(1 + 6Kn)kH} \left[\frac{1}{2}(\frac{1}{4} + Kn)y^2 - \frac{y^4}{12H^2}\right] + f(x)y + g(x) , \tag{dd}$$

where $f(x)$ and $g(x)$ are "constants" of integration. Boundary condition (w) gives

$$f(x) = 0 \, .$$

Solution (dd) becomes

$$T(x, y) = \frac{12 q_s''}{(1 + 6Kn)kH} \left[\frac{1}{2}(\frac{1}{4} + Kn)y^2 - \frac{y^4}{12H^2}\right] + g(x) . \tag{11.58}$$

Boundary condition (x) is automatically satisfied and thus will not yield $g(x)$. To proceed, $g(x)$ will be determined by evaluating the mean temperature T_m using two methods. In the first method, (11.52) is integrated between the inlet of the channel, $x = 0$, and an arbitrary location x

$$\int_{T_{mi}}^{T_m} dT_m = \frac{2q_s''}{\rho c_p u_m H} \int_0^x dx,$$

where

$$T_m(0) = T_{mi}. \tag{11.59}$$

Evaluating the integrals

$$T_m(x) = \frac{2q_s''}{\rho c_p u_m H} x + T_{mi}. \tag{11.60}$$

In the second method, T_m is evaluated using its definition in (11.43). Substituting (11.30) and (11.58) into (11.43)

$$T_m(x) = \frac{-\dfrac{H}{8\mu}\dfrac{dp}{dx}\displaystyle\int_0^{H/2}\left[1+4Kn-4\dfrac{y^2}{H^2}\right]\left\{\dfrac{12q_s''}{(1+6Kn)kH}\left[\dfrac{1}{2}(\dfrac{1}{4}+Kn)y^2-\dfrac{y^4}{12H^2}\right]+g(x)\right\}dy}{-\dfrac{H}{8\mu}\dfrac{dp}{dx}\displaystyle\int_0^{H/2}\left[1+4Kn-4\dfrac{y^2}{H^2}\right]dy}.$$

Evaluating the integrals

$$T_m(x) = \frac{3q_s'' H}{k(1+6Kn)^2}\left[(Kn)^2 + \frac{13}{40}Kn + \frac{13}{560}\right] + g(x). \tag{11.61}$$

Equating (11.60) and (11.61) gives $g(x)$

$$g(x) = T_{mi} + \frac{2q_s''}{\rho c_p u_m H}x - \frac{3q_s'' H}{k(1+6Kn)^2}\left[(Kn)^2 + \frac{13}{40}Kn + \frac{13}{560}\right]. \tag{11.62}$$

Surface temperature $T_s(x, H/2)$ is determined by substituting the temperature solution (11.58) into (11.42)

$$T_s(x) = \frac{3q_s''H}{k(1+6Kn)}\left[\frac{1}{2}Kn + \frac{5}{48}\right] + \frac{2\gamma}{\gamma+1}\frac{q_s''H}{kPr}Kn + g(x). \quad (11.63)$$

Substituting (11.61) and (11.63) into (v), gives the Nusselt number

$$Nu = \frac{2}{\dfrac{3}{(1+6Kn)}\left\{\dfrac{1}{2}Kn + \dfrac{5}{48} - \dfrac{1}{(1+6Kn)}\left[(Kn)^2 + \dfrac{13}{40}Kn + \dfrac{13}{560}\right]\right\} + \dfrac{2\gamma}{\gamma+1}\dfrac{1}{Pr}Kn}.$$

$$(11.64)$$

Using (11.64), the Nusselt number variation with Knudsen number for air, with $\gamma = 1.4$ and $Pr = 0.7$, is plotted in Fig. 11.11. The following remarks are made:

(1) The Knudsen number in (11.64) is a function of local pressure. Since pressure varies along the channel, it follows that the Nusselt number varies with distance x. This is contrary to the no-slip macro-channels case where the Nusselt number is constant.

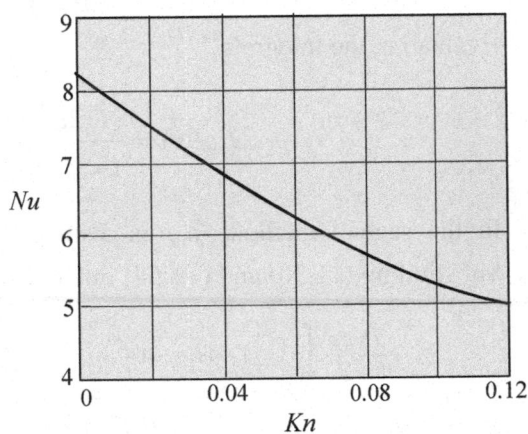

Fig. 11.9 Nusselt number for air flow between parallel plates at uniform surface heat flux for air, $= 1.4$, $Pr = 0.7$, $\sigma_u = \sigma_T = 1$

(2) Unlike macrochannels, the Nusselt number depends on the fluid, as indicated by Pr and γ in (11.64).

(3) The effect of temperature jump on the Nusselt number is represented by the last term in the denominator of (11.64).

(4) The corresponding no-slip Nusselt number for macrochannel flow, Nu_o, is determined by setting $Kn = 0$ in (11.64)

$$Nu_o = \frac{140}{17} = 8.235. \quad (11.65)$$

This is in agreement with the value given in Table 6.2.

(5) Rarefaction and compressibility have the effect of decreasing the Nusselt number. Depending on the Knudsen number, using the no-slip solution, (11.65), can significantly overestimate the Nusselt number.

Example 11.2: Microchannel Heat Exchanger: Uniform Surface Flux

Rectangular microchannels are used to remove heat from a device at uniform surface heat flux. The height, width, and length of each channel are $H = 1.26\,\mu\mathrm{m}$, $W = 90\,\mu\mathrm{m}$, *and* $L = 10$ mm, *respectively. Using air at* $T_i = 20°\mathrm{C}$ *as the coolant fluid, determine the mass flow rate and the variation of Nusselt number along the channel. Assume steady state fully developed conditions. Inlet and outlet pressure are:*

$p_i = 210\,kPa = 210{,}000\,\mathrm{kg/s^2 - m}$

$p_o = 105\,kPa = 105{,}000\,\mathrm{kg/s^2 - m}$

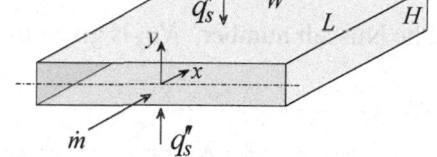

(1) Observations. (i) The problem can be modeled as pressure driven Poiseuille flow between two parallel plates with uniform surface flux. (ii) Assuming fully developed velocity and temperature, the analysis of Section 11.6.3 gives the mass flow rate and Nusselt number. (iii) The Nusselt number depends on the Knudsen number, Kn. Since Kn varies along the channel due to pressure variation, it follows that pressure distribution along the channel must be determined.

(2) Problem Definition. Determine the flow and temperature fields for fully developed Poiseuille flow.

(3) Solution Plan. Apply the results of Section 11.6.3 for the mass flow rate, pressure distribution, and Nusselt number.

(4) Plan Execution.

(i) **Assumptions.** (1) Steady state, (2) laminar flow, (3) two-dimensional (no variation along the width W), (4) slip flow regime ($0.001 < Kn < 0.1$), (5) ideal gas, (6) constant viscosity, conductivity and specific heats, (7) negligible lateral variation of density and pressure, (8) the accommodation coefficients are equal to unity, $\sigma_u = \sigma_T = 1.0$, (9) uniform surface flux, (10) negligible axial conduction, and (11) no gravity.

(ii) **Analysis.** Assuming isothermal flow, the results of Section 11.6.3 give the mass flow rate as

$$m = \frac{1}{24} \frac{W H^3 p_o^2}{\mu L R T_o} \left[\frac{p_i^2}{p_o^2} - 1 + 12 Kn_o \left(\frac{p_i}{p_o} - 1 \right) \right]. \qquad (11.39)$$

The Knudsen number at the exit, Kn_o is

$$Kn_o = \frac{\lambda(p_o)}{H} = \frac{\mu}{H p_o} \sqrt{\frac{\pi}{2} R T_o}, \qquad (11.34)$$

where the temperature T_o at the outlet is assumed to be the same as inlet temperature and the viscosity μ is based on inlet temperature.

The Nusselt number, Nu, is given by

$$Nu = \frac{2}{\dfrac{3}{(1+6Kn)} \left\{ \dfrac{1}{2} Kn + \dfrac{5}{48} - \dfrac{1}{(1+6Kn)} \left[(Kn)^2 + \dfrac{13}{40} Kn + \dfrac{13}{560} \right] \right\} + \dfrac{2\gamma}{\gamma+1} \dfrac{1}{Pr} Kn}$$

$$(11.64)$$

The local Knudsen number, Kn, depends on the local pressure $p(x)$ according to

$$Kn = \frac{\lambda}{H} = \frac{\mu}{H p} \sqrt{\frac{\pi}{2} R T}. \qquad (11.33)$$

Equation (11.35) gives $p(x)$

$$\frac{p(x)}{p_o} = -6 Kn_o + \sqrt{ \left[6 Kn_o + \frac{p_i}{p_o} \right]^2 + \left[(1 - \frac{p_i^2}{p_o^2}) + 12 Kn_o (1 - \frac{p_i}{p_o}) \right] \frac{x}{L} }.$$

$$(11.35)$$

Thus, (11.35) is used to determine $p(x)$, (11.33) to determine $Kn(x)$, and (11.64) to determine the variation of the Nusselt number along the channel.

(iii) Computations. Air properties are determined at 20°C. To compute $p(x)$, $Kn(x)$, and Nu, the following data are used

$H = 1.26 \, \mu$m

$L = 10$ mm

$p_i = 210 \times 10^3 \, \text{kg/s}^2 - \text{m}$

$$p_o = 105 \times 10^3 \text{ kg/s}^2 - \text{m}$$
$$Pr = 0.713$$
$$R = 287 \text{ J/kg} - \text{K} = 287 \text{ m}^2/\text{s}^2 - \text{K}$$
$$T \cong T_i \cong T_o = 20 \text{ °C}$$
$$W = 90 \text{ μm}$$
$$\gamma = 1.4$$
$$\mu = 18.17 \times 10^{-6} \text{ kg/s} - \text{m}$$

Substituting into (11.34)

$$Kn_o = \frac{18.17 \times 10^{-6} (\text{kg/s} - \text{m})}{1.26 \times 10^{-6} (\text{m}) 105 \times 10^3 (\text{kg/s}^2 - \text{m})} \sqrt{\frac{\pi}{2} 287 (\text{m}^2/\text{s}^2 - \text{K})(293.15)(\text{K})} = 0.05$$

Using (11.311) and noting that $p_i / p_o = 2$

$$m = \frac{1}{24} \frac{90 \times 10^{-6} (\text{m})(90 \times 10^{-6})^3 (\text{m}^3)(105 \times 10^3)^2 (\text{kg}^2/\text{s}^4 - \text{m}^2)}{18.17 \times 10^{-6} (\text{kg/s} - \text{m})0.01(\text{m})287(\text{m}^2/\text{s}^2 - \text{K})293.15(\text{K})} \left[(2)^2 - 1 + 12 \times 0.05(2-1)\right]$$

$$m = 19.476 \times 10^{-12} \text{ kg/s}$$

Axial pressure variation is obtain from (11.35)

$$\frac{p(x)}{p_o} = -6 \times 0.05 + \sqrt{(6 \times 0.05 + 2)^2 + \left[1 - (2)^2 + 12 \times 0.05(1-2)\right]\frac{x}{L}} \, ,$$

$$\frac{p(x)}{p_o} = -0.3 + \sqrt{5.29 - 3.6\frac{x}{L}} \, . \tag{a}$$

Equation (a) is used to tabulate pressure variation with x/L. Equations (11.33) and (11.64) are used to compute the corresponding Knudsen and Nusselt numbers.

x/L	p/p_o	Kn	Nu
0	2.000	0.0250	7.333
0.2	1.838	0.0272	7.259
0.4	1.662	0.0301	7.163
0.6	1.469	0.0340	7.035
0.8	1.252	0.0399	6.850
1.0	1.000	0.0500	6.549

(iii) **Checking.** *Dimensional check*: Units for equations (11.33), (11.35), (11.39), and (11.64) are consistent.

Limiting check: No-slip macrochannel Nusselt number is obtained by setting $Kn = 0$ in (11.64). This gives $Nu = 8.235$. This agrees with the value given in Table 6.2.

(5) Comments. (i) To examine the effect of rarefaction and compressibility on the mass flow rate, equation (11.41) is used to calculate m / m_o :

$$\frac{m}{m_o} = \frac{1}{2} \left[\frac{p_i}{p_o} + 1 + 12\, Kn_o \right] = \frac{1}{2} (2 + 1 + 12 \times 0.05) = 1.8 .$$

Thus, neglecting rarefaction and compressibility will underestimate the m by 44%.

(ii) No-slip Nusselt number for fully developed Poiseuille flow between parallel plates with uniform surface heat flux is $Nu = 8.235$. Thus, no-slip theory overestimates the Nusselt number if applied to microchannels.

(iii) It should be noted that the equations used to compute m, $p(x)$, and Nu are based on the assumptions of isothermal conditions in the determination of the flow field. This is a reasonable approximation for typical applications.

11.6.4 Fully Developed Poiseuille Channel Flow: Uniform Surface Temperature [14]

The uniform surface flux of Section 11.6.3 is now repeated with the plates maintained at uniform surface temperature T_s, as shown in Fig. 11.10. We invoke all the assumptions made in solving the uniform flux case.   Since the flow field is assumed independent of temperature, it follows that the solution to the velocity, pressure, and mass flow rate is unaffected by changes in thermal boundary conditions.   Thus, Equation (11.30) for the axial velocity

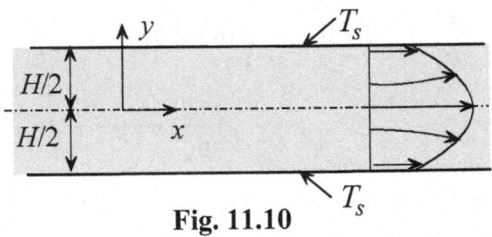

Fig. 11.10

u, (11.35) for pressure variation $p(x) / p_o$, (11.39) for mass flow rate m, and the energy equation, (11.44), are applicable to this case. However, thermal boundary condition at the surface must be changed. Therefore, a new solution to the temperature distribution and Nusselt number must be determined. This change in boundary condition makes it necessary to use a different mathematical approach to obtain a solution. The solution and results detailed in [14] will be followed and summarized here.

Temperature Distribution and Nusselt Number. Using Newton's law of cooling, the Nusselt number for this case is given by

$$Nu = \frac{2Hh}{k} = \frac{-2H}{T_m(x) - T_s} \frac{\partial T(x, H/2)}{\partial y}.$$ (11.66a)

Thus, the problem becomes one of determining the temperature distribution $T(x, y)$ and the mean temperature $T_m(x)$. One approach to this problem is to solve the more general case of Graetz channel entrance problem and specialize it to the fully developed case at $x \to \infty$ [14]. This requires solving a partial differential equation. Although axial conduction was neglected in the uniform heat flux condition, it will be included in this analysis [14]. Thus energy equation (11.44) is modified to include axial conduction

$$\rho c_p u \frac{\partial T}{\partial x} = k \left(\frac{\partial^2 T}{\partial x^2} + \frac{\partial^2 T}{\partial y^2} \right).$$ (11.67a)

The boundary and inlet conditions are

$$\frac{\partial T(x,0)}{\partial y} = 0,$$ (11.68a)

$$T(x, H/2) = T_s - \frac{2\gamma}{\gamma + 1} \frac{H}{Pr} Kn \frac{\partial T(x, H/2)}{\partial y},$$ (11.69a)

$$T(0, y) = T_i,$$ (11.70a)

$$T(\infty, y) = T_s.$$ (11.71a)

The normalized axial velocity is given by (11.56)

$$\frac{u}{u_m} = \frac{6}{1 + 6Kn} \left[\frac{1}{4} + Kn - \frac{y^2}{H^2} \right].$$ (11.56)

Equations (11.66a)-(11.71a) are expressed in dimensionless form using the following dimensionless variables

$$\theta = \frac{T - T_s}{T_i - T_s}, \quad \xi = \frac{x}{H\,RePr}, \quad \eta = \frac{y}{H}, \quad Re = \frac{2\rho u_m H}{\mu}, \quad Pe = RePr.$$ (11.72)

Using (11.56) and (11.72), equations (11.66a)-(11.71a) are transformed to

$$Nu = -\frac{2}{\theta_m}\frac{\partial\theta(\xi,\eta/2)}{\partial\eta},\qquad(11.66)$$

$$\frac{6}{1+6Kn}\left(\frac{1}{4}+Kn-\eta^2\right)\frac{\partial\theta}{\partial\xi} = \frac{1}{(Pe)^2}\frac{\partial^2\theta}{\partial\xi^2}+\frac{\partial^2\theta}{\partial\eta^2},\qquad(11.67)$$

$$\frac{\partial\theta(\xi,0)}{\partial\eta} = 0,\qquad.(11.68)$$

$$\theta(\xi,1/2) = -\frac{2\gamma}{\gamma+1}\frac{1}{Pr}Kn\frac{\partial\theta(\xi,1/2)}{\partial\eta},\qquad(11.69)$$

$$\theta(0,\eta) = 1,\qquad(11.70)$$

$$\theta(\infty,\eta) = 0.\qquad(11.71)$$

In Section 6.6.2, the criterion for neglecting axial conduction is given as

$$Pe = PrRe \geq 100,$$

where Pe is the Peclet number. Because the Reynolds number is usually small in microchannels, the Peclet number may not be large enough to justify neglecting axial conduction. By including the axial conduction term in (11.67), the effect of Peclet number on the Nusselt number can be evaluated.

This problem was solved using the method of separation of variables [14]. The solution is specialized to the fully developed temperature case far away from the inlet. The variation of Nusselt

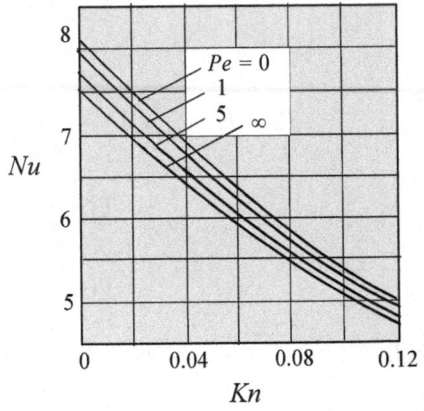

Fig. 11.11 Nusselt number for flow between parallel plates at uniform surface temperature for air, $Pr = 0.7$, $\gamma = 1.4$, $\sigma_u = \sigma_T = 1$, [14]

number with Knudsen number for air at various values of the Peclet number is shown in Fig. 11.11. Examination of Fig. 11.11 leads to the following conclusions:

(1) The Nusselt number decreases as the Knudsen number is increased. Thus using no-slip results to determine microchannel Nusselt number can significantly overestimate its value.

(2) Axial conduction increases the Nusselt number. However, the increase diminishes as the Knudsen number increases.

(3) The limiting case of no-slip ($Kn = 0$) and negligible axial conduction ($Pe = \infty$) gives

$$Nu_o = 7.5407 . \tag{11.73}$$

This is in agreement with the value given in Table 6.2.

(4) If axial conduction is taken into consideration at $Kn = 0$, the Nusselt number increases to $Nu_o = 8.1174$. Thus the maximum error in neglecting axial conduction is 7.1%.

(5) With the Nusselt number known, the heat transfer rate, q_s, is determined following the analysis of Section 6.5

$$q_s = m c_p [T_m (x) - T_{mi}] , \tag{6.14}$$

where the local mean temperature is given by

$$T_m (x) = T_s + (T_{mi} - T_s) \exp[-\frac{P\overline{h}}{m c_p} x] . \tag{6.13}$$

The average heat transfer coefficient, $\overline{h}$, is determine numerically using (6.12)

$$\overline{h} = \frac{1}{x} \int_0^x h(x)dx . \tag{6.12}$$

Example 11.3: Microchannel Heat Exchanger: Uniform Surface Temperature

Repeat Example 11.2 with the channel surface maintained at uniform temperature T_s.

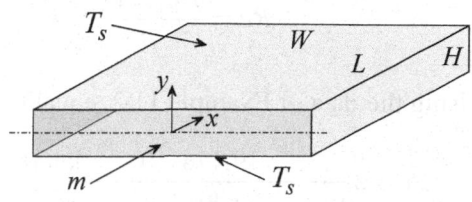

(1) Observations. (i) Since the flow field is assumed independent of temperature, it follows that the velocity, mass flow rate and pressure distribution of Example 11.2 are applicable to this case. (ii) The variation of the Nusselt number with Knudsen number for air is shown in Fig. 11.11. (iii) The determination of Knudsen number as a function of distance along the channel and Fig. 11.11 establish the variation of Nusselt number. (iv) The use of Fig. 11.11 requires the determination of the Peclet number.

(2) **Problem Definition.** Determine the Nusselt number corresponding to each value of Knudsen number of Example 11.2.

(3) **Solution Plan.** Use the tabulated data of Knudsen number and pressure in Example 11.2, compute the Peclet number, and use Fig. 11.11 to determine the Nusselt number variation along the channel.

(4) **Plan Execution.**

(i) **Assumptions.** (1) Steady state, (2) laminar flow, (3) two-dimensional (no variation along the width W), (4) slip flow regime (0.001 < Kn < 0.1), (5) ideal gas, (6) constant viscosity, conductivity and specific heats, (7) negligible lateral variation of density and pressure, (8) the accommodation coefficients are assumed equal to unity, $\sigma_u = \sigma_T = 1.0$, (9) negligible dissipation, (10) uniform surface temperature, and (11) negligible gravity.

(ii) **Analysis and Computations.** Since the velocity and pressure distribution of Example 11.2 are based on the assumption of isothermal conditions, the variation of Knudsen number and pressure with axial distance x/L for uniform surface flux is the same as that for uniform surface temperature. Thus, the tabulated results of Example 11.2 will be used with Fig. 11.11 to determine the variation of Nusselt number with axial distance. The Peclet number is defined as

$$Pe = Re\,Pr\,, \tag{a}$$

where the Reynolds number is given by

$$Re = \frac{\rho u_m 2H}{\mu} = \frac{(m/HW)2H}{\mu} = \frac{2m}{\mu W}\,. \tag{b}$$

Using the data of Example 11.2, equation (b) gives

$$Re = 2\frac{19.476 \times 10^{-12}\,(\text{kg/s})}{18.17 \times 10^{-6}\,(\text{kg/s - m})\,90 \times 10^{-6}\,(\text{m})} = 0.0238$$

Using (a)

$$Pe = 0.02382 \times 0.713 = 0.01698$$

Examination of Fig. 11.11 shows a small change in the Nusselt number as Pe is increased from zero to unity. Thus, we assume that the curve corresponding to $Pe = 0$ applies to this example. The variation of pressure, Knudsen number and Nusselt number with axial distance x/L is tabulated. Also tabulated is the Nusselt number corresponding to negligible axial conduction ($Pe = \infty$).

x/L	p/p_o	Kn	with conduction Nu	no conduction Nu
0	2.000	0.0250	7.38	6.83
0.2	1.838	0.0272	7.31	6.77
0.4	1.662	0.0301	7.22	6.70
0.6	1.469	0.0340	7.12	6.60
0.8	1.252	0.0399	6.91	6.41
1.0	1.000	0.0500	6.65	6.18

(5) Comments. (i) Taking into consideration axial conduction, the no-slip Nusselt number for fully developed Poiseuille flow between parallel plates with uniform surface temperature is $Nu = 8.1174$. The tabulated values of Nusselt numbers are lower due to rarefaction and compressibility. Thus, no-slip theory overestimates the Nusselt number.

(ii) Neglecting axial conduction underestimates the Nusselt number by less than 10%.

(iii) The effect of axial conduction is to shift the values of the Nusselt number for constant surface temperature closer to those for constant surface flux of Example 11.2.

11.6.5 Fully Developed Poiseuille Flow in Microtubes: Uniform Surface Flux [20]

We consider now Poiseuille flow in microtubes at uniform surface heat flux. This problem is identical to Poiseuille flow between parallel plates at uniform flux presented in Section 11.6.3. Fig. 11.12 shows a tube of radius r_o with surface heat flux q_s''. Velocity and temperature are assumed fully developed. Inlet and outlet pressures are p_i and p_o, respectively. We wish to determine the following:

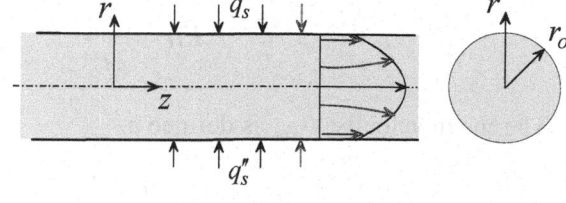

Fig. 11.12

(1) Velocity distribution
(2) Nusselt number

As with pressure driven flow between parallel plates, rarefaction and compressibility alter the familiar flow and heat transfer characteristics of macro tubes. Slip velocity and temperature jump result in axial velocity variation, lateral velocity component, non-parallel streamlines, and non-linear pressure gradient.

Assumptions. We apply the assumptions made in the analysis of Poiseuille flow between parallel plates (see Section 11.6.3).

Flow Field.   Following the analysis of Section 11.6.3, the axial component of the Navier-Stokes equations for constant viscosity, compressible flow in cylindrical coordinates simplifies to

$$\frac{1}{r}\frac{\partial}{\partial r}(r\frac{\partial v_z}{\partial r}) = \frac{1}{\mu}\frac{\partial p}{\partial z}, \tag{a}$$

where $v_z(r,z)$ is the axial velocity. Assuming symmetry and setting $\sigma_u = 1$, the two boundary conditions on v_z are

$$\frac{\partial v_z(0,z)}{\partial r} = 0, \tag{b}$$

$$v_z(r_o,z) = -\lambda\frac{\partial v_z(r_o,z)}{\partial r}. \tag{c}$$

Integration of (a) and application of boundary conditions (b) and (c) give (Problem 11.13):

$$v_z = -\frac{r_o^2}{4\mu}\frac{dp}{dz}\left[1+4Kn-\frac{r^2}{r_o^2}\right]. \tag{11.74}$$

The Knudsen number for tube flow, *Kn*, is defined as

$$Kn = \frac{\lambda}{2r_o}. \tag{11.75}$$

The mean velocity v_{zm} is defined as

$$v_{zm} = \frac{1}{\pi r_o^2}\int_0^{r_o} 2\pi r v_z\, dr .$$

Substituting (11.74) into the above, gives

$$v_{zm} = -\frac{r_o^2}{8\mu}\frac{dp}{dz}(1+8Kn).$$ (11.76)

Combining (11.74) and (11.76)

$$\frac{v_z}{v_{zm}} = 2\frac{1+4Kn-(r/r_o)^2}{1+8Kn}.$$ (11.77)

Note that the Knudsen number varies with pressure along the channel. It represents the effect of rarefaction on the axial velocity. Following the derivation of Section 11.6.3 for the analogous problem of Poiseuille flow between parallel plates, axial pressures distribution is given by (Problem 11.14):

$$\frac{p(z)}{p_o} = -8Kn_o + \sqrt{\left[8Kn_o + \frac{p_i}{p_o}\right]^2 + \left[(1-\frac{p_i^2}{p_o^2})+16Kn_o(1-\frac{p_i}{p_o})\right]\frac{z}{L}}.$$ (11.78)

Using (11.76) and (11.78), and assuming $T \cong T_o$, gives the mass flow rate (Problem 11.15):

$$m = \frac{\pi}{16}\frac{r_o^4 p_o^2}{\mu LRT_o}\left[\frac{p_i^2}{p_o^2}-1+16 Kn_o(\frac{p_i}{p_o}-1)\right].$$ (11.79a)

The corresponding mass flow rate for incompressible no-slip (macroscopic) flow is given by

$$m_o = \frac{\pi}{8}\frac{r_o^4 p_o^2}{\mu LRT}(\frac{p_i}{p_o}-1).$$ (11.79b)

Note that since (11.79a) accounts for rarefaction and compressibility, setting $Kn_o = 0$ in (11.79a) does not reduce to the incompressible no-slip case of (11.79b).

Nusselt Number. Following the analysis of Section 11.6.3, the Nusselt number is defined as

$$Nu = \frac{2r_o h}{k}.$$ (d)

The heat transfer coefficient h for uniform surface flux q_s''' is

$$h = \frac{q_s''}{T_s - T_m}.$$

Substituting into (d)

$$Nu = \frac{2r_o q_s''}{k(T_s - T_m)}, \tag{e}$$

where T_s is tube surface temperature determined from temperature jump condition (11.11)

$$T_s = T(r_o, z) + \frac{2\gamma}{1+\gamma} \frac{\lambda}{Pr} \frac{\partial T(r_o, z)}{\partial r}. \tag{f}$$

The mean temperature T_m for tube flow is given by

$$T_m = \frac{\int_0^{r_o} v_z T r \, dr}{\int_0^{r_o} v_z r \, dr}. \tag{11.80}$$

Thus, the solution to the temperature distribution is needed for the determination of the Nusselt number. Based on the assumptions made, energy equation (2.24) simplifies to

$$\rho c_p v_z \frac{\partial T}{\partial z} = \frac{k}{r} \frac{\partial}{\partial r} \left(r \frac{\partial T}{\partial r} \right). \tag{11.81}$$

The boundary conditions are

$$\frac{\partial T(0, z)}{\partial r} = 0, \tag{g}$$

$$k \frac{\partial T(r_o, z)}{\partial r} = q_s''. \tag{h}$$

We introduce the definition of fully developed temperature profile

$$\phi = \frac{T(r_o, z) - T(r, z)}{T(r_o, z) - T_m(z)}. \tag{11.82}$$

For fully developed temperature, ϕ is assumed independent of z. That is

$$\phi = \phi(r). \tag{11.83}$$

Thus

$$\frac{\partial \phi}{\partial z} = 0. \tag{11.84}$$

Equations (11.82) and (11.84) give

$$\frac{\partial \phi}{\partial z} = \frac{\partial}{\partial z}\left[\frac{T(r_o,z) - T(r,z)}{T(r_o,z) - T_m(z)}\right] = 0.$$

Expanding and using the definition of ϕ in (11.82)

$$\frac{dT(r_o,z)}{dz} - \frac{\partial T}{\partial z} - \phi(r)\left[\frac{dT(r_o,z)}{dz} - \frac{dT_m(z)}{dz}\right] = 0. \tag{11.85}$$

The relationship between the three temperature gradients $\partial T(r,z)/\partial z$, $dT(r_o,z)/dz$, and $dT_m(z)dz$ will be determined. The heat transfer coefficient h, is given by

$$h = \frac{-k\dfrac{\partial T(r_o,z)}{\partial r}}{T_m(z) - T_s(z)}, \tag{i}$$

where $T_s(z)$ is given in (11.11). Temperature gradient in (i) is obtained from (11.81)

$$T(r,z) = T(r_o,z) - [T(r_o,z) - T_m(z)]\phi.$$

Differentiating the above and evaluating the derivative at $r = r_o$

$$\frac{\partial T(r_o,z)}{\partial r} = -[T(r_o,z) - T_m(z)]\frac{d\phi(r_o)}{dr}. \tag{j}$$

Substituting (j) into (i)

$$h = -\frac{k[T(r_o,z) - T_m(z)]}{T_s(z) - T_m(z)}\frac{d\phi(r_o)}{dr}. \tag{k}$$

Newton's law of cooling gives another equation for h

$$h = \frac{q_s''}{T_s(z) - T_m(z)} .$$

Equating the above with (k) and rearranging

$$T(r_o, z) - T_m(z) = -\frac{q_s''}{k \dfrac{d\phi(r_o)}{dr}} = \text{constant.} \qquad (11.86)$$

Differentiating

$$\frac{\partial T(r_o, z)}{\partial z} - \frac{\partial T_m(z)}{\partial z} = 0 .$$

Combining this with (11.85), gives

$$\frac{dT(r_o, z)}{dz} = \frac{dT_m(z)}{dz} = \frac{\partial T}{\partial z} . \qquad (11.87)$$

Equation (11.87) will be used to replace $\partial T / \partial z$ in partial differential equation (11.81) with dT_m / dz. Applying conservation of energy to the element dx in Fig. 11.13 gives

$$2\pi r_o q_s'' dz + mc_p T_m = mc_p \left[T_m + \frac{dT_m}{dz} dz \right] .$$

Simplifying

$$\frac{dT_m}{dz} = \frac{2\pi r_o q_s''}{mc_p} . \qquad (1)$$

However,

$$m = \rho \pi r_o^2 v_{zm} , \qquad (m)$$

where v_{zm} is the mean axial velocity.

Substituting (m) into (1)

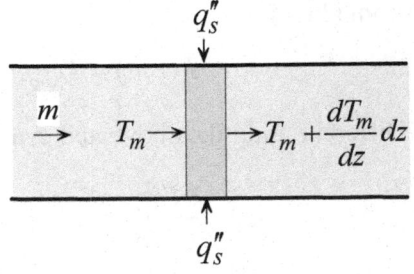

Fig. 11.13

$$\frac{dT_m}{dz} = \frac{2q_s''}{\rho c_p r_o v_{zm}}.$$ (11.88)

Substituting (11.88) into (11.87)

$$\frac{dT(r_o, z)}{dz} = \frac{dT_m(z)}{dz} = \frac{\partial T}{\partial z} = \frac{2q_s''}{\rho c_p r_o v_{zm}}.$$ (11.89)

Equation (11.89) shows that $T(r, z)$, $T_m(z)$ and $T_s(z)$ vary linearly with axial distance z. Substituting (11.89) into (11.81)

$$\frac{\partial}{\partial r}\left(r \frac{\partial T}{\partial r}\right) = \frac{2q_s''}{kr_o} \frac{v_z}{v_{zm}} r.$$ (11.90)

Equation (11.77) is used to eliminate v_z / v_{zm} in the above

$$\frac{\partial}{\partial r}\left(r \frac{\partial T}{\partial r}\right) = \frac{4}{1 + 8Kn} \frac{q_s''}{kr_o}\left[1 + 4Kn - \frac{r^2}{r_o^2}\right] r.$$ (11.91)

Integrating twice

$$T(r, z) = \frac{q_s''}{(1 + 8Kn)kr_o}\left[(1 + 4Kn)r^2 - \frac{1}{4}\frac{r^4}{r_o^2}\right] + f(z)y + g(z).$$ (n)

Boundary condition (g) gives

$$f(z) = 0.$$

Solution (n) becomes

$$T(r, z) = \frac{q_s''}{(1 + 8Kn)kr_o}\left[(1 + 4Kn)r^2 - \frac{1}{4}\frac{r^4}{r_o^2}\right] + g(z).$$ (11.92)

Boundary condition (h) is automatically satisfied. To determine $g(z)$ the mean temperature T_m is evaluated using two methods. In the first method, (11.88) is integrated between the inlet of the tube, $z = 0$, and an arbitrary location z

$$\int_{T_{mi}}^{T_m} dT_m = \frac{2q_s''}{\rho c_p v_{zm} r_o} \int_0^z dz,$$

where

$$T_m(0) = T_{mi}. \tag{11.93}$$

Evaluating the integrals

$$T_m = \frac{2q_s''}{\rho c_p v_{zm} r_o} z + T_{mi}. \tag{11.94}$$

In the second method, T_m is evaluated using its definition in (11.80). Substituting (11.74) and (11.92) into (11.80) and simplifying

$$T_m = \frac{\int_0^{r_o} \left[1 + 4Kn - \frac{r^2}{r_o^2}\right]\left\{\frac{4q_s''}{(1+8Kn)kr_o}\left[\left(\frac{1}{4} + Kn\right)r^2 - \frac{r^4}{16r_o^2}\right] + g(z)\right\} r \, dr}{\int_0^{r_o} \left[1 + 4Kn - \frac{r^2}{r_o^2}\right] r \, dr}.$$

Evaluating the integrals, gives

$$T_m = \frac{q_s'' r_o}{k(1+8Kn)^2}\left[16Kn^2 + \frac{14}{3}Kn + \frac{7}{24}\right] + g(z). \tag{11.95}$$

Equating (11.94) and (11.95), gives $g(z)$

$$g(z) = T_{mi} + \frac{2q_s''}{\rho c_p r_o v_{zm}} z - \frac{q_s'' r_o}{k(1+8Kn)^2}\left[16Kn^2 + \frac{14}{3}Kn + \frac{7}{24}\right]. \tag{11.96}$$

Surface temperature $T_s(r_o, z)$ is determined using (f) and (11.92)

$$T_s(r_o, z) = \frac{4q_s'' r_o}{k(1+8Kn)}\left[Kn + \frac{3}{16}\right] + \frac{4\gamma}{\gamma+1}\frac{q_s'' r_o}{kPr}Kn + g(z). \tag{11.97}$$

Finally, the Nusselt number is determined by substituting (11.95) and (11.97) into (e)

$$Nu = \cfrac{2}{\cfrac{4}{(1+8Kn)}\left(Kn+\cfrac{3}{16}\right)-\cfrac{1}{(1+8Kn)^2}\left[16Kn^2+\cfrac{14}{3}Kn+\cfrac{7}{24}\right]+\cfrac{4\gamma}{\gamma+1}\cfrac{1}{Pr}Kn}.$$

$$(11.98)$$

Using (11.98), the Nusselt number variation with Knudsen number for air, with $\gamma = 1.4$ and $Pr = 0.7$, is plotted in Fig. 11.14. The effect of rarefaction and compressibility is to decrease the Nusselt number.

As with Poiseuille flow between parallel plates, the Nusselt number for fully developed flow depends on the fluid and varies with distance along the channel. The variation of Nu with respect to z in (11.98) is implicit in terms of the Knudsen number, which is a function of pressure. The variation of pressure with axial distance can be determined following the procedure of Section 11.6.3.

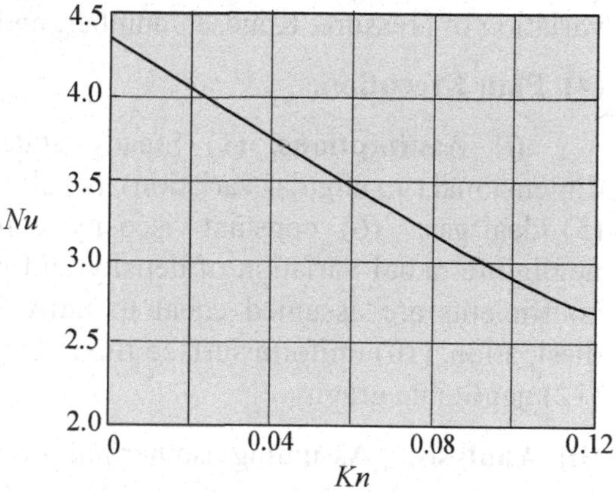

Fig. 11.14 Nusselt number for air flow through tubes at uniform surface heat flux for air, $= 1.4, Pr = 0.7, \sigma_u = \sigma_T = 1, [17]$

The corresponding no-slip Nusselt number, Nu_o, is obtained by setting $Kn = 0$ in (11.98)

$$Nu_o = \frac{48}{11} = 4.364. \qquad (11.99)$$

This agrees with equation (6.55) of the macro tube analysis of Chapter 6.

Example 11.4: Microtube Heat Exchanger: Uniform Surface Flux

Microtubes of radius $r_o = 0.786\,\mu m$ and length $L = 10\,mm$ are used to heat air at $T_i = 20°C$. Assume uniform surface temperature and fully developed conditions; determine the axial variation of Nusselt number. Inlet and outlet pressures are:

$p_i = 315\,kPa = 315{,}000\,kg/s^2 - m$

$p_o = 105\,kPa = 105{,}000\,kg/s^2 - m$

(1) Observations. (i) This is a pressure driven Poiseuille flow problem through a tube. (ii) Axial Nusselt number variation is given in equation (11.98) in terms of the local Knudsen number, Kn. Local Knudsen number depends on local pressure. It follows that pressure distribution along a tube must be determined. (iii) Pressure distribution is given by equation (11.78).

(2) Problem Definition. Determine the axial pressure distribution in a microtube for fully developed Poiseuille flow.

(3) Solution Plan. Use the results of Section 11.6.5 to compute axial variation of pressure, Knudsen number, and Nusselt number.

(4) Plan Execution.

    (i) Assumptions. (1) Steady state, (2) laminar flow, (3) two-dimensional (no angular variation), (4) slip flow regime $(0.001 < Kn < 0.1)$, (5) ideal gas, (6) constant viscosity, conductivity and specific heats, (7) negligible radial variation of density and pressure, (8) the accommodation coefficients are assumed equal to unity, $\sigma_u = \sigma_T = 1.0$, (9) negligible dissipation, (10) uniform surface flux, (11) negligible axial conduction, and (12) negligible gravity.

    (ii) Analysis. Assuming isothermal flow, The Nusselt number, Nu, is given by

$$Nu = \frac{2}{\dfrac{4}{(1+8Kn)}\left(Kn+\dfrac{3}{16}\right) - \dfrac{1}{(1+8Kn)^2}\left[16Kn^2 + \dfrac{14}{3}Kn + \dfrac{7}{24}\right] + \dfrac{4\gamma}{\gamma+1}\dfrac{1}{Pr}Kn} .$$

(11.98)

The local Knudsen number, Kn, depends on the local pressure $p(z)$ according to

$$Kn = \frac{\lambda}{2r_o} = \frac{\mu}{2r_o p}\sqrt{\frac{\pi}{2}RT} ,$$

(a)

$$\frac{p(z)}{p_o} = -8Kn_o + \sqrt{\left[8Kn_o + \frac{p_i}{p_o}\right]^2 + \left[(1-\frac{p_i^2}{p_o^2}) + 16Kn_o(1-\frac{p_i}{p_o})\right]\frac{z}{L}} .$$

(11.78)

    (iii) Computations. Equation (11.78) is used to determine the axial variation of pressure. Equation (a) gives the corresponding Knudsen numbers. The Nusselt number is determined using (11.98). Air properties are evaluated at $20°C$. Computations are based on the following data;

$$r_o = 0.786 \, \mu m$$
$$L = 10 \text{ mm}$$
$$p_i = 315 \times 10^3 \text{ kg/s}^2 - \text{m}$$
$$p_o = 105 \times 10^3 \text{ kg/s}^2 - \text{m}$$
$$Pr = 0.713$$
$$R = 287 \text{ J/kg} - \text{K} = 287 \text{ m}^2/\text{s}^2 - \text{K}$$
$$T \cong T_i \cong T_o = 20 \, °C$$
$$\gamma = 1.4$$
$$\mu = 18.17 \times 10^{-6} \text{ kg/s} - \text{m}$$

Substituting into (a)

$$Kn_o = \frac{18.17 \times 10^{-6} \, (\text{kg/s} - \text{m})}{2 \times 0.786 \times 10^{-6} (\text{m}) 105 \times 10^3 \, (\text{kg/s}^2 - \text{m})} \sqrt{\frac{\pi}{2} 287 (\text{m}^2/\text{s}^2 - \text{K})(293.15)(\text{K})}$$
$$= 0.04$$

Axial pressure variation is obtain from (11.78)

$$\frac{p(z)}{p_o} = -8 \times 0.04 + \sqrt{(8 \times 0.04 + 3)^2 + \left[1 - (3)^2 + 16 \times 0.05(1 - 3)\right]\frac{z}{L}}$$

This gives

$$\frac{p(x)}{p_o} = -0.32 + \sqrt{11.0224 - 9.28\frac{z}{L}} \, . \tag{b}$$

Equations (a), (b), and (11.98) are used to tabulate pressure, Knudsen number and Nusselt number at various values of z/L.

z/L	p/p_o	Kn	Nu
0	3.000	0.0133	4.182
0.2	2.708	0.0148	4.161
0.4	2.384	0.0168	4.130
0.6	2.016	0.0199	4.083
0.8	1.577	0.0254	3.997
1.0	1.000	0.0400	3.766

(iii) **Checking.** *Dimensional check*: Computations confirmed that pressure ratio, Knudsen number and Nusselt number are dimensionless.

Limiting check: No slip macrochannel Nusselt number is obtained by setting $Kn = 0$ in (11.98), giving $Nu = 4.364$. This is close to 4.182 at $z/L = 0$. The difference is due to compressibility effect.

(5) Comments. No-slip Nusselt number for fully developed Poiseuille flow through tubes at uniform surface heat flux is $Nu = 4.364$. The tabulated values for this example show that no-slip theory overestimates the Nusselt number if applied to microchannels.

11.6.6 Fully Developed Poiseuille Flow in Microtubes: Uniform Surface Temperature [14]

The uniform surface flux of Section 11.6.5 is repeated with the tube maintained at uniform surface temperature T_s, as shown in Fig. 11.15. The assumptions made in the solution of the uniform flux condition are applied to this case. The flow field

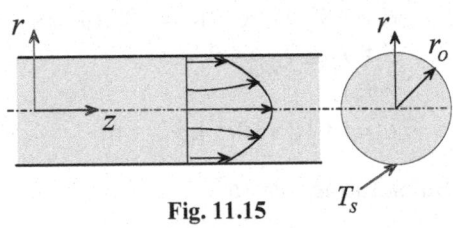

Fig. 11.15

solution is identical for the two cases since it is based on isothermal flow condition. Changing the boundary condition from uniform flux to uniform temperature requires using a different mathematical approach to obtain a solution. Results obtained in [14] are summarized here.

Temperature Distribution and Nusselt Number. The Nusselt number is given by

$$Nu = \frac{2r_o h}{k} = \frac{-2r_o}{T_m(z) - T_s} \frac{\partial T(r_o, z)}{\partial r}. \tag{11.100a}$$

This requires the determination of the temperature distribution $T(r,z)$ and the mean temperature $T_m(z)$. Following the analysis of Section 11.6.4, the solution is based on the limiting case of Graetz tube entrance problem. This approach requires solving a partial differential equation taking into consideration axial conduction. Energy equation (11.81) is modified to include axial conduction

$$\rho c_p v_z \frac{\partial T}{\partial z} = \frac{k}{r} \frac{\partial}{\partial r}\left(r \frac{\partial T}{\partial r}\right) + k\left(\frac{\partial^2 T}{\partial z^2}\right). \tag{11.101a}$$

The boundary and inlet conditions are

$$\frac{\partial T(0,z)}{\partial r} = 0, \tag{11.102a}$$

$$T(r_o,z) = T_s - \frac{2\gamma}{\gamma+1}\frac{2r_o}{Pr}Kn\frac{\partial T(r_o,z)}{\partial r}, \tag{11.103a}$$

$$T(r,0) = T_i, \tag{11.104a}$$

$$T(r,\infty) = T_s. \tag{11.105a}$$

The normalized axial velocity is given by (11.76)

$$\frac{v_z}{v_{zm}} = 2\frac{1+4Kn-(r/r_o)^2}{1+8Kn}. \tag{11.77}$$

Equations (11.100a)-(11.104a) are expressed in dimensionless form using the following dimensionless variables

$$\theta = \frac{T-T_s}{T_i-T_s}, \quad \xi = \frac{z}{2r_o RePr}, \quad R = \frac{r}{r_o}, \quad Re = \frac{2\rho u_m r_o}{\mu}, \quad Pe = RePr.$$
$$\tag{11.106}$$

Using (11.77) and (11.106), equations (11.100a)-(11.105a) are transformed to

$$Nu = -\frac{2}{\theta_m}\frac{\partial\theta(1,\xi)}{\partial R}, \tag{11.100}$$

$$\frac{1+4Kn-R^2}{2(2+16Kn)}\frac{\partial\theta}{\partial\xi} = \frac{1}{R}\frac{\partial}{\partial R}\left(R\frac{\partial\theta}{\partial R}\right) + \frac{1}{(2Pe)^2}\frac{\partial^2\theta}{\partial\xi^2}, \tag{11.101}$$

$$\frac{\partial\theta(0.\xi)}{\partial R} = 0, \tag{11.102}$$

$$\theta(1,\xi) = -\frac{2\gamma}{\gamma+1}\frac{Kn}{Pr}\frac{\partial\theta(1,\xi)}{\partial R}, \tag{11.103}$$

$$\theta(R,0) = 1, \tag{11.104}$$

$$\theta(R,\infty) = 0. \tag{11.105}$$

Equation (11.100) shows that axial conduction becomes important at low Peclet numbers. This problem was solved using the method of separation of variables [14]. The infinite series solution is truncated to determine the Nusselt number for the fully developed case. Fig. 11.16 shows the effect of Peclet and Knudsen numbers on the Nusselt number. Neglecting axial conduction corresponds to $Pe = \infty$. According to Fig. 11.16, axial conduction increases the Nusselt number while rarefaction decreases it. Although axial conduction

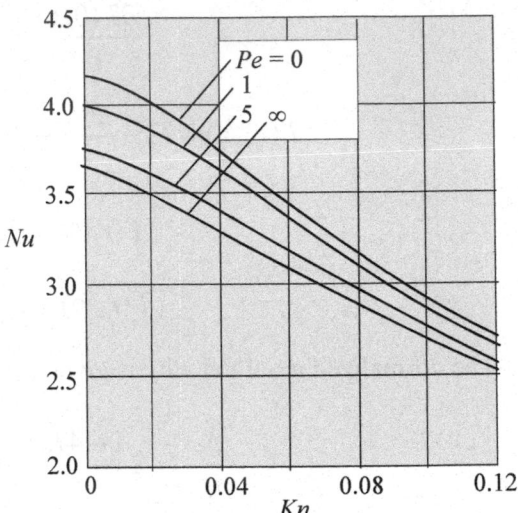

Fig. 11.16 Nusselt number for flow through tubes at uniform surface temperature for air, $Pr = 0.7$, $\gamma = 1.4$, $\sigma_u = \sigma_T = 1$, [14]

increases the Nusselt number, its effect diminishes as the Knudsen number increases. The maximum increase corresponds to the no-slip condition of $Kn = 0$. Fig. 11.16 gives the limiting case of no-slip and negligible axial conduction $(Pe = \infty)$ as

$$Nu_o = 3.657 . \qquad (11.107)$$

This agrees with equation (6.511) obtained in Chapter 6. If axial conduction is taken into consideration at $Kn = 0$, the Nusselt number increases to $Nu_o = 4.175$. Thus the maximum error in neglecting axial conduction is 12.4%.

REFERENCES

[1]   Gad-el-Hak, M., Flow Physics, in *The MEMS Handbook*, M. Gad-el-Hak, ed., CRC Press, 2005.

[2]   Kennard, E.H., *Kinetic Theory of Gases*, McGraw-Hill, New York, 1938.

[3]   Tuckerman, D.B. and Pease, R.F.W., "High Performance Heat Sinking for VLSI," IEEE Electron Dev. Let, EDL-2, 1981, pp.126-129.

[4]   Gad-el-Hak, "Differences between Liquid and Gas Flows at the Microscale," Paper No. ICMM2005-75117, Proceedings of ICMM 2005, Third International Conference on Microchannels and Minichannels, June 13-15, 2005, Toronto, Canada.

[5]   Papautsky, I, Ameel, T. and Frazier, A.B., "A Review of Laminar Single-Phase Flow in microchannels," Proceedings of the 2001 ASME International Mechanical Engineering Congress and Exposition, November 11-16, 2001, New York, NY.

[6]   Steinke, M.E. and Kandlikar, S.G., "Single-Phase Liquid Friction Factors in Microchannels," Paper No. ICMM2005-75112, Proceedings of ICMM 2005, Third International Conference on Microchannels and Minichannels, June 13-15, 2005, Toronto, Canada.

[7]   Morini, G.L., "Single-Phase Convective Heat Transfer in Microchannels: A Review of Experimental Results," International Journal of Thermal Sciences, Vol. 43, 2004 pp. 631-651.

[8]   Guo, Z.Y. and Li, Z. X., "Size Effect on Single-Phase Channel Flow and Heat Transfer at Microscale," International Journal of Heat and Fluid Flow, Vol. 24, 2003, pp. 284-298.

[11]   Steinke, M.E. and Kandlikar, S.G., "Single-Phase Liquid Heat Transfer in Microchannels," Paper No. ICMM2005-75114, Proceedings of ICMM 2005, Third International Conference on Microchannels and Minichannels, June 13-15, 2005, Toronto, Canada.

[10]   Karniadakis, G., Beskok, A. and Aluru, N., Microflows and Nano-flows, Fundamentals and Simulation, Springer, 2005.

[11]   Guo, Z.Y. and Wu, X.B., "Further Study on Compressibility Effect on the Gas Flow and Heat Transfer in a Microtube," Microscale Thermophysical Engineering, Vol. 2, 1998, pp. 111-120.

[12]   Kavehpour, H.P., Faghri, M. and Asako, Y., "Effects of Compressibility and Rarefaction on Gaseous Flows in Microchannels," Numerical Heat Transfer, Part A, Vol. 32, 1997, pp. 677-696.

[13]   Guo, Z.Y. and Wu, X.B., "Compressibility Effect on the Gas Flow and Heat Transfer in a Microtube," International Journal of Heat and Mass Transfer, Vol. 40, 1997, pp. 3251-3254.

[14] Hadjiconstantinou, N.G. and Simek, O., "Constant-Wall-Temperature Nusselt Number in Micro and Nano-Channels," Journal of Heat Transfer, Vol. 124, 2002, pp. 356-364.

[15]   Zohar, Y., *Heat Convection in Micro Ducts*, Kluwer Academic Publishers, Norwell, Massachusetts, 2003.

[16] Hadjiconstantinou, N.G., "Validation of a Second-Order Slip Model for Transition-Range Gaseous Flows," Paper No. ICMM2004-2344, Second International Conference on Microchannels and Minichannels, June 17-19, 2004, Rochester, NY.

[17]   Arkilic, E.B., Schmidt, M.A., and Breuer, K.S., "Gaseous Slip Flow in Microchannels," in *Rarefied Gas Dynamics*: Proceedings of the 111th International Symposium, eds. Harvey, J. and Lord, G. pp. 347-353, Oxford, 1995.

[18]   Hadjiconstantinou, N.G., "Convective Heat Transfer in Micro and Nano Channels: Nusselt Number beyond Slip Flow," HTD-Vol. 366-2, Proceedings of the ASME Heat Transfer Division-2000, pp. 13-23, ASME 2000.

[19]   Arkilic, E.B., Schmidt, M.A., and Breuer, K.S., "Gaseous Slip Flow in Long Microchannels," Journal of Microelecromechanical Systems, Vol. 6, 1997, pp. 167-178.

[20] Sparrow, E.M. and Lin, S.H., "Laminar Heat Transfer in Tubers Under Slip-Flow Conditions, ASME Journal of Heat Transfer, Vol. 84, 1962, pp.363-369.

PROBLEMS

11.1 The speed of sound, c, in an ideal gas is given by

$$c = \sqrt{\gamma RT} \, ,$$

where γ is the specific heat ratio, R is gas constant and T is temperature. Show that

$$Kn = \sqrt{\frac{\pi}{2} \gamma} \, \frac{M}{Re} \, ,$$

where M is mach number defined as

$$M = V / c \, .$$

11.2   Reported discrepancies in experimental data on the fiction factor f are partially attributed to errors in measurements. One of the key quantities needed to calculate f is channel diameter D. Show that

$$f \propto D^5 .$$

11.3   Consider shear driven Couette flow between parallel plates separated by a distance H. The lower plate is stationary while the upper plate moves with a velocity u_s. Assume that no heat is conducted through the lower plate and that the upper plate is maintained at uniform temperature T_s. Taking into consideration dissipation, velocity slip and temperature jump, determine the Nusselt number. Assume steady state ideal gas flow.

11.4   A large plate moves with constant velocity u_s parallel to a stationary plate separated by a distance H. An ideal gas fills the channel formed by the plates. The stationary plate is at temperature T_o and the moving plate is at temperature T_s. Assume laminar flow and take into consideration dissipation and velocity slip and temperature jump:

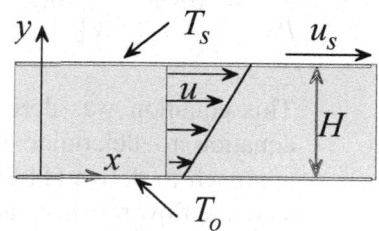

(a) Show that the temperature distribution is given by

$$T = T_o + \frac{\mu u_s^2}{2k(1+2Kn)^2}\left[\frac{2\gamma}{\gamma+1}\frac{Kn}{Pr}+\frac{y}{H}-\frac{y^2}{H^2}\right]+\frac{T_s-T_o}{1+2\dfrac{2\gamma}{\gamma+1}\dfrac{Kn}{Pr}}\left[\frac{2\gamma}{\gamma+1}\frac{Kn}{Pr}+\frac{y}{H}\right].$$

(b) Determine the heat flux at the plates.

11.5   Consider Couette flow between two parallel plates separated by a distance H. The lower plate moves with velocity u_{s1} and the upper plate moves in the opposite direction with velocity u_{s2}. The channel is filled with ideal gas. Assume velocity slip conditions, determine the mass flow rate. Under what condition will the net flow rate be zero?

11.6   Determine the frictional heat generated by the fluid in Example 11.1.

11.7   Consider   shear   driven
        Couette   flow   between
        parallel plates. The upper
        plate moves with velocity
        u_s and is maintained at
        uniform temperature T_s.
        The lower plate is heated

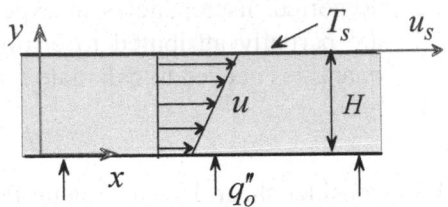

with uniform flux q_o''. The fluid between the two plates is an ideal
gas. Taking into consideration velocity slip, temperature jump, and
dissipation, determine the temperature of the lower plate.

11.8   Pressure distribution in Poiseuille flow between parallel plates is
        given by

$$\frac{p(x)}{p_o} = -6Kn_o + \sqrt{\left[6Kn_o + \frac{p_i}{p_o}\right]^2 + \left[(1 - \frac{p_i^2}{p_o^2}) + 12Kn_o(1 - \frac{p_i}{p_o})\right]\frac{x}{L}} .$$

(11.35)

This equation was derived in Section 11.6.3 using the continuity
equation to determine the y-velocity component v. An alternate
approach to derive (11.35) is based on the condition that for steady
state the flow rate is invariant with axial distance x. That is

$$\frac{dm}{dx} = \frac{d}{dx}\left[2W\int_0^{H/2} \rho u\, dy\right] = 0 .$$

where W is channel width. Derive (11.35) using this approach.

11.9   One of the factors affecting mass flow rate through microchannels is
        channel height H. To examine this effect, consider air flow through
        two microchannels. Both channels have the same length, inlet
        pressure and temperature and outlet pressure. The height of one
        channel is double that of the other. Compute the mass flow ratio for
        the following: $H_1 = 5\,\mu m$, $H_2 = 10\,\mu m$, $T_i = 30\,^\circ C$, $p_i = 420$ kPa,
        $p_o = 105$ kPa.

11.10 A micro heat exchanger consists of rectangular channels of height
        $H = 25\,\mu m$, width $W = 600\,\mu m$, and length $L = 10$ mm. Air enters
        the channels at temperature $T_i = 20\,^\circ C$ and pressure $p_i = 420$ kPa.
        The outlet pressure is $p_o = 105$ kPa. The air is heated with uniform
        surface heat flux $q_s'' = 1100$ W/m^2. Taking into consideration

velocity slip and temperature jump, assume fully developed conditions and compute the following:

(a) Mass flow rate, m.

(b) Mean outlet temperature, T_{mo}.

(c) Heat transfer coefficient at the outlet, $h(L)$.

(d) Surface temperature at the outlet, $T_s(L)$.

11.11 Rectangular microchannels are used to remove heat from a device at uniform surface heat flux. The height, width, and length of each channel are $H = 6.29\,\mu m$, $W = 90\,\mu m$, and $L = 10$ mm, respectively. Using air at $T_i = 20°C$ as the coolant fluid, determine the mass flow rate and the variation of Nusselt number along the channel. Inlet and outlet pressures are $p_i = 410$ kPa and $p_o = 105$ kPa. Assume steady state fully developed slip flow and temperature jump conditions.

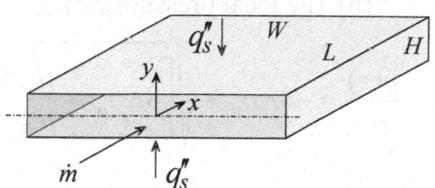

11.12 A micro heat exchanger consists of rectangular channels of height $H = 6.7\,\mu m$, width $W = 400\,\mu m$, and length $L = 8$ mm. Air enters the channels at temperature $T_i = 30\,°C$ and pressure $p_i = 510$ kPa. The outlet pressure is $p_o = 102$ kPa. Channel surface is at uniform temperature $T_s = 50°C$. Assume fully developed flow and temperature, compute:

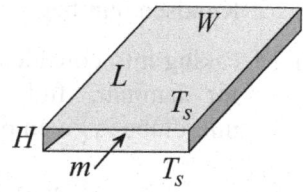

(a) Mass flow rate, m.

(b) Heat transfer coefficient at the inlet, $h(0)$, and outlet, $h(L)$.

(c) Mean outlet temperature, T_{mo}.

(d) Surface heat flux at the outlet, $q_s''(L)$.

11.13 Consider isothermal Poiseuille flow of gas in a microtube of radius r_o. Taking into consideration velocity slip, show that the axial velocity is given by

$$v_z = -\frac{r_o^2}{4\mu}\frac{dp}{dz}\left[1 + 4Kn - \frac{r^2}{r_o^2}\right]. \qquad (11.74)$$

11.14 Consider fully developed isothermal Poiseuille flow through a microtube. Follow the analysis of Section 11.6.3 and use the continuity equation in cylindrical coordinates to derive the following:

(a) The radial velocity component v_r

$$v_r = \frac{r_o^3}{4\mu}\frac{1}{p}\frac{\partial}{\partial z}\left\{p\frac{dp}{dz}\left[\frac{1}{2}\frac{r}{r_o} - \frac{1}{4}\frac{r^3}{r_o^3} + 2\frac{r}{r_o}Kn(p)\right]\right\},$$

where $Kn(p)$ is the local Knudsen number.

(b) The local pressure $p(z)$

$$\frac{p(z)}{p_o} = -8Kn_o + \sqrt{\left[8Kn_o + \frac{p_i}{p_o}\right]^2 + \left[(1-\frac{p_i^2}{p_o^2}) + 16\,Kn_o\,(1-\frac{p_i}{p_o})\right]\frac{z}{L}},$$

$$(11.78)$$

where p_i is inlet pressure, p_o outlet pressure, and Kn_o is the outlet Knudsen number.

11.15 Taking into consideration velocity slip, show that the mass flow rate for laminar, fully developed isothermal Poiseuille flow in a microtube is given by

$$m = \frac{\pi}{16}\frac{r_o^4 p_o^2}{\mu LRT}\left[\frac{p_i^2}{p_o^2} - 1 + 16\,Kn_o\,(\frac{p_i}{p_o} - 1)\right]. \qquad (11.711a)$$

11.16 Pressure distribution for fully developed Poiseuille flow through tubes is given by

$$\frac{p(z)}{p_o} = -8Kn_o + \sqrt{\left[8Kn_o + \frac{p_i}{p_o}\right]^2 + \left[(1-\frac{p_i^2}{p_o^2}) + 16\,Kn_o\,(1-\frac{p_i}{p_o})\right]\frac{z}{L}}.$$

$$(11.78)$$

Derive this equation using the condition that, for steady state, the mass flow rate is invariant with axial distance z. That is

$$\frac{dm}{dz} = \frac{d}{dz}\left[2\pi\int_0^{r_o}\rho v_z r\,dr\right] = 0.$$

11.17 Air is heated in a microtube of radius $r_o = 5\,\mu m$ and length $L = 2\,mm$. Inlet temperature and pressure are $T_i = 20\,^\circ C$ and $p_i = 600\,kPa$. Outlet pressure is $p_o = 100\,kPa$. Uniform surface flux, $q''_s = 1500$ W/m^2, is used to heat the air. Taking into consideration velocity slip and temperature jump and assuming fully developed flow and temperature, compute:

(a) Mass flow rate, m.

(b) Mean outlet temperature, T_{mo}.

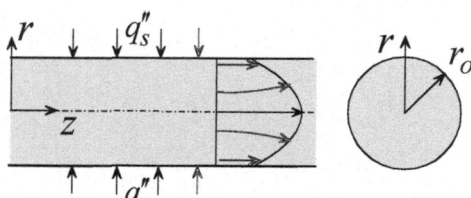

(c) Heat transfer coefficient at the outlet, $h(L)$.

(d) Surface temperature at the outlet, $T_s(L)$.

11.18 Determine the axial variation of the Nusselt number and heat transfer coefficient of the microtube in Problem 11.17.

11.19 A micro heat exchanger uses microtubes of radius $r_o = 3\,\mu m$ and length $L = 6\,mm$. Inlet air temperature and pressure are $T_i = 20\,^\circ C$ and pressure $p_i = 600\,kPa$.

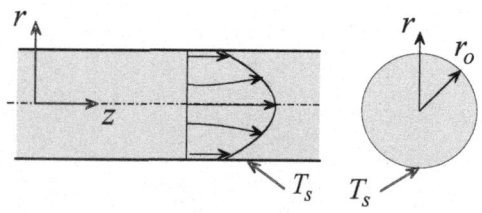

Outlet pressure is $p_o = 100\,kPa$. Each tube is maintained at uniform surface tempera-ture $T_s = 60\,^\circ C$. Taking into consideration velocity slip and temperature jump and assuming fully developed flow and temperature, determine the following:

(a) Heat transfer coefficient at the inlet, $h(0)$, and outlet, $h(L)$.

(b) Mean outlet temperature T_{mo}.

11.20 Air enters a microtube at temperature $T_i = 20\,^\circ C$, and pressure $p_i = 600\,kPa$. Outlet pressure is $p_o = 100\,kPa$. Tube radius is $r_o = 1\,\mu m$ and length is $L = 6\,mm$. The surface is maintained at uniform temperature $T_s = 40\,^\circ C$. Taking into consideration velocity slip and temperature jump and assuming fully developed conditions, determine the variation along the tube of the following:

(a) Nusselt number, $Nu(z)$.

(b) Heat transfer coefficient, $h(z)$.

(c) Mean temperature, $T_m(z)$.

APPENDIX A

CONSERVATION OF ENERGY: THE ENERGY EQUATION

The derivation of energy equation (2.15) in Section 2.6 is presented in detail. We consider the element $dxdydz$ in Fig. A.1 and apply conservation of energy (first law of thermodynamics). We assume continuum and neglect nuclear, electromagnetic and radiation energy transfer. Our starting point is equation (2.14) [1]:

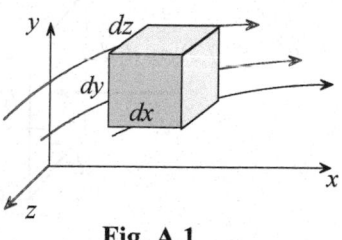

Fig. A.1

A Rate of change of internal and kinetic energy of element	$=$	**B** Net rate of internal and kinetic energy transport by convection
$+$	**C** Net rate of heat added by conduction	$-$ **D** Net rate of work done by element on surroundings

(2.14)

Note that *net rate* in equation (2.14) refers to rate of energy added minus rate of energy removed. We will formulate expressions for each term in equation (2.14).

(1) A = Rate of change of internal and kinetic energy of element

The material inside the element has internal and kinetic energy. Let

$\hat{u}$ = internal energy per unit mass
V = magnitude of velocity

Thus

$$A = \frac{\partial}{\partial t}\left[\rho\,(\hat{u} + V^2 / 2)\right]dxdydz .$$

(A-1)

(2) B = Net rate of internal and kinetic energy transport by convection

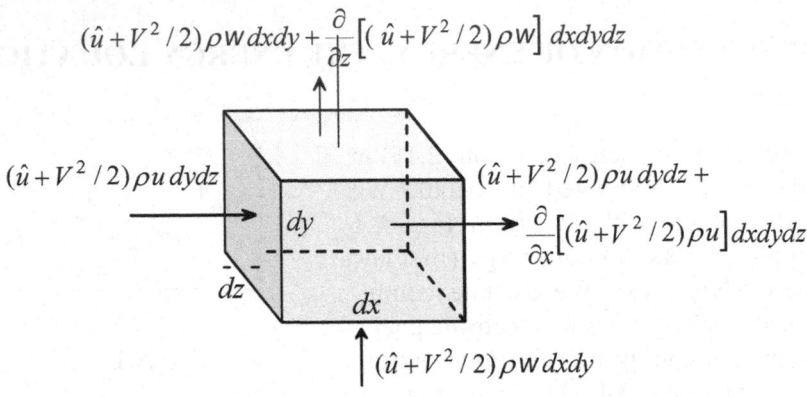

Fig. A.2

Mass flow through the element transports kinetic and thermal energy. Fig. A.2 shows energy convected in the x and y-directions only. Not shown is energy carried in the z-direction. To understand the components of energy transport shown in Fig. A.2, we examine the rate of energy entering the $dydz$ surface. Mass flow rate through this area is $\rho u\, dydz$. When this is multiplied by internal and kinetic energy per unit mass, $(\hat{u} + V^2/2)$, gives the rate of energy entering $dydz$ due to mass flow $(\hat{u} + V^2/2)\rho u\, dydz$. Similar expressions are obtained for the energy transported through all sides. Using the components shown in Fig. A.2 and including energy transfer in the z-direction (not shown) we obtain

$$
\begin{aligned}
\mathbf{B} = {} & (\hat{u} + V^2/2)\,\rho u\, dydz + (\hat{u} + V^2/2)\,v\, dxdz + (\hat{u} + V^2/2)\,\rho w dxdy \\
& - (\hat{u} + V^2/2)\,\rho u\, dydz - \frac{\partial}{\partial x}\Big[(\hat{u} + V^2/2)\,\rho u\Big] dxdydz \\
& - (\hat{u} + V^2/2)\,\rho v\, dxdz - \frac{\partial}{\partial y}\Big[(\hat{u} + V^2/2)\,\rho v\Big] dxdydz \\
& - (\hat{u} + V^2/2)\,\rho w\, dxdy - \frac{\partial}{\partial z}\Big[(\hat{u} + V^2/2)\,\rho w\Big] dxdydz .
\end{aligned}
$$

Simplifying

$$
\mathbf{B} = -\left\{ \frac{\partial}{\partial x}\Big[(\hat{u} + V^2/2)\,\rho u\Big] + \frac{\partial}{\partial y}\Big[(\hat{u} + V^2/2)\,\rho v\Big] + \frac{\partial}{\partial z}\Big[(\hat{u} + V^2/2)\,\rho w\Big] \right\} \times dxdydz .
$$

Making use of the definition of divergence (1.19) the above becomes

$$\mathbf{B} = - \left\{ \nabla \bullet \left[(\hat{u} + V^2/2) \rho \vec{V} \right] \right\} dxdydz.$$ (A-2)

(3) C = Net rate of heat addition by conduction

Let

$q'' = $ heat flux $=$ rate of heat conduction per unit area

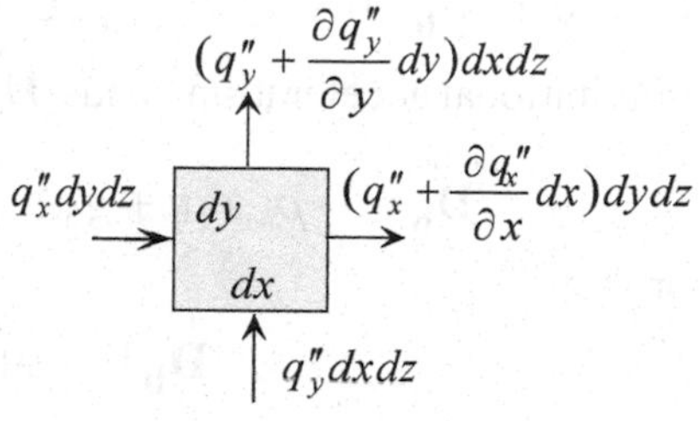

Fig. A.3 shows the z-plane of the element $dxdydz$. Taking into consideration conduction in the z-direction, the net energy conducted through the element is given by

$$ (q''_y + \frac{\partial q''_y}{\partial y} dy)dxdz $$

$$ q''_x dydz \qquad (q''_x + \frac{\partial q''_x}{\partial x} dx)dydz $$

$$ q''_y dxdz $$

Fig. A.3

$$C = q''_x dydz + q''_y dxdz + q''_z dxdy - (q''_x + \frac{\partial q''_x}{\partial x} dx)dydz$$

$$ - (q''_y + \frac{\partial q''_y}{\partial y} dy)dxdz - (q''_z + \frac{\partial q''_z}{\partial z} dz)dxdy.$$

Simplifying

$$C = - \left[\frac{\partial q''_x}{\partial x} + \frac{\partial q''_y}{\partial y} + \frac{\partial q''_z}{\partial z} \right] dxdydz.$$

Introducing the definition of divergence

$$C = -(\nabla \bullet \vec{q''})\, dxdydz.$$ (A-3)

(4) D = Net rate of work done by the element on the surroundings

Rate of work is defined as force $\times$ velocity. Thus

Rate of work = force × velocity.

Work done by the element on the surroundings is negative because it represents energy loss. We thus examine all forces acting on the element and their corresponding velocities. As we have done previously in the formulation of the equations of motion, we consider body and surface forces. Thus

$$D = D_b + D_s, \qquad (A\text{-}4)$$

where

D_b = Net rate of work done by body forces on the surroundings

D_s = Net rate of work done by surface forces on the surroundings

Consider D_b first. Let g_x, g_y and g_z be the three components of gravitational acceleration. Thus D_b is given by

$$D_b = -\rho(g_x u + g_y v + g_z w)\,dxdydz\,,$$

or

$$D_b = -\rho(\vec{V} \cdot \vec{g}) \qquad (A\text{-}5)$$

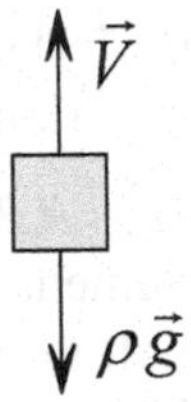

Fig. A.4

To further clarify the negative sign consider a fluid particle being raised vertically, as shown Fig. A.4. If the particle is being raised, work is being done on it, and yet the dot product $(\vec{V} \cdot \vec{g})$ is negative; hence the negative sign being added to the work term to make the work positive.

Next we formulate an equation for rate of work done by surface stresses D_s. Fig. A.5 shows an element with some of the surface stresses. For the purpose of clarity, only stresses on two faces are shown. Each stress is associated with a velocity component. The product of stress, surface area and velocity represents rate of work done. Summing all such products, we obtain

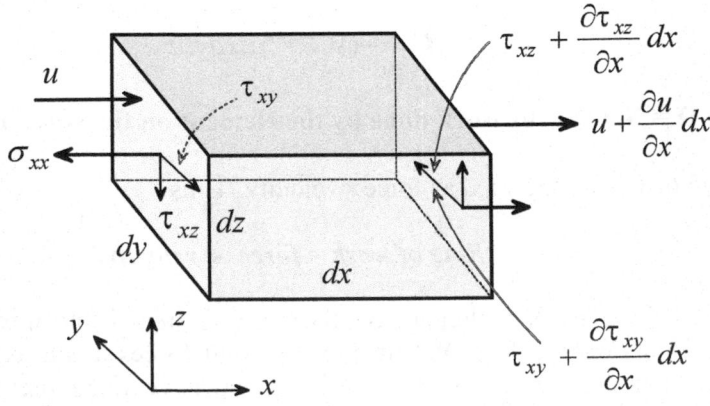

Fig. A.5

$$\mathbf{D_s} = -\left(u + \frac{\partial u}{\partial x}dx\right)\left(\sigma_{xx} + \frac{\partial \sigma_{xx}}{\partial x}dx\right)dydz - u(-\sigma_{xx})dydz$$

$$-\left(w + \frac{\partial w}{\partial x}dx\right)\left(\tau_{xz} + \frac{\partial \tau_{xz}}{\partial x}dx\right)dydz - w(-\tau_{xz})dydz$$

$$-\left(v + \frac{\partial v}{\partial x}dx\right)\left(\tau_{xy} + \frac{\partial \tau_{xy}}{\partial x}dx\right)dydz - v(-\tau_{xy})dydz$$

$$-\left(w + \frac{\partial w}{\partial z}dz\right)\left(\sigma_{zz} + \frac{\partial \sigma_{zz}}{\partial z}dz\right)dxdy - w(-\sigma_{zz})dxdy$$

$$-\left(v + \frac{\partial v}{\partial z}dz\right)\left(\tau_{zy} + \frac{\partial \tau_{zy}}{\partial z}dz\right)dxdy - v(-\tau_{zy})dxdy$$

$$-\left(v + \frac{\partial v}{\partial y}dy\right)\left(\sigma_{yy} + \frac{\partial \sigma_{yy}}{\partial y}dy\right)dxdz - v(-\sigma_{yy})dxdz$$

$$-\left(u + \frac{\partial u}{\partial y}dy\right)\left(\tau_{yx} + \frac{\partial \tau_{yx}}{\partial y}dy\right)dxdz - u(-\tau_{yx})dxdz$$

$$-\left(w + \frac{\partial w}{\partial y}dy\right)\left(\tau_{yz} + \frac{\partial \tau_{yz}}{\partial y}dy\right)dxdz - w(-\tau_{yz})dxdz$$

$$-\left(u + \frac{\partial u}{\partial z}dz\right)\left(\tau_{zx} + \frac{\partial \tau_{zx}}{\partial z}dz\right)dxdy - u(-\tau_{zx})dxdy.$$

Note that negative sign indicates work is done by element on the surroundings. Neglecting higher order terms the above simplifies to

$$\mathbf{D_s} = -\left\{ u\left(\frac{\partial \sigma_{xx}}{\partial x} + \frac{\partial \tau_{yx}}{\partial y} + \frac{\partial \tau_{zx}}{\partial z}\right) + v\left(\frac{\partial \tau_{xy}}{\partial x} + \frac{\partial \sigma_{yy}}{\partial y} + \frac{\partial \tau_{zy}}{\partial z}\right) + \right.$$

$$w\left(\frac{\partial \tau_{xz}}{\partial x} + \frac{\partial \tau_{yy}}{\partial y} + \frac{\partial \sigma_{zz}}{\partial z}\right) + \left(\sigma_{xx}\frac{\partial u}{\partial x} + \tau_{yx}\frac{\partial u}{\partial y} + \tau_{zx}\frac{\partial u}{\partial z}\right) +$$

$$\left.\left(\tau_{xy}\frac{\partial v}{\partial x} + \sigma_{yy}\frac{\partial v}{\partial y} + \tau_{zy}\frac{\partial v}{\partial z}\right) + \left(\tau_{xz}\frac{\partial w}{\partial x} + \tau_{yz}\frac{\partial w}{\partial y} + \sigma_{zz}\frac{\partial w}{\partial z}\right)\right\}dxdydz.$$

$$(A-6)$$

Substituting (A-5) and (A-6) into (A-4)

$$\mathbf{D} = -\rho\left(\vec{V}\cdot\vec{g}\right)dxdydz - \left[\frac{\partial}{\partial x}(u\sigma_{xx} + v\tau_{xy} + w\tau_{xz}) + \right.$$

$$\left. \frac{\partial}{\partial y}(u\tau_{yx} + v\sigma_{yy} + w\tau_{yz}) + \frac{\partial}{\partial z}(u\tau_{zx} + v\tau_{zy} + w\sigma_{zz})\right]dxdydz. \text{ (A - 7)}$$

Substituting (A-1), (A-2), (A-3) and (A-7) into (2.14)

$$\frac{\partial}{\partial t}\left[\rho\left(\hat{u} + \frac{1}{2}V^2\right)\right] = -\nabla\cdot\left[\left(\hat{u} + \frac{1}{2}V^2\right)\rho\vec{V}\right] - \nabla\cdot\vec{q}'' + \rho\left(\vec{V}\cdot\vec{g}\right) +$$

$$\frac{\partial}{\partial x}(u\sigma_{xx} + v\tau_{xy} + w\tau_{xz}) + \frac{\partial}{\partial y}(u\tau_{yx} + v\sigma_{yy} + w\tau_{yz})$$

$$+ \frac{\partial}{\partial z}(u\tau_{zx} + v\tau_{zy} + w\sigma_{zz}). \tag{A - 8}$$

Note that equation (A-8) contains the nine normal and shearing stresses that appear in the formulation of the momentum equations (2.6). We will now use (2.6) to simplify (A-8). Multiplying equations (2.6a), (2.6b) and (2.6c) by the velocity components u, v and w, respectively, and adding the resulting three equations, we obtain

$$\rho\left(u\frac{Du}{Dt} + v\frac{Dv}{Dt} + w\frac{Dw}{Dt}\right) = \rho\left(ug_x + vg_y + wg_z\right)$$

$$+ u\left(\frac{\partial\sigma_{xx}}{\partial x} + \frac{\partial\tau_{yx}}{\partial y} + \frac{\partial\tau_{zx}}{\partial z}\right) + v\left(\frac{\partial\tau_{xy}}{\partial x} + \frac{\partial\sigma_{yy}}{\partial y} + \frac{\partial\tau_{zy}}{\partial z}\right) + w\left(\frac{\partial\tau_{xz}}{\partial x} + \frac{\partial\tau_{yz}}{\partial y} + \frac{\partial\sigma_{zz}}{\partial z}\right).$$

$$\tag{A-9}$$

However,

$$\rho\left(u\frac{Du}{Dt} + v\frac{Dv}{Dt} + w\frac{Dw}{Dt}\right) = \frac{\rho}{2}\frac{DV^2}{Dt}, \tag{A-10}$$

and

$$\left(ug_x + vg_y + wg_z\right) = \vec{V}\cdot\vec{g} \tag{A-11}$$

Substituting (A-10) and (A-11) into (A-9)

$$\frac{\rho}{2}\frac{D\vec{V}^2}{Dt} = \rho\vec{V}\cdot\vec{g} + u\left(\frac{\partial\sigma_{xx}}{\partial x} + \frac{\partial\tau_{yx}}{\partial y} + \frac{\partial\tau_{zx}}{\partial z}\right) + v\left(\frac{\partial\tau_{xy}}{\partial x} + \frac{\partial\sigma_{yy}}{\partial y} + \frac{\partial\tau_{zy}}{\partial z}\right)$$

$$+ w\left(\frac{\partial\tau_{xz}}{\partial x} + \frac{\partial\tau_{yz}}{\partial y} + \frac{\partial\sigma_{zz}}{\partial z}\right)$$

$$. \qquad (A\text{-}12)$$

Returning to (A-8), the first and second terms are rewritten as follows

$$\frac{\partial}{\partial t}\left[\rho\left(\hat{u}+\frac{1}{2}V^2\right)\right] = \left(\hat{u}+\frac{1}{2}V^2\right)\frac{\partial\rho}{\partial t} + \rho\frac{\partial}{\partial t}\left(\hat{u}+\frac{1}{2}V^2\right), \qquad (A\text{-}13)$$

$$\nabla\cdot\left[\left(\hat{u}+\frac{1}{2}V^2\right)\rho\vec{V}\right] = \left(\hat{u}+\frac{1}{2}V^2\right)\nabla\cdot\rho\vec{V} + \rho\vec{V}\cdot\nabla\left(\hat{u}+\frac{1}{2}V^2\right). \qquad (A\text{-}14)$$

Substituting (A-12), (A-13) and (A-14) into (A-8)

$$-\left(\hat{u}+\frac{1}{2}V^2\right)\overbrace{\left(\frac{\partial\rho}{\partial t}+\nabla\cdot\rho\vec{V}\right)}^{0} - \rho\overbrace{\left[\frac{\partial}{\partial t}\left(\hat{u}+\frac{1}{2}V^2\right)+\vec{V}\cdot\nabla\left(\hat{u}+\frac{1}{2}V^2\right)\right]}^{\frac{D}{Dt}\left(\hat{u}+\frac{1}{2}V^2\right)} - \nabla\cdot\vec{q}''$$

$$+\frac{\rho}{2}\frac{DV^2}{Dt} + \left(\sigma_{xx}\frac{\partial u}{\partial x} + \tau_{yx}\frac{\partial u}{\partial y} + \tau_{zx}\frac{\partial u}{\partial z}\right) + \left(\tau_{xy}\frac{\partial v}{\partial x} + \sigma_{yy}\frac{\partial v}{\partial y} + \tau_{yz}\frac{\partial v}{\partial z}\right)$$

$$+\left(\tau_{xz}\frac{\partial w}{\partial x} + \tau_{yz}\frac{\partial w}{\partial y} + \sigma_{zz}\frac{\partial w}{\partial z}\right) = 0.$$

The above equation simplifies to

$$\rho\frac{D\hat{u}}{Dt} = -\nabla\cdot\vec{q}'' + \left(\sigma_{xx}\frac{\partial u}{\partial x} + \tau_{yx}\frac{\partial u}{\partial y} + \tau_{zy}\frac{\partial u}{\partial z}\right) + \left(\tau_{xy}\frac{\partial v}{\partial x} + \sigma_{yy}\frac{\partial v}{\partial y} + \tau_{zy}\frac{\partial v}{\partial z}\right)$$

$$+\left(\tau_{xz}\frac{\partial w}{\partial x} + \tau_{yz}\frac{\partial w}{\partial y} + \sigma_{zz}\frac{\partial w}{\partial z}\right). \qquad (A\text{-}15)$$

Equation (A-15) is based on the principle of conservation of energy. In addition, conservation of mass and momentum were used. Note that the only assumptions made so far are: continuum and negligible nuclear, electromagnetic and radiation energy transfer. We next introduce constitutive equations to express the heat flux $\vec{q''}$ in terms of the temperature field, and the normal and shearing stresses in terms of the velocity field. For the former we use Fourier's law (1.8) and for the latter we apply Newtonian approximation (2.7). Application of Fourier's law (1.8) gives the heat flux in the n-direction as

$$q_n'' = -k_n \frac{\partial T}{\partial n},$$ (A-16)

where k_n is thermal conductivity in the n-direction. Assuming isotropic material, we write

$$k_n = k_x = k_y = k_z = k.$$ (A-17)

Using the operator ∇, equation (A-16) is expressed as

$$\vec{q''} = -k\nabla T.$$ (A-18)

Substituting (A-18) and (2.7) into (A-15) and rearranging, we obtain

$$-\rho \frac{D\hat{u}}{Dt} = -\nabla \cdot k\nabla T - p\nabla \cdot \vec{V} + \mu \Phi,$$ (A-18)

where $\hat{u}$ internal energy and Φ is the dissipation function defined as

$$\Phi = 2\left[\left(\frac{\partial u}{\partial x}\right)^2 + \left(\frac{\partial v}{\partial y}\right)^2 + \left(\frac{\partial w}{\partial z}\right)^2\right] + \left[\frac{\partial v}{\partial x} + \frac{\partial u}{\partial y}\right]^2 + \left[\frac{\partial u}{\partial z} + \frac{\partial w}{\partial x}\right]^2$$

$$+ \left[\frac{\partial w}{\partial y} + \frac{\partial v}{\partial z}\right]^2 - \frac{2}{3}\left[\frac{\partial u}{\partial x} + \frac{\partial v}{\partial y} + \frac{\partial w}{\partial z}\right]^2.$$ (A-19)

Equation (A-18) is based on the following assumptions: (1) continuum, (2) negligible nuclear, electromagnetic and radiation energy, (3) isotropic material, and (4) Newtonian fluid.

The next step is to express (A-18) first in terms of enthalpy and then in terms of temperature. Starting with the definition of enthalpy h

$$\hat{h} = \hat{u} + \frac{P}{\rho}.$$ (A-20)

Differentiating (A-20)

$$\frac{D\hat{h}}{Dt} = \frac{D\hat{u}}{Dt} + \frac{1}{\rho}\frac{Dp}{Dt} - \frac{P}{\rho^2}\frac{D\rho}{Dt}. \tag{A-21}$$

Substituting (A-21) into (A-18)

$$\rho\frac{D\hat{h}}{Dt} = \nabla \cdot k\nabla T + \frac{Dp}{Dt} + \mu\Phi - \frac{p}{\rho}\overbrace{\left(\frac{D\rho}{Dt} + \rho\nabla \cdot \vec{V}\right)}^{0}. \tag{A-22}$$

Application of the continuity equation (2.2c) to (A-22) eliminates the last two terms. Thus (A-22) simplifies to

$$\rho\frac{D\hat{h}}{Dt} = \nabla \cdot k\nabla T + \frac{Dp}{Dt} + \mu\Phi. \tag{A-23}$$

We next express enthalpy in (A-23) in terms of temperature using the following thermodynamic relation [2]

$$d\hat{h} = c_p dT + \frac{1}{\rho}(1 - \beta T)dp, \tag{A-24}$$

where β is the coefficient of thermal expansion, defined as

$$\beta = -\frac{1}{\rho}\left(\frac{\partial\rho}{\partial T}\right)_p. \tag{A-25}$$

Taking the total derivative of (A-24)

$$\frac{D\hat{h}}{Dt} = c_p\frac{DT}{Dt} + \frac{1}{\rho}(1 - \beta T)\frac{Dp}{Dt}. \tag{A-26}$$

Substituting (A-26) into (A-23)

$$\rho c_p\frac{DT}{Dt} = \nabla \cdot k\nabla T + \beta T\frac{Dp}{Dt} + \mu\Phi. \tag{2.15}$$

REFERENCES

[1] Bird, R.B., W.E. Stewart and E.N. Lightfoot, *Transport Phenomena*, John Wiley & Sons, 1960.

[2] Van Wylen, G. J. and R.E. Sonntag, *Fundamentals of Classical Thermodynamics*, 2ndd ed., John Wiley & Sons, 1973.

APPENDIX B: POHLHAUSEN'S SOLUTION

The transformed energy equation is

$$\frac{d^2\theta}{d\eta^2}+\frac{Pr}{2}f(\eta)\frac{d\theta}{d\eta}=0.\qquad(4.61)$$

The boundary conditions are

$$\theta(0)=0,\qquad\qquad(4.62\text{a})$$
$$\theta(\infty)=1,\qquad\qquad(4.62\text{b})$$
$$\theta(\infty)=1.\qquad\qquad(4.62\text{c})$$

Note that boundary conditions (4.60b) and (4.60c) coalesce into a single condition, as shown in (4.62b) and (4.62c). Equation (4.61) is solved by first separating the variables as

$$\frac{d\left(\dfrac{d\theta}{d\eta}\right)}{\dfrac{d\theta}{d\eta}}=-\frac{Pr}{2}f(\eta)d\eta.$$

Integrating the above from $\eta=0$ to η

$$\int_0^\eta\frac{d\left(\dfrac{d\theta}{d\eta}\right)}{\dfrac{d\theta}{d\eta}}=-\frac{Pr}{2}\int_0^\eta f(\eta)d\eta.$$

Evaluating the integral on the left-hand-side

$$\ln\frac{\dfrac{d\theta}{d\eta}}{\dfrac{d\theta(0)}{d\eta}}=-\frac{Pr}{2}\int_0^\eta f(\eta)d\eta.$$

Taking the anti log of the above

$$\frac{d\theta}{d\eta} = \frac{d\theta(0)}{d\eta}\exp\left[-\frac{Pr}{2}\int_0^\eta f(\eta)d\eta\right].$$

Integrating again from η to $\eta = \infty$ and using boundary condition (4.62b)

$$\int_\eta^\infty d\theta = \frac{d\theta(0)}{d\eta}\int_\eta^\infty \exp\left[-\frac{Pr}{2}\int_0^\eta f(\eta)d\eta\right]d\eta.$$

This gives

$$\theta(\eta) = 1 - \frac{d\theta(0)}{d\eta}\int_\eta^\infty \exp\left[-\frac{Pr}{2}\int_0^\eta f(\eta)d\eta\right]d\eta. \qquad (a)$$

The constant $d\theta(0)/d\eta$ in (a) is unknown. It is determined by satisfying boundary condition (4.62a), which gives

$$\frac{d\theta(0)}{d\eta} = \left\{\int_0^\infty \exp\left[-\frac{Pr}{2}\int_0^\eta f(\eta)d\eta\right]d\eta\right\}^{-1}. \qquad (b)$$

Substituting (b) into (a)

$$\theta(\eta) = 1 - \frac{\displaystyle\int_\eta^\infty \exp\left[-\frac{Pr}{2}\int_0^\eta f(\eta)d\eta\right]d\eta}{\displaystyle\int_0^\infty \exp\left[-\frac{Pr}{2}\int_0^\eta f(\eta)d\eta\right]d\eta}. \qquad (c)$$

The integral in (c) can be simplified using the transformed momentum equation (4.44)

$$2\frac{d^3 f}{d\eta^3} + f(\eta)\frac{d^2 f}{d\eta^2} = 0. \qquad (4.44)$$

Solving (4.44) for $f(\eta)$ and integrating

$$-\frac{1}{2}\int_0^\eta f(\eta)\,d\eta = \int_0^\eta \frac{\dfrac{d^3 f}{d\eta^3}}{\dfrac{d^2 f}{d\eta^2}}\,d\eta = \ln\frac{\dfrac{d^2 f}{d\eta^2}}{\dfrac{d^2 f(0)}{d\eta^2}}.$$

Multiplying both sides of the above by Pr and taking the anti log of the above

$$\exp\left[-\frac{Pr}{2}\int_0^\eta f(\eta)d\eta\right] = \frac{\left[\dfrac{d^2 f}{d\eta^2}\right]^{Pr}}{\left[\dfrac{d^2 f(0)}{d\eta^2}\right]^{Pr}}. \tag{d}$$

Substituting (d) into (c) gives

$$\theta(\eta) = 1 - \frac{\displaystyle\int_\eta^\infty \left[\dfrac{d^2 f}{d\eta^2}\right]^{Pr}\,d\eta}{\displaystyle\int_0^\infty \left[\dfrac{d^2 f}{d\eta^2}\right]^{Pr}\,d\eta}. \tag{e}$$

Similarly, substituting (d) into (b) gives the temperature gradient at the wall

$$\frac{d\theta(0)}{d\eta} = \frac{\left[\dfrac{d^2 f(0)}{d\eta^2}\right]^{Pr}}{\displaystyle\int_0^\infty \left[\dfrac{d^2 f}{d\eta^2}\right]^{Pr}\,d\eta}. \tag{4.63}$$

The constant $\dfrac{d^2 f(0)}{d\eta^2}$ in (f) is obtained from Table 4.1

$$\frac{d^2 f(0)}{d\eta^2} = 0.332.$$

Thus (4.63) becomes

$$\frac{d\theta(0)}{d\eta} = \frac{[0.332]^{Pr}}{\int_0^\infty \left[\frac{d^2 f}{d\eta^2}\right]^{Pr} d\eta}.$$ (4.64)

APPENDIX C

LAMINAR BOUNDARY LAYER FLOW OVER SEMI-INFINITE PLATE:
VARIABLE SURFACE TEMPERATURE [1]

Surface temperature varies with distance along the plate according to

$$T_s(x) - T_\infty = Cx^n.$$ (4.72)

Based on the assumptions listed in Section 4.3, temperature distribution is governed by energy equation (4.18)

$$u\frac{\partial T}{\partial x} + v\frac{\partial T}{\partial y} = \alpha\frac{\partial^2 T}{\partial y^2}.$$ (4.18)

The velocity components u and v in (4.18) are given in Blasius solution

$$\frac{u}{V_\infty} = \frac{df}{d\eta},$$ (4.42)

$$\frac{v}{V_\infty} = \frac{1}{2}\sqrt{\frac{v}{V_\infty x}}\left(\eta\frac{df}{d\eta} - f\right).$$ (4.43)

The boundary conditions are:

$$T(x,0) = T_s = T_\infty + Cx^n,$$ (4.74a)

$$T(x,\infty) = T_\infty,$$ (4.74b)

$$T(0,y) = T_\infty.$$ (4.74c)

The solution to (4.18) is obtained by the method of similarity transformation. We define a dimensionless temperature θ as

$$\theta = \frac{T - T_s}{T_\infty - T_s}.$$ (a)

We assume that

$$\theta(x,y) = \theta(\eta),$$ (b)

where

$$\eta = y\sqrt{\frac{V_\infty}{\nu x}} \ . \tag{c}$$

Equation (4.18) is transformed in terms of $\theta(\eta)$ and η. Equation (a) is solved for $T(x,y)$

$$T = T_s + (T_\infty - T_s)\theta \ .$$

Substituting (4.72) in the above

$$T = T_\infty + Cx^n - Cx^n\theta \ . \tag{d}$$

The derivatives in (4.18) are formulated using (b)-(c) and the chain rule:

$$\frac{\partial T}{\partial x} = Cnx^{n-1} - Cnx^{n-1}\theta - Cx^n\frac{\partial\theta}{\partial x} \ .$$

However,

$$\frac{\partial\theta}{\partial x} = \frac{d\theta}{d\eta}\frac{\partial\eta}{\partial x} = -\frac{\eta}{2x}\frac{d\theta}{d\eta} \ .$$

Substituting into the above

$$\frac{\partial T}{\partial x} = Cnx^{n-1} - Cnx^{n-1}\theta + \frac{C}{2}x^{n-1}\eta\frac{d\theta}{d\eta} \ . \tag{e}$$

Similarly

$$\frac{\partial T}{\partial y} = -Cx^n\frac{d\theta}{d\eta}\frac{\partial\eta}{\partial y} = -Cx^n\sqrt{\frac{V_\infty}{\nu x}}\frac{d\theta}{d\eta} \ , \tag{f}$$

$$\frac{\partial^2 T}{\partial y^2} == -Cx^{n-1}\frac{V_\infty}{\nu}\frac{d^2\theta}{d\eta^2} \ . \tag{g}$$

Substituting (4.42), (4.43) and (e)-(f) into (4.18)

$$V_\infty\frac{df}{d\eta}\left[Cnx^{n-1} - Cnx^{n-1}\theta + \frac{C}{2}x^{n-1}\eta\frac{d\theta}{d\eta}\right]$$

$$-\frac{V_\infty}{2}\sqrt{\frac{\nu}{V_\infty x}}\left[\eta\frac{df}{d\eta} - f\right]Cx^n\sqrt{\frac{V_\infty}{\nu x}}\frac{d\theta}{d\eta} = -\alpha\frac{V_\infty}{\nu x}\frac{d^2\theta}{d\eta^2} \ .$$

This simplifies to

$$\frac{d^2\theta}{d\eta^2} + nPr\frac{df}{d\eta}(1-\theta) + \frac{Pr}{2}f(\eta)\frac{d\theta}{d\eta} = 0, \qquad (4.75)$$

where

$$Pr = \frac{\nu}{\alpha}.$$

REFERENCE

[1] Oosthuizen, P.H. and D. Naylor, *Introduction to Convection Heat Transfer Analysis*, McGraw-Hill, 1999.

APPENDIX D

THE VON KÁRMÁN MOMENTUM AND HEAT TRANSFER ANALOGY

The von Kármán Analogy is an extension of the Reynolds and Prandtl-Taylor Analogies that include a buffer layer between the viscous sublayer and outer layer. The viscous sublayer is described similar to (8.91):

$$T_s - \overline{T}_{b1} = \frac{q_o''}{\tau_o c_p} Pr \, \overline{u}_{b1} \,, \tag{a}$$

where the subscript $b1$ defines the lower edge of the buffer layer (which is also the edge of the viscous sublayer). As in the Prandtl-Taylor Analogy, a value of $u^+ = y^+ \approx 5$ is chosen to approximate the edge of the viscous sublayer, leading to equation (8.95),

$$\frac{\overline{u}_1}{V_\infty} = 5\sqrt{\frac{C_f}{2}} \,. \tag{8.95}$$

Combining the above with (a) gives

$$T_s - \overline{T}_{b1} = \frac{q_o''}{\tau_o c_p} Pr \, (5) V_\infty \sqrt{\frac{C_f}{2}} \,. \tag{D-1}$$

For the buffer layer, von Kármán approximated the shape of the velocity profile as

$$u^+ = 5.0 \ln y^+ - 3.05 \,, \tag{D-2}$$

which extends from $y^+ \approx 5$ to $y^+ \approx 30$. This region is a transitional one, where the flow field is dominated by molecular diffusion at one end, and by turbulent diffusion at the other. Unfortunately, the terms $(v + \varepsilon_M)$ and $(\alpha + \varepsilon_H)$ cannot be assumed to be equal in the buffer region; their ratio will vary from $v/\alpha \, (= Pr)$ to $\varepsilon_M / \varepsilon_H \, (\approx 1)$. Strictly speaking, then, the momentum and heat transfer equations are not analogous in this region.

This region requires a different approach from the others. First, we can develop an expression ε_M as follows. From the velocity profile (D-2), the partial derivative of u^+ is

$$\frac{\partial u^+}{\partial y^+} = \frac{5}{y^+} . \tag{a}$$

We can substitute this into the transformed Couette flow assumption, equation (8.52), to obtain

$$\varepsilon_M = \nu \left(\frac{y^+}{5} - 1 \right) . \tag{b}$$

Noting that $\varepsilon_M = \varepsilon_H$, the ratio $(\nu + \varepsilon_M)/(\alpha + \varepsilon_H)$ can be expressed as

$$\frac{(\nu + \varepsilon_m)}{(\alpha + \varepsilon_h)} = \frac{y^+}{y^+ + 5(1/Pr - 1)} .$$

Substituting the above into (8.86), we obtain

$$\frac{q''}{\tau} = -c_p \frac{y^+ + 5(1/Pr - 1)}{y^+} \frac{\partial \bar{T} / \partial y}{\partial \bar{u} / \partial y} . \tag{D-3}$$

We can separate the derivatives and integrate with respect to y^+ if we make two more adjustments to the equation: the first is to substitute variables y^+ for y, u^+ for $\bar{u}$ in the derivatives in (D-3). The second is to assume that q''/τ is constant; a reasonable assumption, since we saw that the ratio is in fact constant in the other regions. Thus (D-3) becomes

$$-\frac{q_o'' u^*}{\tau_o c_p} \frac{y^+}{y^+ + 5(1/Pr - 1)} \frac{\partial u^+}{\partial y^+} = \frac{\partial \bar{T}}{\partial y^+} . \tag{c}$$

Finally, substituting $u^* = V_\infty \sqrt{C_f / 2}$ in (D.3), and integrating both sides between $y^+ = 5$ to $y^+ = 30$, we obtain

$$\bar{T}_{b2} - \bar{T}_{b1} = -\frac{q_o'' V_\infty \sqrt{C_f / 2}}{\tau_o c_p} 5 \ln(5Pr + 1) , \tag{D-4}$$

where $\overline{T}_{b1}$ and $\overline{T}_{b2}$ are the temperatures at each end of the buffer layer.

The outer layer is described similar to that of the Prandtl Analogy,

$$\overline{T}_{b2} - T_\infty = \frac{q_o''}{\tau_o c_p}(V_\infty - \overline{u}_{b2}) \ . \tag{D-5}$$

Now, to obtain an expression for $\overline{u}_{b2}$, we can use the velocity profile assumed for the buffer layer, equation. (D-2). Assuming that the buffer layer ends at $y^+ \approx 30$,

$$u_{b2}^+ \approx 5\ln(30) - 3 \ , \tag{d}$$

or, from the definition of u^+.

$$\overline{u}_{b2} \approx (5\ln 30 - 3)V_\infty \sqrt{C_f / 2} \ . \tag{e}$$

Substituting these expressions into (D-5) gives[*]

$$\overline{T}_{b2} - T_\infty = \frac{q_o''}{\tau_o c_p}V_\infty\left[1 - (5\ln 30 - 3)\sqrt{\frac{C_f}{2}}\right] \ . \tag{D-6}$$

Finally, adding (D-1), (D-4), and (D-6), and rearranging, the analogy becomes

$$St_x \equiv \frac{Nu_x}{Re_x Pr} = \frac{C_f / 2}{1 + 5\sqrt{\dfrac{C_f}{2}}\left\{\left(Pr - \dfrac{3}{5}\right) + \ln\left[\dfrac{5Pr + 1}{30}\right]\right\}} \ , \tag{D-7}$$

This result is not significantly different, numerically, from von Kármán's result, which is

---

[*] von Kármán used a slightly different approximation for $\overline{u}_{b2}$, approximating it as $\overline{u}_{b2} = \overline{u}_{b1} + (\overline{u}_{b2} - \overline{u}_{b1})$. The velocity $\overline{u}_{b1}$ was evaluated using (8.93), while $\overline{u}_{b2} - \overline{u}_{b1} \approx 5\ln(30/5)$ from the buffer layer velocity profile. The approximation used here yields virtually the same result.

$$St_x = \frac{C_f / 2}{1 + 5\sqrt{\dfrac{C_f}{2}}\left\{(Pr - 1) + \ln\left[\dfrac{5Pr + 1}{6}\right]\right\}} . \qquad (8.97)$$

This expression gives the local heat flux for a flat plate with unheated starting length of x_0 .

REFERENCES

[1] von Kármán, T., "The Analogy between Fluid Friction and Heat Transfer," *Trans. ASME*, Vol. 61, 1938, pp. 705-710.

[2] Kakaç, S. and Yener, Y., *Convective Heat Transfer*, 2nd Ed., CRC Press, Boca Raton, 1995.

APPENDIX E

TURBULENT HEAT TRANSFER FROM A FLAT PLATE WITH UNHEATED STARTING LENGTH [1,2]

Consider a flat plate in turbulent flow, as depicted in Fig. 8.19.

The velocity and temperature profiles are estimated using the $1/7^{th}$ power law,

$$\frac{\bar{u}}{V_\infty} = \left(\frac{y}{\delta}\right)^{1/7} \tag{8.65}$$

and

$$\frac{\bar{T} - T_s}{T_\infty - T_s} = \left(\frac{y}{\delta_t}\right)^{1/7}. \tag{8.114}$$

For this development, we will assume $\alpha = \nu$ $(Pr = 1)$ and $\varepsilon_H = \varepsilon_M (Pr_t = 1)$.

The energy integral equation is as follows for constant-property, incompressible flow over an impermeable flat plate:

$$-\alpha \frac{\partial T(x,0)}{\partial y} = \frac{d}{dx} \int_0^{\delta_t(x)} \bar{u}(\bar{T} - T_\infty) dy, \tag{5.7}$$

Substituting the $1/7^{th}$-law velocity and temperature profiles into right-hand side of (5.7), and evaluating the integral, we obtain

$$-\alpha \frac{\partial T(x,0)}{\partial y} = V_\infty (T_s - T_\infty) \frac{d}{dx} \left(\frac{7}{72} \frac{\delta_t^{8/7}}{\delta^{1/7}}\right). \tag{E-1}$$

The problem with this expression is that we can't use the $1/7^{th}$ power law temperature profile to evaluate the left-hand of (E-1). This is because the gradient becomes undefined (that is, goes to infinity) at $y = 0$. Recall from

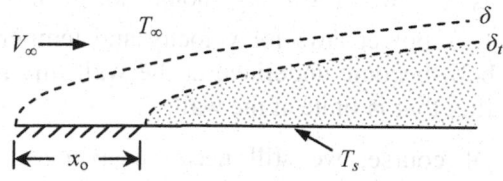

Fig. 8.19

Chapter 8 that we faced the same problem evaluating the wall shear in the momentum integral method. As a result, we have to develop some other means of estimating the wall heat transfer, one that does not result in an unrealistic value for the heat flux at the wall.

It turns out that if we look at the heat flux anywhere in the boundary layer, and not simply at the wall, we can find an approach to the problem that works. Consider the definition of the apparent heat flux,

$$\frac{q''(x,y)}{\rho c_p} = -(\alpha + \varepsilon_H)\frac{\partial \overline{T}}{\partial y} . \tag{8.41}$$

Since we are assuming $\alpha = \nu$ and $\varepsilon_H = \varepsilon_M$, we can write (8.41) as

$$\frac{q''(x,y)}{\rho c_p} = -(\nu + \varepsilon_M)\frac{\partial \overline{T}}{\partial y} . \tag{a}$$

If we substitute the definition of the apparent shear stress, equation (8.40), for the term $(\nu + \varepsilon_M)$ into the above,

$$\frac{q''(x,y)}{\rho c_p} = -\frac{\tau(x,y)}{\rho}\frac{\partial \overline{T}/\partial y}{\partial \overline{u}/\partial y} . \tag{E-2}$$

If we then substitute the $1/7^{\text{th}}$ power law velocity and temperature profiles into this expression and simplify, we obtain

$$\frac{q''(x,y)}{\rho c_p} = \frac{\tau(x,y)}{\rho}\left(\frac{\delta}{\delta_t}\right)^{1/7}\frac{(T_s - T_\infty)}{V_\infty} . \tag{E-3}$$

Notice that the variable y does not appear in the above expression – it cancels out. The result is fortunate, because with y missing from (E-3), the heat flux does not become undefined (or zero, which is equally unrealistic) as $y \to 0$. The irony should not go unnoticed: even though we used the $1/7^{\text{th}}$ power laws for velocity and temperature in the development, which both become undefined at the wall, the resulting expression yields a heat flux that is finite at the wall.

Of course, we still need an expression for the shear stress $\tau(x,y)$ anywhere in the boundary layer, which we have not attempted to model before. To model the shear stress, we can invoke an integral momentum equation, like equation (5.5),

$$\nu \frac{\partial \overline{u}(x,0)}{\partial y} = V_\infty \frac{d}{dx} \int_0^{\delta(x)} \overline{u}\,dy - \frac{d}{dx} \int_0^{\delta(x)} \overline{u}^2\,dy \ . \tag{5.5}$$

Notice that this equation is integrated from $y = 0$, and yields an expression for shear stress at the wall. The problem with this is that we want an expression for shear stress at any y location. A more general integral equation requires a new derivation.

(i) Integral Momentum Equation for General y Location

The development parallels the original development in Section 5.6, and so many of the details are left as an exercise. We will limit the derivation to incompressible flow over an impermeable flat plate.

First, conservation of mass is applied to the element shown in Fig. E.1. The derivation is identical to the original development, and yields the following expression for the differential mass flow at the boundary layer edge,

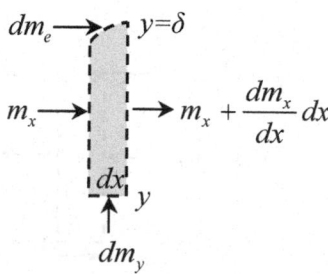

Fig. E.1

$$dm_e = \frac{d}{dx}\left[\int_y^{\delta(x)} \rho \overline{u}\,dy \right] dx - \rho \overline{v}(y)dx \ . \tag{E-4}$$

Application of the momentum theorem in the x-direction to the element depicted in Fig. E.2 gives

$$\sum F_x = M_x\,(\text{out}) - M_x\,(\text{in}) \ . \tag{b}$$

What is different about this derivation is that x-momentum enters through the bottom of the element by way of the y-component of velocity. Applying the forces and momentum fluxes of Fig. E.2 to the conservation equation (b) gives

$$p(\delta - y) + \left(p + \frac{dp}{2}\right)d\delta - p(\delta - y) - \frac{d}{dx}\left[p(\delta - y)\right]dx - \tau dx$$

$$= \left(M_x + \frac{dM_x}{dx}dx\right) - M_x - V_\infty dm_e - \overline{u}(\rho \overline{v}dx) \ . \tag{c}$$

The momentum flux is

$$M_x = \int_y^{\delta(x)} \rho \bar{u}^2 \, dy \,.$$

(d)

Substituting (d) and (E-4) into (c), neglecting higher-order terms, and simplifying,

$$-\delta \frac{dp}{dx} - \tau(x,y) = \frac{d}{dx} \int_y^{\delta(x)} \rho \bar{u}^2 \, dy - V_\infty \frac{d}{dx} \int_y^{\delta(x)} \rho \bar{u} \, dy + \rho \left[V_\infty - \bar{u}(y) \right] \bar{v}(y)$$

(E-5)

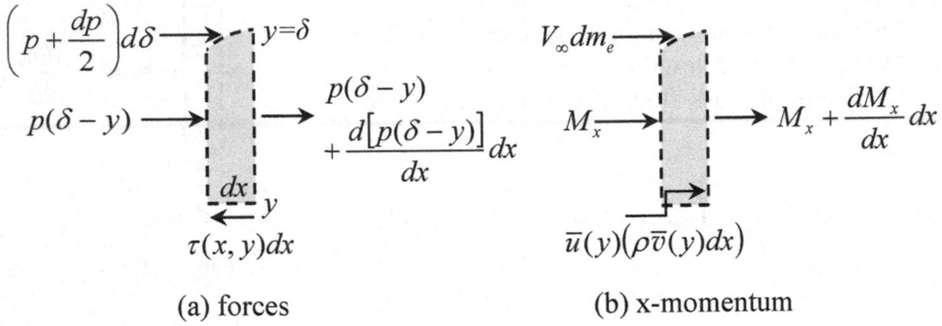

(a) forces   (b) x-momentum

Fig. E.2

We need to solve (E-5) for the shear stress, $\tau(x,y)$. For a flat plate, dp/dx is zero, and the $1/7^{th}$ power law gives us an estimate for the velocity profile, $\bar{u}(y)$. However, we still need an expression for the y component velocity, $\bar{v}(y)$.

(ii) Evaluation on of y velocity component

We can determine an expression for $\bar{v}(y)$ by employing the differential conservation of mass, which for incompressible, two-dimensional flow is

$$\frac{\partial \overline{u}}{\partial x} + \frac{\partial \overline{v}}{\partial y} = 0 \ . \tag{E-6}$$

Rearranging and integrating from 0 to y,

$$\int_0^y \frac{\partial \overline{v}}{\partial y} dy = -\int_0^y \frac{\partial \overline{u}}{\partial x} dy \ , \tag{e}$$

or $\qquad\qquad \overline{v}(y) - \overline{v}(0) = -\int_0^y \frac{\partial \overline{u}}{\partial x} dy \ . \tag{f}$

We can evaluate the integral on the right side by invoking Leibniz's rule for differentiating under the integral sign: for an integrand $f(x,y)$ with limits of integration $a(x)$ and $b(x)$,

$$\frac{d}{dx} \int_a^b f(x,y) dy = \int_a^b \frac{\partial}{\partial x} f(x,y) dy + f(b,x)\frac{db}{dx} - f(a,x)\frac{da}{dx} . \tag{E-7}$$

For our purposes, $f \to \overline{u}$, $a \to 0$, and $b \to y$. Since both limits of integration are not functions of x, then (e) becomes simply

$$\overline{v}(y) = -\frac{d}{dx} \int_0^y \overline{u} dy \ . \tag{g}$$

Substituting the $1/7^{\text{th}}$ power law (8.65), into (f), and evaluating, it can be shown that

$$\overline{v}(y) = \frac{V_\infty}{8} \left(\frac{y}{\delta}\right)^{8/7} \frac{d\delta}{dx} \ . \tag{E-8}$$

(iii) Evaluation of Shear Stress $\tau(x,y)$

Expressing equation (E-5) in terms of the shear stress,

$$\tau(x,y) = -\frac{d}{dx} \int_y^{\delta(x)} \rho \overline{u}^2 dy + V_\infty \frac{d}{dx} \int_y^{\delta(x)} \rho \overline{u} dy - \rho [V_\infty - \overline{u}(y)] \overline{v}(y) \tag{E-9}$$

.

Substituting the velocity expressions (8.65) and (E-9), it can be shown that the shear stress can be expressed as

$$\frac{\tau(x, y)}{\rho} = \frac{7}{72} V_\infty^2 \left[1 - \left(\frac{y}{\delta} \right)^{9/7} \right] \frac{d\delta}{dx} . \tag{E-10}$$

Note that at $y = 0$, we find that the wall stress is

$$\frac{\tau_o}{\rho} = \frac{7}{72} V_\infty^2 \frac{d\delta}{dx},$$

Which is exactly what we found in the Prandtl-von Karman integral solution (8.68). In fact, we can express (E-10) as the following ratio:

$$\frac{\tau(x, y)}{\tau_o} = 1 - \left(\frac{y}{\delta} \right)^{9/7} . \tag{E-11}$$

Equation (E-11) is more convenient than (E-10) in that we have already solved for the wall shear stress in the Prandtl-von Karman integral solution. This will become convenient as we move forward with the derivation.

(iii) Evaluation of Heat Flux $q''(x, y)$

With an expression for the shear stress $\tau(x, y)$ in hand, we can substitute it into (E-3) to obtain an expression for the heat flux $q''(x, y)$. The result is

$$\frac{q''(x, y)}{\rho c_p} = \left(\frac{\delta}{\delta_t} \right)^{1/7} \frac{(T_s - T_\infty)}{V_\infty} \frac{\tau_o}{\rho} \left[1 - \left(\frac{y}{\delta} \right)^{9/7} \right], \tag{h}$$

or, from the definition of friction factor,

$$\frac{q''(x, y)}{\rho c_p} = \left(\frac{\delta}{\delta_t} \right)^{1/7} \frac{(T_s - T_\infty)}{V_\infty} \left(V_\infty^2 \frac{C_f}{2} \right) \left[1 - \left(\frac{y}{\delta} \right)^{9/7} \right]. \tag{i}$$

Finally, we can evaluate this expression at the wall ($y = 0$), which gives

$$\frac{q_o''}{\rho c_p} = \left(\frac{\delta}{\delta_t} \right)^{1/7} (T_s - T_\infty) V_\infty \frac{C_f}{2} . \tag{E-12}$$

(iv) Solution of Integral Energy Equation

Finally, we can substitute our expression for the heat flux (E-12) into the integral energy equation, (E-1). Doing so, and simplifying, we obtain

$$\left(\frac{\delta}{\delta_t}\right)^{1/7}\left(\frac{C_f}{2}\right) = \frac{d}{dx}\left(\frac{7}{72}\frac{\delta_t^{8/7}}{\delta^{1/7}}\right). \tag{E-13}$$

To evaluate this expression, we note that the Prandtl-von Karman integral solution gives us a solution for the friction factor,

$$\frac{C_f}{2} = \frac{0.02968}{\mathrm{Re}_x^{1/5}}. \tag{8.71}$$

Substituting this into (E-13) gives

$$\frac{d}{dx}\left(\frac{7}{72}\frac{\delta_t^{8/7}}{\delta^{1/7}}\right) = 0.02968\left(\frac{\delta}{\delta_t}\right)^{1/7}\left(\frac{V_\infty x}{v}\right)^{-1/5}. \tag{E-14}$$

We will find it easier to evaluate (E-14) if we define $\xi = \delta_t / \delta$. Equation (E-14) can be written as

$$\frac{7}{72}\frac{d}{dx}\left(\delta\xi^{8/7}\right) = 0.02968(\xi)^{-1/7}\left(\frac{V_\infty x}{v}\right)^{-1/5}. \tag{j}$$

Applying the product rule to the derivative, (i) can be written as

$$\delta\frac{8}{7}\xi^{1/7}\frac{d\xi}{dx} + \xi^{8/7}\frac{d\delta}{dx} = 0.02968\left(\frac{72}{7}\right)\xi^{-1/7}\left(\frac{V_\infty x}{v}\right)^{-1/5},$$

or, multiplying both sides by $\xi^{1/7}$,

$$\delta\frac{8}{7}\xi^{2/7}\frac{d\xi}{dx} + \xi^{9/7}\frac{d\delta}{dx} = 0.02968\left(\frac{72}{7}\right)\left(\frac{V_\infty x}{v}\right)^{-1/5}. \tag{k}$$

We can again invoke the Prandtl-von Karman solution, and substitute (8.67) for the velocity boundary layer:

$$\frac{\delta}{x} = \frac{0.3816}{\mathrm{Re}_x^{1/5}}. \tag{8.70}$$

Doing this, we obtain

$$\frac{8}{7}(0.3816)\left(\frac{V_{\infty}x}{v}\right)^{-1/5} x\xi^{2/7}\frac{d\xi}{dx} + \xi^{9/7}(0.3816)\frac{4}{5}\left(\frac{V_{\infty}x}{v}\right)^{-1/5}$$

$$= 0.02968\left(\frac{72}{7}\right)\left(\frac{V_{\infty}x}{v}\right)^{-1/5} ,$$

which reduces to

$$\frac{8}{7}\xi^{2/7}\frac{d\xi}{dx} + \xi^{9/7}\frac{4}{5}x^{-1} = \frac{0.02968}{0.3816}\left(\frac{72}{7}\right)x^{-1} . \tag{1}$$

Now, by the product rule,

$$\xi^{2/7}\frac{d\xi}{dx} = \frac{7}{9}\frac{d\xi^{9/7}}{dx} ,$$

So (1) becomes

$$\frac{8}{9}\frac{d\xi^{9/7}}{dx} + \xi^{9/7}\frac{4}{5}x^{-1} = \frac{0.02968}{0.3816}\left(\frac{72}{7}\right)x^{-1} , \tag{m}$$

or

$$\frac{d\xi^{9/7}}{dx} + \frac{9}{10}\xi^{9/7}x^{-1} = \frac{0.02968}{0.3816}\left(\frac{72}{7}\right)\left(\frac{9}{8}\right)x^{-1} . \tag{E-15}$$

To solve this equation, we first note that (E-15) is of the form

$$\frac{dY}{dx} + \frac{AY}{x} = \frac{B}{x} ,$$

where $Y = \xi^{9/7}$ and $A = B = 0.9$. This type of equation can be solved by use of an integrating factor, the result being

$$\left(\frac{\delta_t}{\delta}\right)^{9/7} = \frac{10}{9}B + \frac{C}{x^{9/10}} , \tag{n}$$

where C is a constant of integration. To evaluate this constant, we can invoke the boundary condition $\delta_t = 0$ at $x = x_0$. This gives

$$C = -\frac{10}{9}Bx_0^{9/10} , \tag{o}$$

and so (n) becomes

$$\left(\frac{\delta_t}{\delta}\right)^{9/7} = \frac{10}{9}B - \frac{10}{9}B\left(\frac{x_o}{x}\right)^{9/10},$$

or, with $B = 0.9$,

$$\frac{\delta_t}{\delta} = \left[1 - \left(\frac{x_o}{x}\right)^{9/10}\right]^{7/9}. \qquad \text{(E-16)}$$

Finally, we can use this in our expression for the heat flux (E-12),

$$\frac{q_o''}{\rho c_p} = (T_s - T_\infty)V_\infty \frac{C_f}{2}\left[1 - \left(\frac{x_o}{x}\right)^{9/10}\right]^{-1/9}, \qquad \text{(p)}$$

And rearranging, we can show that

$$\frac{q_o''}{\rho c_p V_\infty (T_s - T_\infty)} \equiv St_x = \frac{C_f}{2}\left[1 - \left(\frac{x_o}{x}\right)^{9/10}\right]^{-1/9} \qquad \text{(8.121)}$$

This expression gives the local heat flux for a flat plate with unheated starting length of x_o.

REFERENCES

[1] Burmeister, L.C., *Convective Heat Transfer*, 2nd Ed., John Wiley and Sons, New York, 1993.

[2] Kays, W.M., Crawford, M.E., and Weigand, B., *Convective Heat and Mass Transfer*, 4th Ed., McGraw-Hill, Boston, 2005.

APPENDIX F: Properties of Air at Atmospheric Pressure

T	C_p	ρ	μ	ν	k	Pr
°C	J/kg-°C	kg/m^3	kg/s-m	m^2/s	W/m-°C	
-40	1006.0	1.5141	15.17×10^{-6}	10.02×10^{-6}	0.02086	0.731
-30	1005.8	1.4518	15.69×10^{-6}	10.81×10^{-6}	0.02168	0.728
-20	1005.7	1.3944	16.20×10^{-6}	11.62×10^{-6}	0.02249	0.724
-10	1005.6	1.3414	16.71×10^{-6}	12.46×10^{-6}	0.02329	0.721
0	1005.7	1.2923	17.20×10^{-6}	13.31×10^{-6}	0.02408	0.718
10	1005.8	1.2467	17.69×10^{-6}	14.19×10^{-6}	0.02487	0.716
20	1006.1	1.2042	18.17×10^{-6}	15.09×10^{-6}	0.02564	0.713
30	1006.4	1.1644	18.65×10^{-6}	16.01×10^{-6}	0.02638	0.712
40	1006.8	1.1273	19.11×10^{-6}	16.96×10^{-6}	0.02710	0.710
50	1007.4	1.0924	19.57×10^{-6}	17.92×10^{-6}	0.02781	0.709
60	1008.0	1.0596	20.03×10^{-6}	18.90×10^{-6}	0.02852	0.708
70	1008.7	1.0287	20.47×10^{-6}	19.90×10^{-6}	0.02922	0.707
80	1009.5	0.9996	20.92×10^{-6}	20.92×10^{-6}	0.02991	0.706
90	1010.3	0.9721	21.35×10^{-6}	21.96×10^{-6}	0.03059	0.705
100	1011.3	0.9460	21.78×10^{-6}	23.02×10^{-6}	0.03127	0.704
110	1012.3	0.9213	22.20×10^{-6}	24.10×10^{-6}	0.03194	0.704
120	1013.4	0.8979	22.62×10^{-6}	25.19×10^{-6}	0.03261	0.703
130	1014.6	0.8756	23.03×10^{-6}	26.31×10^{-6}	0.03328	0.702
140	1015.9	0.8544	23.44×10^{-6}	27.44×10^{-6}	0.03394	0.702
150	1017.2	0.8342	23.84×10^{-6}	28.58×10^{-6}	0.03459	0.701
160	1018.6	0.8150	24.24×10^{-6}	29.75×10^{-6}	0.03525	0.701
170	1020.1	0.7966	24.63×10^{-6}	30.93×10^{-6}	0.03589	0.700
180	1021.7	0.7790	25.03×10^{-6}	32.13×10^{-6}	0.03654	0.700
190	1023.3	0.7622	25.41×10^{-6}	33.34×10^{-6}	0.03718	0.699
200	1025.0	0.7461	25.79×10^{-6}	34.57×10^{-6}	0.03781	0.699
210	1026.8	0.7306	26.17×10^{-6}	35.82×10^{-6}	0.03845	0.699
220	1028.6	0.7158	26.54×10^{-6}	37.08×10^{-6}	0.03908	0.699
230	1030.5	0.7016	26.91×10^{-6}	38.36×10^{-6}	0.03971	0.698
240	1032.4	0.6879	27.27×10^{-6}	39.65×10^{-6}	0.04033	0.698
250	1034.4	0.6748	27.64×10^{-6}	40.96×10^{-6}	0.04095	0.698
260	1036.5	0.6621	27.99×10^{-6}	42.28×10^{-6}	0.04157	0.698
270	1038.6	0.6499	28.35×10^{-6}	43.62×10^{-6}	0.04218	0.698
280	1040.7	0.6382	28.70×10^{-6}	44.97×10^{-6}	0.04279	0.698
290	1042.9	0.6268	29.05×10^{-6}	46.34×10^{-6}	0.04340	0.698
300	1045.2	0.6159	29.39×10^{-6}	47.72×10^{-6}	0.04401	0.698
310	1047.5	0.6053	29.73×10^{-6}	49.12×10^{-6}	0.04461	0.698
320	1049.9	0.5951	30.07×10^{-6}	50.53×10^{-6}	0.04521	0.698
330	1052.3	0.5853	30.41×10^{-6}	51.95×10^{-6}	0.04584	0.698
340	1054.4	0.5757	30.74×10^{-6}	53.39×10^{-6}	0.04638	0.699
350	1056.8	0.5665	31.07×10^{-6}	54.85×10^{-6}	0.04692	0.700

APPENDIX G: Properties of Saturated Water

T	C_p	ρ	μ	v	k	α	β	Pr
°C	J/kg-°C	kg/m³	kg/s-m	m²/s	W/m-°C	m²/s	1/K	
0	4218	999.8	1.791×10^{-3}	1.792×10^{-6}	0.5619	1.332×10^{-7}	-0.0853×10^{-3}	13.45
5	4203	1000.0	1.520×10^{-3}	1.520×10^{-6}	0.5723	1.362×10^{-7}	0.0052×10^{-3}	11.16
10	4193	999.8	1.308×10^{-3}	1.308×10^{-6}	0.5820	1.389×10^{-7}	0.0821×10^{-3}	9.42
15	4187	999.2	1.139×10^{-3}	1.140×10^{-6}	0.5911	1.413×10^{-7}	0.148×10^{-3}	8.07
20	4182	998.3	1.003×10^{-3}	1.004×10^{-6}	0.5996	1.436×10^{-7}	0.207×10^{-3}	6.99
25	4180	997.1	0.8908×10^{-3}	0.8933×10^{-6}	0.6076	1.458×10^{-7}	0.259×10^{-3}	6.13
30	4180	995.7	0.7978×10^{-3}	0.8012×10^{-6}	0.6150	1.478×10^{-7}	0.306×10^{-3}	5.42
35	4179	994.1	0.7196×10^{-3}	0.7238×10^{-6}	0.6221	1.497×10^{-7}	0.349×10^{-3}	4.83
40	4179	992.3	0.6531×10^{-3}	0.6582×10^{-6}	0.6286	1.516×10^{-7}	0.389×10^{-3}	4.34
45	4182	990.2	0.5962×10^{-3}	0.6021×10^{-6}	0.6347	1.533×10^{-7}	0.427×10^{-3}	3.93
50	4182	988.0	0.5471×10^{-3}	0.5537×10^{-6}	0.6405	1.550×10^{-7}	0.462×10^{-3}	3.57
55	4184	985.7	0.5043×10^{-3}	0.5116×10^{-6}	0.6458	1.566×10^{-7}	0.496×10^{-3}	3.27
60	4186	983.1	0.4668×10^{-3}	0.4748×10^{-6}	0.6507	1.581×10^{-7}	0.529×10^{-3}	3.00
65	4187	980.5	0.4338×10^{-3}	0.4424×10^{-6}	0.6553	1.596×10^{-7}	0.560×10^{-3}	2.77
70	4191	977.7	0.4044×10^{-3}	0.4137×10^{-6}	0.6594	1.609×10^{-7}	0.590×10^{-3}	2.57
75	4191	974.7	0.3783×10^{-3}	0.3881×10^{-6}	0.6633	1.624×10^{-7}	0.619×10^{-3}	2.39
80	4195	971.6	0.3550×10^{-3}	0.3653×10^{-6}	0.6668	1.636×10^{-7}	0.647×10^{-3}	2.23
85	4201	968.4	0.3339×10^{-3}	0.3448×10^{-6}	0.6699	1.647×10^{-7}	0.675×10^{-3}	2.09
90	4203	965.1	0.3150×10^{-3}	0.3264×10^{-6}	0.6727	1.659×10^{-7}	0.702×10^{-3}	1.97
95	4210	961.7	0.2978×10^{-3}	0.3097×10^{-6}	0.6753	1.668×10^{-7}	0.728×10^{-3}	1.86
100	4215	958.1	0.2822×10^{-3}	0.2945×10^{-6}	0.6775	1.677×10^{-7}	0.755×10^{-3}	1.76
120	4246	942.8	0.2321×10^{-3}	0.2461×10^{-6}	0.6833	1.707×10^{-7}	0.859×10^{-3}	1.44
140	4282	925.9	0.1961×10^{-3}	0.2118×10^{-6}	0.6845	1.727×10^{-7}	0.966×10^{-3}	1.23
160	4339	907.3	0.1695×10^{-3}	0.1869×10^{-6}	0.6815	1.731×10^{-7}	1.084×10^{-3}	1.08
180	4411	886.9	0.1494×10^{-3}	0.1684×10^{-6}	0.6745	1.724×10^{-7}	1.216×10^{-3}	0.98
200	4498	864.7	0.1336×10^{-3}	0.1545×10^{-6}	0.6634	1.706×10^{-7}	1.372×10^{-3}	0.91
220	4608	840.4	0.1210×10^{-3}	0.1439×10^{-6}	0.6483	1.674×10^{-7}	1.563×10^{-3}	0.86
240	4770	813.6	0.1105×10^{-3}	0.1358×10^{-6}	0.6292	1.622×10^{-7}	1.806×10^{-3}	0.84
260	4991	783.9	0.1015×10^{-3}	0.1295×10^{-6}	0.6059	1.549×10^{-7}	2.130×10^{-3}	0.84
280	5294	750.5	0.0934×10^{-3}	0.1245×10^{-6}	0.5780	1.455×10^{-7}	2.589×10^{-3}	0.86

Index